Tropical Forestry

For other volumes of the series, go to
www.springer.com/series/5439

Sheona Shackleton • Charlie Shackleton •
Patricia Shanley
Editors

Non-Timber Forest Products in the Global Context

With 45 Figures and 12 Tables

Springer

Editors
Dr. Sheona Shackleton
Rhodes University
Dept. Environmental Science
6140 Grahamstown
South Africa
s.shackleton@ru.ac.za

Prof. Charlie Shackleton
Rhodes University
Dept. Environmental Science
6140 Grahamstown
South Africa
c.shackleton@ru.ac.za

Dr. Patricia Shanley
Center for International
Forestry Research (CIFOR)
Bogor
BOCBD
Indonesia
p.shanley@cgiar.org

ISSN 1614-9785
ISBN 978-3-642-17982-2 e-ISBN 978-3-642-17983-9
DOI 10.1007/978-3-642-17983-9
Springer Heidelberg Dordrecht London New York

Library of Congress Control Number: 2011925367

Cover photo: ready2shoot/Fotolia.com

Cover design: deblik, Heidelberg, Germany

Printed on acid-free paper

Springer is part of Springer Science+Business Media (www.springer.com)

Foreword

In the mid-1980s, I lived in the Peruvian Amazon and conducted research on the native fruits of the region. My work was focused primarily on the ecology and sustainable harvesting of three forest taxa, but I also interacted quite a bit with local collectors, middlemen, and vendors in the sprawling Belen market of Iquitos. The great majority of the fruits sold in the Iquitos market during those years were wild harvested. In trying to figure out where all of this material was coming from, I discovered several interesting things about the local fruit trade. Ribereño communities upriver knew the location and fruiting times of wild populations of all the commercial fruit species in the vicinity; they visited these populations every year and harvested commercial quantities of fruit, they figured out ways to get the fruit to market, they haggled with buyers about the price, and they usually made some money doing this.

Forest fruits were still characterised as minor forest products at this point in time, but to many of the villagers that I worked with along the Ucayali River, these non-timber forest products (NTFPs) were very important to their livelihood, by no means minor. I wrote a couple of papers about the ecology and management of forest fruits in the Peruvian Amazon and even published a comment about the relative economic value of non-timber resources in a prestigious scientific journal (Peters et al. 1989). One might say that I was one of the early adopters in the NTFP sector.

And then the momentum started to build, and NTFPs became a cause célébre, and fruits, and nuts, and latex, and indigenous people, and rubber tappers started appearing everywhere. Through it all, there seemed to be an unstated assumption that somehow the NTFP acronym – as well as the people that collect them and the markets through which they are sold – represented a distinct and relatively homogenous category of products and processes. Things got confusing. Is the fragrant heartwood of the *Aquilaria* tree, obtained by tree felling in the Asian tropics, a timber or a non-timber resource? Is palizada, the pole-sized stems logged commercially from the Selva Maya for use in house construction, an NTFP?

Subsequent research that I have conducted in other parts of the world have consistently highlighted the differences, rather than the similarities, in the ways that communities collect, manage, and market NTFPs. In some cases, the collection of NTFPs provides a reliable source of income, plays an important cultural role in the community, and offers a convincing reason to keep forests as forests, rather than converting them to pastures, oil palm plantations or estate crops. At the other extreme, some communities exploit NTFPs to the point of depletion, are enslaved by compromises to local markets, or are impoverished by existing regulations governing the collection and sale of these resources. In spite of how much one might want NTFPs to be a predictable, well-defined commodity group, there is actually a great amount of contextual chaos associated with these resources. This does not negate the potential of NTFPs but certainly does argue against blanket prescriptions and standardised governance.

There are as many different NTFP systems as there are non-timber forest products. The Peruvian case was my first peek behind the curtain of this wonderful people and plants show. Every interaction between a community and an NTFP, the positive ones as well as the negative ones, can teach participants, researchers, and onlookers something about sustainable resource use. The editors of the present volume are to be congratulated for embracing this diversity and for weaving together the many dimensions of NTFPs into such a comprehensive, informative overview.

Kate E. Tode Curator of Botany Charles M. Peters
The New York Botanical Garden, Bronx NY, USA

Reference

Peters CM, Gentry A, Mendelsohn R (1989) Valuation of a tropical forest in Peruvian Amazonia. Nature 339:655–657

Preface and Acknowledgments

The idea to write this book on non-timber forest products (NTFPs) first came from Anette Lindqvist of Springer-Verlag. Springer was interested in adding such a volume to their tropical forest series. After circulation amongst several NTFP researchers, Patricia Shanley and I, with input from Bruce Campbell, discussed the proposal at length in 2006 at the Center for International Forestry Research (CIFOR). I was, at the time, spending 6 months at CIFOR as a visiting scientist, and Patricia was working there. We agreed to take on the challenge of compiling this book, providing that we could extend the discussion beyond just tropical forests to a global review that drew on work from the humid and dry tropics, temperate regions, and from both the developing and developed world. We also stipulated that it needed to be an edited volume with contributions from leading researchers and commentators in the field. Work began in earnest in 2007, when the editing team was expanded to include Charlie Shackleton. Three years later, we were able to submit this book to the publishers. It represents the collective output of 43 authors of chapters and boxes, each with extensive and unique insights and experience in researching and working with NTFPs. It is a truly interdisciplinary volume, with authors coming from sociological, anthropological, economic, ecological, environmental science, policy, botanical, and geographical backgrounds, but all comfortable working beyond conventional disciplinary boundaries.

Our aim for this book was to provide a comprehensive, global synthesis of current knowledge on the potential and challenges associated with the multiple roles, use, management, and marketing of non-timber forest-products (NTFPs) across the world. The time is opportune for such a synthesis as the last two and half decades have seen much research and policy effort around NTFPs and many questions. This book explores the evolution of sentiments regarding the potential of NTFPs in promoting options for sustainable multipurpose forest management, income generation, and poverty alleviation. Based on critical analysis of the debates and discourses, it employs a systems approach to present a balanced and realistic perspective on the benefits and challenges associated with NTFP use and management within local livelihoods and landscapes, supporting this with case examples from both the southern and northern hemispheres. Authors attempted to include and give equal weight to the varied and complex social, economic, and ecological dimensions of NTFPs. In summary, the purpose of this book is to:

- Provide a global review of the multiple roles of NTFPs.
- Provide an up-to-date critical analysis of the debates and discourses surrounding NTFPs.
- Ensure a strong academic base, but also provide practical information so as to be of use to a wider audience.
- Cover the sector in an open and questioning manner that is more provocative than the standard text book.
- Provide solid case study material.
- Highlight the many complexities of the sector.

In addition to the authors, another important group of people contributed to this book by providing invaluable commentary and feedback on draft chapters. This input has been essential in ensuring quality. Each chapter has been read by at least two, sometimes three, external reviewers as well as at least one of the editors. We gratefully acknowledge these willing reviewers, namely Bruce Campbell, Campbell Plowden, Darcy Mitchell, Frances (Jack) Putz, Freerk Weirsum, Carol Colfer, Jenne de Beer, Johan Poulsen, Kamalit Bawa, Mary Menton, Robert Voeks, Roger Leakey, Sarah Laird, Soul Shava, Steven Siebert, Uma Shaanker, and Wayne Twine.

All authors contributed to this book because of their belief in the necessity for such a volume and received no payment. The editors are grateful to CIFOR and Rhodes University for time to work on the book. Charlie Shackleton was able to undertake much of the editing of chapters during his sabbatical in 2009/2010 funded by Rhodes University. I used a small portion of my National Research Foundation (NRF) research grant in the finalisation of this book. Chapter 8 is based on an article by Manuel Guariguata and colleagues published by Elsevier in Forest Ecology and Management Volume 259. All views are authors' own.

Grahamstown, South Africa
February 2011

Sheona Shackleton

Contents

Contributors

Arild Angelsen Center for International Forestry Research (CIFOR), Bogor, Indonesia; Department of Economics and Resource Management, Norwegian University of Life Sciences (UMB), 1430 Ås, Norway, arild.angelsen@umb.no

Yildiz Aumeeruddy-Thomas Centre d'Ecologie Foncionnelle et Evolutive, Université de Montpellier II, UMR 5175 Montpellier, France, cdelang@cuhk.edu.hk

Alejandro Casas Centro de Investigaciones en Ecosistemas, Universidad Nacional Autónoma de México, Morelia, Mexico, acasas@oikos.unam.mx

Michelle Cocks Institute for Social and Economic Research (ISER), Rhodes University, P.O. Box 94, Grahamstown 6140, Eastern Cape, South Africa, m.cocks@ru.ac.za

Wendy Cocksedge Centre for Non-Timber Resources, Royal Roads University, 2005 Sooke Road, Victoria, BC, Canada V9B 5Y2, Wendy.cocksedge@royalroads.ca

Priscilla Cooke St Clair Department of Economics, Pacific Lutheran University, Tacoma, WA 98447-0003, USA, stclairp@plu.edu

Peter Cronkleton Center for International Forestry Research (CIFOR), P.O. Box 0113, BOCBD Bogor 16000, Indonesia, p.cronkleton@cgiar.org

Anthony B. Cunningham Centre for International Forestry Research, Bogor, Indonesia; School of Plant Biology, University of Western Australia, 35 Stirling Highway, Crawley 6009, Australia, tonyc05@bigpond.net.au

Annika Dahlberg Department of Geography and Quaternary Geology, Stockholm University, Stockholm 10691, Sweden, annika.dahlberg@natgeo.su.se

Jenne de Beer Darrell Posey Field Fellow, International Society for Ethnobiology, Manila, Philippines, ntfp7@yahoo.co.uk

Claudio O. Delang Department of Geography and Resource Management, The Chinese University of Hong Kong, Shatin, N.T., Hong Kong, cdelang@cuhk.edu.hk

Tony Dold Botany Department, Selmar Schonland Herbarium (GRA), Rhodes University, P.O. Box 94, Grahamstown 6140, Eastern Cape, South Africa, t.dold@ru.ac.za

Carmen García-Fernández Biodiversity International, Via dei Tre Denari 472/a, 00057 Maccarese, Rome, Italy, Carmen.GARCIA-FERNANDEZ@ec.europa.eu

Suresh Kumar Ghimire Central Department of Botany, Tribhuvan University, Kirtipur, Katmandu, Nepal, Sureshkghimire@yahoo.com

Manuel R. Guariguata Center for International Forestry Research (CIFOR), P.O. Box 0113, BOCBD Bogor 16000, Indonesia, m.guariguata@cgiar.org

Susan Harrison-Sanchez Life Cycle Associates LLC, 985 Portola Valley Road, Portola Valley, CA 94028, USA, sanchez@lifecycleassociates.com

Cristina Herrero-Jauregui Departamento de Ecología, Facultad de Biología, Universidad Complutense de Madrid, Madrid 28040, España, crherrero@bio.ucm.es

Verina Ingram Centre for International Forestry Research (CIFOR), P.O. Box 0113, BOCBD Bogor 16000, Indonesia, v.ingram@cgiar.org

Ramesh Kannan ATREE, Royal Enclave, Sri Rampura, Jakkur Post, Bangalore, 560064 Karnataka, India, Rameshkannan.h@gmail.com

Koen Kusters Amsterdam Institute of Social Science Research, University of Amsterdam, Nieuwe Prinsengracht 130, 1018 VZ, Amsterdam, The Netherlands, k.kusters@uva.nl

Sarah A. Laird People and Plants International, P.O. Box 251, Bristol, VT 05443, USA, sarahlaird@aol.com

Citlalli López Centro de Investigaciones Tropicales, Universidad Veracruzana, Calle Araucarias s/n, Interior de la Ex-hacienda Lucas Martin, Colonia Periodistas, C.P. 91019, Apartado Postal 252, Xalapa, Veracruz, México, citlalli_lb@yahoo.com

Dylan McGarry Department of Environmental Science, Rhodes University, Grahamstown 6140, South Africa, armadylan@gmail.com

Doyle McKey Centre d'Ecologie Foncionnelle et Evolutive, Université de Montpellier II, 5175 Montpellier, France, doyle.mckey@cefe.cnrs.fr

Rebecca J. McLain Institute for Culture and Ecology, P.O. Box 6681, Portland, OR 97214, USA, mclain@ifcae.org

Sharif Ahmed Mukul Department of Forestry and Environmental Science, Shahjalal University of Science and Technology, Sylhet 3114, Bangladesh, sharif_a_mukul@yahoo.com

Robert Nasi Center for International Forestry Research (CIFOR), P.O. Box 0113, BOCBD, Bogor 16000, Indonesia, r.nasi@cgiar.org

Ousseynou Ndoye Food and Agriculture Organisation of the United Nations (FAO), P.O. Box 281, Yaoundé, Cameroon, ousseynou.ndoye@fao.org

Fiona Paumgarten School of Animal, Plant and Environmental Sciences, University of the Witwatersrand, Johannesburg, South Africa; Center for International Forestry Research, Lusaka, Zambia, fi.paumgarten@gmail.com

Alan Pierce Independent researcher, 1061 Mountainview Road, Duxbury, VT, USA, arp@sover.net

Mirjam A.F. Ros-Tonen Amsterdam Institute of Social Science Research, University of Amsterdam, Nieuwe Prinsengracht 130, 1018 VZ, Amsterdam, The Netherlands, m.a.f.ros-tonen@uva.nl

Yvonne M. Scherrer Program Sustainability Research, University of Basel, Klingelbergstrasse 50, CH 4056, Basel, Switzerland, y.scherrer@pantek.org

Charlie Shackleton Department of Environmental Science, Rhodes University, Grahamstown 6140, South Africa, c.shackleton@ru.ac.za

Sheona Shackleton Department of Environmental Science, Rhodes University, Grahamstown 6140, South Africa, s.shackleton@ru.ac.za

Patricia Shanley Center for International Forestry Research and Woods and Wayside International, 211 Ridgeview Road, Princeton, NJ 08540, USA, p.shanley@cgiar.org

Douglas Sheil Institute of Tropical Forest Conservation (ITFC), Bwindi Impenetrable National Park (at Ruhija), P.O. Box 44, Kabale, Uganda; Center for International Forestry Research, Bogor, Indonesia, dsheil@cgiar.org

Erin Sills Department of Forestry & Environmental Resources, North Carolina State University, Raleigh, NC 27695–8008, USA, sills@ncsu.edu

Terry C. H. Sunderland Center for International Forestry Research (CIFOR), P.O. Box 0113, BOCBD, Bogor 16000, Indonesia, t.sunderland@cgiar.org

Andrew Taber Center for International Forestry Research (CIFOR), P.O. Box 0113, BOCBD, Bogor 16000, Indonesia, a.taber@cgiar.org

Tamara Ticktin Department of Botany, University of Hawai'i, Honolulu, HI USA, ticktin@hawaii.edu

Lucrezia Tincani Centre for Development, Environment and Policy, SOAS, University of London, 36 Gordon Square, London, WC1H 0PD, UK, lucrezia.tincani@soas.ac.uk

Nathalie van Vliet Post Doctoral researcher, Department of Geography and Geology, University of Copenhagen, Oster Voldgade, 101350 Copenhagen K, Denmark, vanvlietnathalie@yahoo.com

Nathan Vogt ACT, Indiana University, 179 Rua de Obidos, Edificio Montreal, Apto. 802, Cidade Velha, Belém 66020-440, Brasil, ndvogt@gmail.com

Rachel Wynberg Environmental Evaluation Unit, University of Cape Town, Rondebosch 7700, South Africa, Rachel@iafrica.com

Part I
Introduction: Non-timber Forest Products in the Global Context

Chapter 1
Non-timber Forest Products: Concept and Definitions

Charlie Shackleton, Claudio O. Delang, Sheona Shackleton, and Patricia Shanley

Abstract Since the potential value and role of non-timber forest products (NTFPs) was first mooted in the 1980s, there has been a tremendous escalation in research, practice, and policy interest across numerous disciplines including conservation, livelihood studies, economics, forestry, and anthropology. Inevitably this has resulted in altered and evolving definitions of what is, or is not, an NFTP. This chapter discusses the various attributes of an NTFP and proposes a working definition that could be applicable across disciplines. The final section of the chapter provides a brief overview of the contributions to this book.

1.1 Introduction

Over the last two decades, the importance of non-timber forest products (NTFPs) to rural livelihoods, income generation, local economies, and in some instances forest conservation has become increasingly recognised and appreciated in both the research and policy sectors. So much so, that the FAO, in its regular compilation on the *State of the World's Forests*, has prompted national governments to collect and report data pertaining to the extraction rates and sustainability of NTFP harvests. Yet this is no small task, being complicated at a number of levels.

C. Shackleton (✉) and S. Shackleton
Department of Environmental Science, Rhodes University, Grahamstown 6140, South Africa
e-mail: c.shackleton@ru.ac.za; s.shackleton@ru.ac.za

C.O. Delang
Department of Geography and Resource Management, The Chinese University of Hong Kong, Shatin, N.T., Hong Kong
e-mail: cdelang@cuhk.edu.hk

P. Shanley
Center for International Forestry Research, Bogor, Indonesia
e-mail: p.shanley@cgiar.org

S. Shackleton et al. (eds.), *Non-Timber Forest Products in the Global Context*,
Tropical Forestry 7, DOI 10.1007/978-3-642-17983-9_1,

- First is the complication resulting from the sheer number of different species and types of NTFPs; ranging from large and whole plants and animals to smaller parts such as fruits, leaves, flowers, seeds, roots and bulbs, bark, honey, insects, resins, horns, skins, and many more.
- Second is the tremendous variation in volumes extracted, used, and traded; from just a small handful of a particular product during times of need (such as bark from a medicinal tree (Fig. 1.1) to hundreds of thousands of tons of product on an annual basis (such as fuelwood, some fruit species, fibre products).
- Third, enumeration is further made more difficult because much of what is extracted is done so by rural communities far from infrastructure and formal recording or census agencies.
- Fourth, the assessment of sustainability of extraction is almost impossible because there is insufficient biological knowledge, understanding, and monitoring of most NTFP species with respect to their growth and mortality rates, productivity, reproduction, and responses to harvesting.
- Last is the debate, confusion, and inconsistencies in what constitutes an NTFP and what does not, which determines what types and species of products should be recorded.

Consequently, the estimates reported in the *State of the World's Forests* are extremely unreliable or totally erroneous [e.g., see the comment by Shackleton (2009) of the reporting of fruit NTFPs for South Africa]. It is this last dimension that is the subject of this chapter and sets the basis for interpretation of the details, facts, and figures presented in the rest of the book.

In considering what is an NTFP, some authors have suggested it may be easier to rather deal with what it is not (Neumann and Hirsch 2000; Belcher 2003). However, whilst defining what are not NTFPs may seem relatively straightforward, one finds that ambiguity persists on both sides of the ontological coin, and thus we have not opted for that route.

The most intuitive and widely known definition comes from the landmark work by De Beer and McDermott (1989) on the economic value of NTFPs in southeast Asia. They defined NTFPs as encompassing "all biological materials other than timber which are extracted from forests for human use" (De Beer and McDermott 1989). In line with this definition, most researchers agree that NTFPs include floral products such as grasses, roots, flowers, fruits, and bamboo (Fig. 1.2), which people use for a variety of purposes (e.g., as food for themselves and their domesticated animals, as medicinal plants, as ornaments, and as raw material for tools), as well as faunal products such as insects, birds, fish, or game (Figs. 1.1 – 1.4). However, some have excluded fodder for livestock (Rajchal 2008).

However, considerable debate and inconsistencies arise when considering the woody portion of plants, such as the stems, branches, and bark of trees or shrubs. At one end of the continuum, some have argued, such as the FAO, that all timber products should be excluded from the definition of NTFPs, such that they recommend use of the alternative term of "nonwood forest product" (NWFP) instead of

Fig. 1.1 Traditional healer collecting bark from *Ozoroa* sp. for medicinal purposes, Bushbuckridge, South Africa (photo: Charlie Shackleton)

Fig. 1.2 Children on the way to the river to collect water, Highland Kammu village in Bo Kaeo province, northern Lao PDR. Each bamboo tube contains 4 or 5 l of water (photo: Claudio Delang)

NTFP. At the opposite end are those who include all wood products as NTFPs, except those extracted and sold by large-scale capital intensive commercial interests (De Beer and McDermott 1989). Thus, wood harvested from forested lands for fuel, construction, tools or crafts, medicines, and carving would be considered NTFPs, provided it is by local subsistence users or small-scale local traders. Several positions can be found between these two ends, especially around the inclusion (or not) of fuelwood as an NTFP, as well as inclusion, or not, of wood products extracted for trade as opposed to simply subsistence consumption. The evolution of wood products traded on local markets to international niche markets, therefore also complicates the definition.

There are some important organisational factors that led to the FAO's introduction of the NWFP concept. The FAO definition emerged from the desire to "build a classification system" (FAO 1999) that would harmonise with its Central Product Classification System, as well as from the fact that the FAO already had a Wood Product division and a separate group that dealt with fuelwood, rather than from the desire to create a category that would help in management, research, and policy (Belcher 2003). That other FAO units (such as its Community Forestry Unit) continue to use the term NTFP underlines the fact that, ultimately, the terminology used depends on the particular research and policy objectives. The FAO's concept is quite useful for statistical recording, particularly in national accounts. But it is arguably less useful for field-based research into livelihoods, value chains, and forest conservation, as local users and managers do not make the fine distinctions that the FAO does, and consequently nor can local researchers. As Belcher (2003: 165) comments, "for those interested in community development, forest conservation, or other aspects of forest management, the distinction between wood and nonwood is neither relevant nor helpful". Because of this, the usage of the term NTFP has gained currency within a progressively expanding field of study.

Despite its complexities, the term NTFP has certainly served to nominally unify contributions from a range of disciplines, which a decade or more ago used a range of terms, such as minor forest products, wild products, secondary forest products, or by-products of forests. However, in bringing together different disciplines, it is inevitable that there are at times subtle differences in how different researchers, regions, or disciplines perceive the term NTFP and what it includes. These different applications of the term have come under heightened scrutiny and the concept has undergone various qualifying modifications and transformations. These changes underscore the propensity of many researchers to problematise the term, quite justifiably, for its inherent conceptual ambiguities. Such problematisation, however, has been relatively unsuccessful in producing conceptually viable theories on the relationships between NTFPs and the major development issues to which their investigation is most relevant, such as property regime transformation, forest and natural resource conservation, social and environmental sustainability, poverty alleviation or poverty traps, among others.

The tendency to dissect and critique the definition of the term NTFP (e.g., Belcher 2003; Rajchal 2008) stems from the fact that the very notion of an NTFP is premised upon other terms whose conceptual foundations are subject to deconstruction. The idea of an NTFP results from the compounding of four other contested ideas, which, in the context of contribution to the notion of an NTFP, can be described as extractive scale, ecosystem value, forests, and domestication.

1.2 Defining Non-timber Forest Products

1.2.1 Non-timber or Nonwood?

Firstly, as acknowledged by De Beer and McDermott (1989), distinguishing between "timber" and "non-timber" is not so easy, especially when taking into account the scale of extraction and the interests served. When exactly do wood products extracted from trees constitute non-timber products? Some commentators suggest that the trunk of a tree should be considered as a timber product, whilst its branches (whether collected from already dead trees and used as fuelwood or whether cut from living trees and used as construction materials) are not. And where does bark fit into this debate and attempt to create a dichotomy? However, as De Beer and McDermott (1989) argue, the part of the tree that is collected is not the only information necessary to determine whether the product is an NTFP. NTFPs are usually collected by local consumers or traders, on a small scale, and when these products are collected on an industrial scale, they should no longer enter into the NTFPs category. As De Beer and McDermott (1989: 24) argue, "large-scale fuelwood harvesting for urban markets and the use of timber in rural house construction [should be excluded from the category of NTFPs]; the former is excluded on the grounds that it violates the criteria of scale and rural use/benefit, whilst the latter must be omitted from consideration if only on the grounds of consistency". Whilst providing a useful rationale, many interpretative contradictions can be found. For example, the supply of urban fuelwood markets may also benefit rural communities if they are part of the supply chain, as is often the case (e.g., McCrary et al. 2005). And making scale as the criterion for inclusion or not is no less problematic than other criteria because how does one define large scale (and therefore excludes some products) as opposed to small scale and thus certain products are included? Scale is relative in spatial and temporal context.

1.2.2 Biological Nature of NTFPs?

Some commentators contest the biological materials requisite for NTFPs. Whilst some early researchers sought to include abiotic materials such as rocks, clay, sand, and water for local use into the NTFP category, others go beyond just ecosystem goods to include services such as watershed services, carbon sequestration functions, and ecotourism capacities (Belcher 2003). These considerations reflect the attempts of researchers to measure as comprehensively as possible the conventionally un- or under-appreciated importance of forests and so strengthen arguments for their conservation and wise use. However, since popularisation of the term "ecosystem goods and services" and their classification into four categories as used by the Millennium Ecosystem Assessment, the arguments are waning for including these services as NTFPs. Cavendish (2000), in a landmark livelihoods study, used the term "environmental income" to include nonbiological landscape assets in his analysis, which also included NTFPs. In general, however, we perceive that the consensus seems to be gravitating towards exclusion of abiotic products because, since they are not renewable resources, they pose very different issues around sustainable use and conservation.

1.2.3 Only from Forests?

Regarding the source of the product, i.e., forests, there are numerous inconsistencies in the NTFP literature concerning what exactly merits the designation, as well as what is a forest (e.g., Putz and Redford 2010). The primary debates here relate to (a) do NTFPs come only from forests or can they also come from other ecosystems? and (b) do they come only from natural lands (forests or otherwise) or can they also be found and extracted from human modified or impacted systems? Returning once again to De Beer and McDermott (1989), they argued that "managed, secondary or degraded forests are sources of non-timber forest products, plantations are not". The rationale behind this is clear, i.e., "NTFPs should be used as elements of in situ systems of conservation" in which "NTFP production . . . is compatible with and can give value to natural forests" (Belcher 2003). To include plantations would be to "undermine the conservation objective" (Belcher 2003). However, as De Beer and McDermott (1989: 25) subsequently admit, most natural forests are, or have been at some point, managed by human populations, and therefore in some cases this "distinction is ultimately as arbitrary as the one between timber and non-timber". Moreover, whilst one need not invoke a conservation argument to maintain plantation forests, there is a host of products that can be harvested from plantations that are useful to local communities for subsistence or for local trade; for example, mushrooms, floricultural and medicinal species, berries, or bushmeat (e.g., Porembski and Biedinger 2001; Clason et al. 2008; Parry et al. 2009).

Perhaps because (a) few untouched natural forests can be found and (b) the need to show value for the conservation of biodiversity at all scales applies not only to forests but also to grasslands, wetlands, scrublands, and the like, many researchers have shifted the focus from the place of origin of the species (the forest) to the conditions surrounding the growth of the species. Consequently, they consider NTFPs as all wild, uncultivated plants that grow (or wildlife that is found) anywhere. Thus, most researchers include in the NTFPs category not only fish and other aquatic species found in streams and lakes, because of these bodies' ecological interconnectedness to forests but also wild, uncultivated flora, or undomesticated fauna that are found in grasslands, or other ecosystems. Additionally, the oft encountered difficulty in demarcating clear boundaries between the forest and the villages in communities that are highly dependent on NTFPs prompts many to include wild plants that grow within the village boundaries and fields. It also raises the question of whether or not NTFPs can be found in urban areas where most (but not all) of the conservation functions of the original forest or natural land have been

Fig. 1.3 Marketing indigenous vegetables in urban market, Vietnam (photo: Terry Sunderland)

Fig. 1.4 Sal (*Shorea robusta*) leaves have numerous cultural and practical uses throughout India (photo: Terry Sunderland)

largely lost. Can collection of fuelwood, wild edible fruits and herbs or medicinal bark from single trees in a street, or patches of wasteland or forests between buildings or along roadsides be deemed as collecting NTFPs? Increasingly the answer seems yes. This is because whilst conservation of landscapes and habitats is important (and the initial rationale for the valuation and recognition of NTFPs), but so too is conservation at the species level. This is particularly pertinent as the bulk of the world's biodiversity is located outside of protected areas. Therefore, if species conservation can also be achieved in modified habitats such as agroforests, fields and villages or urban green spaces, whilst contributing to local wellbeing, then it would fit within the NTFP ambit. Moreover, such use from urban systems or others has livelihood benefits, and so brings us back to the other core dimension of NTFPs, i.e., use and benefit.

1.2.4 Only from the Wild?

Cultivated plants are frequently, but not always, excluded from the NTFPs category, whether they grow in forests or any other biome. The rationale for the exclusion of cultivated products (including those grown from wild seeds) from the NTFPs category lies in the fact that cultivation indicates a human-dominated and managed system and thus the conservation merits of highlighting the value of NTFPs have been undermined. But once again this is not as clear cut as it may seem. Wiersum (1997) explores the significant array of states between a *relatively* intact, wild forest (or grassland, etc.) and a totally transformed one dominated by largely monocultures of commercial agriculture. Thus, what may appear as a wild forest to the external observer may have actually experienced decades or millennia of enrichment planting, protection of key species, suppression of pests or fires, and recruitment of particular species mediated by human disturbance. Are not some of these actions legitimate forms of cultivation? At the opposite end, there are many undomesticated or wild species and individual trees and herbs that occur spontaneously and are retained in human-dominated landscapes, fields, and homesteads (High and Shackleton 2000; Schrekenberg 1999; Gonźalez-Amaro et al. 2009). These were not planted or cultivated in a conventional sense of the word, but are promoted by human presence, either by disturbance, removal of competition or predators, or by indirect additions (such as manure or water) from the farming actions. In both these extremes, considerable value is obtained from these species, much of which has been ignored by officials and researchers; however, they are vital to local livelihoods. Additionally, in both examples, the conservation of specific species is being practiced and should be encouraged. Home gardens and agroforests in many parts of the world are extremely biodiverse and contain many plant species in common with the neighbouring forest lands (e.g., Aguilar-Støen et al. 2009; Sahoo 2009), which therefore also provide habitat to dozens of invertebrate species and food for small mammals and avifauna. The complications escalate when a species that used to be harvested largely from the wild becomes commercially domesticated and grown either in small-holders' forest gardens [e.g., fibre bromeliads in Mexico (Ticktin and Nantel 2004)] or more commercially orientated enterprises akin to monocropping of conventional crops [e.g., plantation for Khayu oil in Indonesia (Budiadi et al. 2005)].

1.2.5 Only Indigenous?

The widely used definition of De Beer and McDermott (1989) does not reflect on whether NTFPs should be indigenous or may also include alien species. For example, many inland fisheries for subsistence communities are based on alien species, as is much collection of fuelwood (e.g., Geesing et al. 2004; de Neergaard et al. 2005). But local collection and use of alien species are not restricted to these types of resources and include craft woods and fibres, wild fruits, edible leaves, or spinaches and medicinal plants (Box 1.1). However, given the rationale of NTFPs also contributing to conservation of the forest through realisation of their value, then it is implicit that de Deer and McDermott referred to indigenous species. Yet, in situations where alien species are used (therefore contributing to livelihoods), and that such use (a) helps keep the alien species in check and (b) relieves the pressure on indigenous species or sites, could that not then be construed as a conservation benefit? If so, then the alien species, if meeting the other criteria to be an NTFP, could be described as an NTFP. Indeed, some conservation agencies have deliberately introduced or made use of alien species specifically for the purpose of diverting harvesting pressures away from conservation areas or threatened species; for example, belts of alien tree species near protected areas (or more widely to relieve deforestation pressures) to provide construction timber and fuelwood for local populations (Struhsaker 1987; Viisteensaari et al. 2000; Kasolo and Temu 2008). However, as with most of the other criteria there are degrees to which this can be taken. For example, the alien species might be a highly aggressive invader, and although its use as an NTFP does check its spread to some localised extent, its overall impact is negative in conservation terms, and if the costs stemming from its invasion continue to escalate, it will ultimately be negative on livelihoods too (Shackleton et al. 2007). But for non- or mildly invasive alien species, it seems clear that any use would be welcomed by conservationists, whilst also contributing to livelihoods and thereby be fitting as an NTFP. Indeed, local users of alien NTFPs are not always aware that they are alien, and do not discriminate between alien and indigenous (Shackleton et al. 2007).

Box 1.1 Alien Invasive Species Used as an NTFP by the Forest-Dependent Communities in Southern India

Ramesh Kannan, Ashoka Trust for Research in Ecology and the Environment (ATREE), Bangalore, India, kannan@atree.org

European colonisation laid the foundation for multiple deliberate transfers and accidental introductions of species in the Old World as well as in the New World. The introduction of alien invasive species in India was largely intentional. Most of them were introduced as ornamental plants by the British during their colonial period. *Lantana trifolia* was introduced by W. Carey in 1807, followed by *Lantana aculeata* by W. Hamilton in 1809 in the Royal Botanical Garden in Calcutta, India. It is commonly known as "The Lantana". H.D. Hooker (1885) mentioned in his book "The Flora of British India" that

(continued)

Fig. 1.5 Lantana tree worshiped by the Palliyar communities in the Palani Hills in the Southern Western Ghats

Lantana was found throughout much of western Deccan Peninsula and Ceylon (Sri Lanka) (Fig. 1.5).

Given its invasive properties, management of *Lantana camara* has been a troublesome conservation and economic issue, both in the agricultural and forested landscapes. In India, quite often there have been ad hoc attempts in clearing *Lantana* along roadsides, manually or by machines. However, since most of the roots are deeply lodged in the soil, these efforts have mostly served to provide only temporary relief and therefore have not contributed significantly to lowering the weed biomass.

The current inability to control the spread of Lantana has impacts beyond just conservation. It may also impact on the supply and accessibility of non-timber forest products (NTFPs) in the forests, which are vital components of the livelihoods of forest-dependent communities. Indeed, use of and trade in NTFPs constitute an important source of livelihood for millions of forest-dependent communities in India. Recent studies in India have shown that traditional income sources from NTFPs could be jeopardised by various reasons, including alien invasive species. On the other hand, in these landscapes, alien invasive species have often been viewed as a resource that can accrue potential economic benefits to aid rural livelihoods. As Ralph Waldo Emerson once stated, "*What is a weed? A plant whose virtues have not been discovered*". Local communities in southeast Asian countries explored the possibility of using locally abundant alien invasive species such as Water

(continued)

hyacinth (*Eichhornia crassipes*) to manufacture furniture. Similarly, communities in Kenya found ways to use invasive *Prosopis* spp in a variety of livelihood activities and needs. In India, some indigenous communities in the Western Ghats successfully adopted the *Lantana* as a key livelihood resource. The Kuravas are the traditional basket weaving community in Sirumalai, Western Ghats. They successfully experimented with and subsequently adopted *Lantana* as a favoured resource for the manufacture of baskets. It is not clear when and why the Koravas shifted from indigenous NTFPs (bamboo) to alien invasive *Lantana* for the construction of furniture. Preliminary studies suggest that the practice of using the *Lantana*, instead of bamboo, might have arisen due to (a) the nonavailability of bamboo, (b) increasing difficulty in accessing bamboo even when available, and (c) the initial investment need to be made to procure bamboo poles. It is estimated that the Koravas of Natham village obtain more than 80% of their cash income from the sale of *Lantana* products. Interestingly, many of these community members did not believe that the *Lantana* is an alien plant. They view *Lantana* as an NTFP, similar to bamboo.

1.2.6 The Nature and Scale of Extraction

As the real and potential value of NTFPs has been documented, it has stimulated significant work by development and government agencies to help capture the value, typically through development of or support to trade networks and markets. This is undoubtedly beneficial to local livelihoods, and perhaps conservation, but serves, yet again, to add another situation that does not conform to the previous conceptions of what is an NTFP. If the definition is restricted to subsistence and local extraction what then happens when a market is found for that particular subsistence resource? Does it cease to be an NTFP? Pragmatists would argue not, as it is the very act of valuing, and adding value to, NTFPs that heralded their initial promise as an important argument for forest conservation and improving human wellbeing. Thus, logically, the higher the value that can be realised from NTFPs, the better the cause of forest conservation will be advanced. Whilst the empirical basis for the relationship between NTFP value and consequent forest conservation can be debated (Crook and Clapp 1998; Shackleton 2001, Chapter 10), it does not negate the logic that development of trade (local or external) in what was previously a subsistence NTFP should not suddenly mean it is no longer an NTFP. It is still a valuable product from the forest additional to any high-value timber. Further complication is added when small-scale markets develop, but not by people resident close to or within the forest, but by urban migrants. For example, Stoian (2005) reported significant contributions of NTFP trade to livelihoods of the urban and peri-urban poor who were new migrants to the urban setting in Bolivia. This is not restricted to developing countries. Kilchling et al. (2009) reported that approximately 60% of urban dwellers in Switzerland visit urban and peri-urban forests weekly or more often, where they collect berries, honey, and products that are typically called NTFPs in rural or developing country contexts, as is further described in Boxes 5.1 and 5.2 in Chapter 5.

1.2.7 Only for Consumptive Uses?

A more recent dimension to debates around definitions of NTFPs is consideration of the use of species and places in forests for nonutilitarian purposes (Cocks and Wiersum 2003, Chapter 5). We highlighted above the dichotomies between subsistence use and trade, but both these imply harvest and consumption. Yet, rural communities also require species in and from natural and semi-natural lands for spiritual and traditional purposes. Some of this may require harvesting of some product from a particular species, but at times not. Individual, family, or group ceremonies or spiritual rejuvenation may occur at specific sites within a forest (e.g., Sithole 2004; Anthwal et al. 2010) (such as sacred groves in India and many countries in sub-Saharan Africa), or by a specific tree or species (Shackleton et al. 2002; Dafni 2006; Sitaramam et al. 2009), which also confers conservation benefits on these sites and species (Campbell 2005; Khan et al. 2008; Page et al. 2009). As yet there are so few studies in this regard that are designed and reported within an NTFP paradigm to require a substantial revision of the prevailing conceptions of NTFPs, but they will grow. A key issue will be that whilst the value of these sacred sites and trees to local wellbeing is undisputed, this value cannot be monetised in a manner that is being promoted for other NTFP products. Consequently, the evaluation of trade-offs and relative returns from different land uses become nigh impossible to compare.

1.2.8 Only for Local Benefit?

Combining debates around the scale of extraction and for consumptive use only, or not, essentially revolves around the increasing consensus that NTFPs are about local benefits; local benefits to wellbeing and local benefits to the environment. This is because it is only through local people capturing the benefits of NTFPs will they contribute to and complement other reasons to conserve habitats and species, which are ever being challenged by changing contexts and often vulnerable circumstances. It is here that conservationists, human geographers, economists, and ecologists seem to have a common vision. Thus, the notions of livelihoods and livelihood benefits have to take central stage in the concept and studies around NTFPs. But we must caution against totally overshadowing the conservation tenet and utility of the NTFP concept.

In focusing on local benefits we are, once again, challenged by the dynamic nature of local livelihoods and constantly changing circumstances. Thus, does an NTFP that used to be collected and used traditionally, but after time is harvested and sold to intermediaries (either by a local entrepreneur or a supportive NGO) who add value and sell it on, remain an NTFP? From a biological sense, it does because it is the same species and part. But from a forest conservation incentive and livelihood perspective it might not; it is now simply another commodity or product, the same as any other. As soon as the local benefit criterion is diminished, the local conservation incentive criterion too gets diminished, and so it would be injudicious to continue to consider it as an NTFP if it is simply the development of another commodity chain. It is on this criterion that we realise that the concept and the reality of an NTFP is not solely biological (as per Sects. 1.2.1–1.2.5) but also socioeconomic. And it is the need to merge these different disciplinary criteria that make a unified definition of NTFPs so elusive. Consequently, we argue that if the commercialisation is developed on the basis of strong local participation and benefit, then it would remain an NTFP. But if the commercialisation model was exploitation for the benefit of one or two elites or external players, at the expense of broader local access or share of the resource, then it would not be an NTFP.

1.3 Seeking a Unified Definition

What is clear from the discussion is that the concept and definition of the term NTFP has evolved and mutated. This is a function of (a) the large array of species and situations studied, (b) the exponentially increasing information and knowledge base, and (c) the growing range of disciplines involved and questions posed during research on NTFPs. As the NTFP research agenda has expanded to include increasingly diverse projects conducted in a greater variety of socio-ecological contexts, so too the very concept of NTFPs has expanded. Rather than reflecting a negative by-product of this form of research, the multitude of dimensions to the concept should be viewed as reflecting the diverse group of researchers involved, ranging from anthropologists to biologists, ecologists to economists, ethnobotanists to geographers, etc. Understandably, within a group representing such a variety of backgrounds, there emerges a wealth of perspectives and disciplinary nuances.

The chapters in this book demonstrate this diversity. However, it is possible that in the near future, a typology of NTFPs may be required. That would move the debate away from what is an NTFP (or not) to one of when and where it is and what sort, and how does one type of NTFP respond differently to another one, or require different management or policy environment. Such a typology could well be useful for developing more predictive conceptions around NTFPs rather than simply descriptions of use and value. We suggest that already there is an emerging dichotomy between research and policy examining (a) the importance and usefulness of previously neglected and underestimated biological resources (NTFPs?) in livelihoods and income generation and (b) mechanisms to promote conservation of reasonably un-impacted species, sites, and ecosystems through assigning and realising a value stream from biological resources whose extraction has relatively small impacts on the species composition, structure, and function of the site or ecosystem. The contributors to this book occupy both sides of this dichotomy and some of the grey areas in between. We have encouraged contributions in chapters and boxes that will display the multitude of definitions of NTFPs.

For the purposes of this book, we have a broader interpretation of what constitutes an NTFP than that of the widely used definition of De Beer and McDermott (1996) (especially in relation to fuelwood), but at the same time emphasise the livelihood benefits to local actors, rather than the very general "for human use" of De Beer and McDermott (although they implicitly meant local actors). The essential ingredients of this working definition include:

- Biological products (i.e., not abiotic products or ecosystem services).
- Wild species (indigenous, naturalised, or alien) which means that the bulk of the total species population is self-replicating without human agency. A small proportion of the total species population may be only recently cultivated or domesticated at a local level, or self-reproducing within human-dominated systems.
- Harvested by humans, and thus fodder consumed by free-ranging animals would be excluded (as it would be accounted for under benefits from agriculture rather than NTFPs), unless it was harvested by humans and transported to the animals to consume.
- Consumptive and nonconsumptive uses.
- Available from any landscapes or ecosystems (including human dominated).
- The broad scale management objectives are set, monitored, and regulated by those on whose land the NTFP occurs.
- Most, if not all, of the benefits from the direct or indirect use accrue to local livelihoods and wellbeing.
- The benefits accruing can act as an incentive to conserve the species or site if the necessary enabling factors and institutions are in place.

In the interim, it is important that researchers and commentators define what they mean by the use of the term when conducting their work. This is of special significance in studies that attempt to quantify amounts, values, or contributions to livelihoods of a range of NTFPs. For example, if some studies include timber and others do not, or if some authors include fodder and others do not, then the results of value to households are not comparable between these two studies. Works that include the full range of biological resources as NTFPs are likely to find higher values or higher importance indices than studies that include only a subset. Readers need to be alerted to such potential incompatibilities by being clearly informed by the authors what resources were considered NTFPs for the purposes of their study. This problem does not apply for sectoral studies on values and market chains for specific resources, e.g., only medicinal plants, honey, or fuelwood.

1.4 Purpose and Structure of the Book

In compiling this book, we sought to meet several different purposes, including:

- Providing an up-to-date critical analysis and thought-provoking perspective on the benefits and challenges associated with NTFP use and management.
- Highlighting the many complexities and disciplinary nuances of the subject.
- Taking a systems view to place NTFPs within the bigger picture with respect to both livelihoods and landscapes.
- Providing a text that would be an integrated and comprehensive resource useful to those researchers and commentators already active in the study and policy aspects of NTFPs, and also for those less familiar with the subject area to gain a thorough overview and insightful understanding of the issues, debates, and complexities pertaining to NTFPs and their management.

The book is comprised of 12 chapters. These have been broadly grouped into four sections, each of which is described below.

Part 1: Introduction – non-timber forest products in the global context.

This section contains two chapters. The first deals with discussions around what NTFPs are, why these have evolved, and how that influences the design of research and policy. The second, led by Erin Sills, provides the reader with a succinct overview of the evolution of the ideas and perceived potentials of NTFPs over the last 2–3 decades. It places the current debates in a historical context and explains how we have come to the recent much more nuanced view.

Part 2: Multiple roles and values of non-timber forest products.

Having set the scene in Part 1, Part 2 follows on with four chapters that together provide an overview of the different roles and values of NTFPs in terms of contributing to human wellbeing. Chapter 3, led by Sheona Shackleton, grapples with the key debate on whether the use of NTFPs can make meaningful contributions to local livelihoods, and thus offer potential for poverty alleviation. This covers the direct-use values, safety net functions, and trade in NTFPs, drawing on case studies from around the world. They conclude that NTFPs are not going to be a panacea for poverty alleviation on a large scale, but in some contexts and circumstances a measurable proportion of households can be lifted out of poverty, or prevented from slipping deeper into poverty, due to the availability and use of NTFPs. In Chapter 4, Tony Cunningham examines examples of scaling up of these local benefits to wider scales and international markets. He demonstrates that the number of people and cash flows involved are huge, although aspects of equity are rarely reported. In moving beyond the local rural areas, one cannot omit the use and marketing of NTFPs in urban areas, which can be significant. The role of cultural norms and beliefs in sustaining demand for NTFPs even in rapidly developing and modernising societies is considered in Chapter 5 by Michelle Cocks and Citlalli Lopez. Examining case studies of traditional brooms in South Africa and traditional bark paper in Mexico, it becomes clear that the cultural dimensions of NTFP demand and debates have been seriously neglected, and moreover, that cultural demand evolves and transforms to take advantage of changing opportunities in a transforming world. The urban and cultural dimensions are considered further by Nathalie van Vliet and co-authors in Chapter 6, using a case study of bushmeat in central and western Africa. Not only are the volumes large, they also argue that demand in urban areas is likely to remain significant due to a host of cultural considerations, economic differentials, and local preferences. In other words, NTFPs are not just small-scale harvests for local consumption in remote rural areas.

Part 3: Systems for sustainable management of non-timber forest products.

This section also comprises of four chapters which seek to summarise debates and offer new perspectives on the tricky aspect of managing NTFPs to capture the values and benefits described in Part 2 whilst simultaneously conserving the habitats and systems in which they are found. In Chapter 7, Tamara Ticktin and Charlie Shackleton start with questioning if, and how, NTFPs can be ecologically sustainably harvested, and what characterises those systems where ecological sustainability has been achieved relative to those where it has not. Unfortunately, however, many empirical studies fail to include all the dimensions, scales, and variables necessary to draw confident conclusions, but examples of both can be found. Thereafter, Manuel Guariguata and colleagues explore in Chapter 8 the old, but still very pressing, debate around whether, and under what circumstances, harvesting systems for high-value tropical timber can be compatible with NTFP systems within the same forest. Although this is possible, there are not many real examples on the ground. This is due to a number of constraints, notable ones of which relate to tenure, appropriate training of forestry practitioners and officials, and the weak knowledge base on the biology and harvesting levels and responses of all but a few NTFP species.

Given the importance of tenure, Mirjam Ros-Tonen and Koen Kusters examine it as a key aspect of governance systems (along with regulations, markets, and partnerships), in more detail in Chapter 9. Governance aspects are vital not only in securing access and rights to use specific NTFPs but also in channelling the benefits to local communities and preventing a deepening of poverty. Consequently, governance covers both human and ecological facets and relates to justice, equity, power relations, and wellbeing. They conclude that the building of partnerships is a critical starting point in developing appropriate governance mechanisms and institutions. In rounding off Part 3, Terry Sunderland and co-authors interrogate (in Chapter 10) whether the rights of people and the rights of the forest biodiversity can be matched. Whilst the value of NTFPs is beyond doubt, does realisation of that value promote forest and species conservation? Drawing of examples principally from tropical forests, they conclude that the two have been rarely met. Rather, due to a number of largely management and economic factors (but also tenure), they conclude that the most common outcome is for harvesters to extract as much value in as short a time as possible, which is usually to the detriment of the species or local habitat. The Holy Grail of balancing the wellbeing of local communities and the forest has yet to be attained, and perhaps due to the nature of most NTFPs systems, is unlikely to be?

Part 4: Building on the opportunities offered by non-timber forest products.

This concluding section has two chapters that attempt to draw from the preceding ones and look towards what next. In Chapter 11, Sarah Laird and colleagues probe the policy issues and what needs to be done to secure the rights to and benefits from NTFPs for poor rural communities. They argue that in most instances, policies around NTFPs are ill informed and ill conceived based on insufficient information or failing to make use of available information. What is more, there is a plethora of policies ostensibly designed for other sectors, but which have major impacts on the sustainability of NTFP harvesting or trade. They conclude with several well-argued policy issues that would be a useful starting point for those operating in the policy and regulation sector, but recognising the context-specific nature of many policy requirements.

In the final chapter (Chapter 12), the editors have grappled with the unenviable task of bringing it all together. To that end, we have recapped the major discussions and debates in the preceding chapters, so as to highlight the complexity of NTFP use and management and to ensure an integrated understanding. We also recognise the rapidly changing world we live in and consider what this means for forests and NTFPs. Lastly we consider some of what we perceive to be the key areas that need investigation or resolution over the coming decade and highlight what we believe to be the major "take home" messages.

References

Aguilar-Støen M, Moe SR, Camargo-Ricalde SL (2009) Home gardens sustain crop diversity and improve farm resilience in Candelaria Loxicha, Oaxaca, Mexico. Hum Ecol 37:55–77

Anthwal A, Gupta N, Sharma A, Anthwal S, Kim K-H (2010) Conserving biodiversity through traditional beliefs in sacred groves in Uttarakhand Himalaya, India. Resour Conserv Recycl 54:962–971

Belcher BM (2003) Comment: what isn't an NTFP? Int Forest Rev 5:161–168

Budiadi KY, Ishii HT, Sabarnurdin MS, Suryanto P (2005) Productivity of kayu putih (Melaleuca leucadendron LINN) tree plantation managed in non-timber forest production systems in Java, Indonesia. Agrofor Syst 64:143–155

Campbell MO (2005) Sacred groves for forest conservation in Ghana's coastal savannas: assessing ecological and social dimensions. Singap J Trop Geogr 26:151–169

Cavendish W (2000) Empirical regularities in the poverty-environment relationship of rural households: evidence from Zimbabwe. World Dev 28:1979–2003

Clason AJ, Lindgren PMF, Sullivan TP (2008) Comparison of potential non-timber forest products in intensively managed young stands and mature/old-growth forests in south-central British Columbia. For Ecol Manage 256:1897–1909

Cocks ML, Wiersum KF (2003) The significance of plant diversity to rural households in Eastern Cape Province of South Africa. For Trees Liveli 13:39–58

Crook C, Clapp RA (1998) Is market-orientated forest conservation a contradiction in terms? Environ Conserv 25:131–145

Dafni A (2006) On the typology and the worship status of sacred trees with a special reference to the Middle East. J Ethnobiol Ethnomed 2:26 [online: http://www.ethnobiomed.com/content/2/1/26]

De Beer JH, McDermott M (1989) The economic value of non-timber forest products in South East Asia. The Netherlands Committee for IUCN, Amsterdam
De Beer JH, McDermott M (1996) The economic value of non-timber forest products in Southeast Asia. Netherlands Committee for IUCN, Amsterdam
De Neergaard A, Saarnak C, Hill T, Khanyile M, Martinez Berzosa A, Birch-Thomsen T (2005) Australian wattle species in the Drakensberg region of South Africa: an invasive alien or a natural resource? Agric Syst 85:216–233
FAO (1999) Towards a harmonised definition for non-wood forest products. Unasylva 50 (198):63–64
Geesing D, Al-Khawlani M, Abba ML (2004) Management of introduced *Prosopis* species: can economic exploitation control an invasive species? Unasylva 55:36–44
Gonźalez-Amaro RM, Martinez-Bernal A, Basurto-Peña F, Vibrans H (2009) Crop and non-crop productivity in a traditional maize agroecosystem of the highland of Mexico. J Ethnobiol Ethnomed 5: online [DOI: 10.1186/1746-4269-5-35]
Gram S (2001) Economic valuation of special forest products: an assessment of methodological shortcomings. Ecol Econ 36:109–117
High C, Shackleton CM (2000) The comparative value of wild and domestic plants in home gardens of a South African rural village. Agrofor Syst 48:141–156
Kasolo WK, Temu AB (2008) Tree species selection for buffer zone agroforestry: the case of Budongo Forest in Uganda. Int Forest Rev 10:52–64
Khan ML, Khumbongmayum AD, Tripathi RS (2008) The sacred groves and their significance in conserving biodiversity: an overview. Int J Ecol Environ Sci 34:277–291
Kilchling P, Hansmann R, Seeland K (2009) Demand for non-timber forest products: surveys of urban consumers and sellers in Switzerland. For Policy Econ 11:294–300
McCrary JK, Walsh B, Hammett AL (2005) Species, sources, seasonality and sustainability of fuelwood commercialisation in Masaya, Nicaragua. For Ecol Manage 205:299–309
Neumann RP, Hirsch E (2000) Commercialisation of non-timber forest products: review and analysis of research. CIFOR, Bogor. 176 pp
Page NV, Quereshi Q, Rawat GS, Kushalappa CG (2009) Plant diversity of sacred forest fragments of Western Ghats: a comparative study of four life forms. Plant Ecol 206:237–250
Pant P (2007) Bamboo for life ...... for people and red panda. A Study on utilitarian value and management of bamboo and its relation with red panda conservation in Maimjhuwa VDC, Ilam, Nepal. MSc Thesis, Wageningen University and Research Center, Wageningen
Parry L, Barlow J, Peres CA (2009) Allocation of hunting effort by Amazonian smallholders: implications for conserving wildlife in mixed-use landscapes. Biol Conserv 142:1777–1786
Phuthego TC, Chanda R (2004) Traditional ecological knowledge and community-based natural resource management: lessons from a Botswana wildlife management area. Appl Geogr 24:57–76
Plotkin MJ, Famolare L (1992) Sustainable harvest and marketing of rain forest products. Island, Washington, DC
Porembski S, Biedinger N (2001) Epiphytic ferns for sale: influence of commercial plant collection on the frequency of *Platycerium stemaria* (Polypodiaceae) in coconut plantations on the southeastern Ivory Coast. Plant Biol 3:72–76
Putz FE, Redford KH (2010) The importance of defining 'forest': tropical forest degradation, deforestation, long-term phase shifts, and further transitions. Biotropica 42:10–20
Rajchal R (2008) Analytical review of the definitions of non-timber forest product. In: Hussain A (ed) Managing non-timber forest products: global perspectives. ICFAI University Press, Hyderabad, pp 3–17
Sahoo UK (2009) Traditional home gardens and livelihood security in North-East India. J Food Agric Environ 7:665–670
Schrekenberg K (1999) Products of a managed landscape: non-timber forest products in the parklands of Bassila region, Benin. Glob Ecol Biogeogr 8:279–289

Shackleton CM (2001) Re-examining local and market orientated use of wild species for the conservation of biodiversity. Environ Conserv 28:70–278
Shackleton CM (2009) Will the real custodian of natural resource management please stand up? S Afr J Sci 105:91–93
Shackleton CM, McGarry D, Fourie S, Gambiza J, Shackleton SE, Fabricius C (2007) Assessing the effects of invasive alien species on rural livelihoods: case examples and a framework from South Africa. Hum Ecol 35:113–127
Shackleton SE, Shackleton CM, Cunningham AB, Lombard C, Sullivan CA, Netshiluvhi TR (2002) Knowledge on *Sclerocarya birrea* subsp. *caffra* with emphasis on its importance as a non-timber forest product in South and southern Africa: a summary. Part 1: taxonomy, ecology and role in rural livelihoods. S Afr For J 194:27–42
Sitaramam V, Jog SR, Tetaki P (2009) Ecology of *Ficus religiosa* accounts for its association with religion. Curr Sci 97:637–640
Sithole B (2004) Sacred groves on dambos in Zimbabwe: are these private resources for rural elites? For Trees Liveli 14:121–136
Stoian D (2005) Making the best of two worlds: rural and peri-urban livelihood options sustained by non timber forest products from the Bolivian Amazon. World Dev 33:1473–1490
Struhsaker TT (1987) Forestry issues and conservation in Uganda. Biol Conserv 39:95–99
Thomas MG, Schumann DR (1994) Income opportunities in special forest products: self-help suggestions for rural entrepreneurs. DIANE, USA Dept of Agriculture, Washington, DC
Ticktin T, Nantel P (2004) Dynamics of harvested populations of the tropical understory herb *Aechmea magdalenae* in old-growth versus secondary forests. Biol Conserv 120:461–470
Viisteensaari J, Johansson S, Kaarakka V, Luukkanen O (2000) Is the alien tree species *Maesopsis eminii* Engl. (Rhamnaceae) a threat to tropical forest conservation in the East Usambaras, Tanzania? Environ Conserv 27:76–81
Wiersum KF (1997) Indigenous exploitation and management of tropical forest resources: an evolutionary continuum in forest-people interactions. Agric Ecosyst Environ 63:1–16

Chapter 2
Evolving Perspectives on Non-timber Forest Products

Erin Sills, Patricia Shanley, Fiona Paumgarten, Jenne de Beer, and Alan Pierce

Abstract Many individual non-timber forest products (NTFPs) were historically mainstream trade commodities, but their diminished importance in international trade after World War II meant that they become almost invisible in forest statistics, management, and policy. They were rediscovered as a category in the late 1980s, provoking high hopes by many, suspicion by some, and a new research agenda on their potential role in the sustainable development of tropical forest regions. This was followed by general disenchantment with NTFPs that dominated the literature and policy discussion at the turn of the century, which in turn gave way to today's more nuanced understanding and policy recommendations, as described in many chapters of this book. We identify four themes in recent literature that serve as guideposts to a realistic and moderate assessment of NTFPs (1) centrality of culture

E. Sills (✉)
Department of Forestry and Environmental Resources, North Carolina State University, Raleigh, NC 27695-8008, USA
and
Center for International Forestry Research, Bogor, Indonesia
e-mail: sills@ncsu.edu

P. Shanley
Center for International Forestry Research, Bogor, Indonesia
e-mail: p.shanley@cgiar.org

F. Paumgarten
School of Animal, Plant and Environmental Sciences, University of the Witwatersrand, Johannesburg, South Africa
and
Center for International Forestry Research, Lusaka, Zambia
e-mail: fi.paumgarten@gmail.com

J. de Beer
Darrell Posey Field Fellow, International Society for Ethnobiology, Manila, Philippines
e-mail: ntfp7@yahoo.co.uk

A. Pierce
Independent Researcher, 1061 Mountainview Road, Duxbury, VT, USA
e-mail: arp@sover.net

S. Shackleton et al. (eds.), *Non-Timber Forest Products in the Global Context*, Tropical Forestry 7, DOI 10.1007/978-3-642-17983-9_2,

and tradition, (2) local and regional markets, (3) value of diversity in and of itself, and (4) continuum of forest management.

2.1 Introduction

Over the past quarter century, the dominant narrative about non-timber forest products (NTFPs) swung from optimism to pessimism about their potential to alleviate poverty and encourage conservation. In this chapter, we first provide historical context, then describe the motivations, assumptions, and operating principles of the optimists and pessimists. Out of this debate, there is emerging a new middle ground of research and policy that focuses on NTFPs that are grounded in cultural traditions, that are traded in local and regional markets, and that are managed in subtle ways across a spectrum of forest types. These NTFPs make up a diverse basket of products that insure and enhance the quality of life of forest users.

2.2 History

Although not always termed NTFPs, such products have been used and traded for centuries. Consequently, their roles and importance in trade and societies have varied through time from key commodities during periods of early colonial conquest to secondary or minor resources, and once again more recently back in the international spotlight.

2.2.1 Mainstream

Historically, many NTFPs were key global commodities and an important component of international trade, driving the fabled spice trade between Asia and Europe, expanding in the colonial period with products such as shea butter (*Vitellaria paradoxa*) and gum Arabic (*Acacia* spp.) from Africa, and feeding the industrial revolution with products such as rubber from the Amazon (*Heavea brasilenses*). The economic importance and often exploitative nature of the international trade in NTFPs are amply documented in case studies of particular products [e.g., Weinstein (1983) on rubber in the Amazon, Hanson (1992) on gum Arabic in West Africa, Peluso (1992) on rattan in Indonesia] and in the history literature (Wolters 1967; Turner and Loewen 1998; Donkin 2003).

2.2.2 Invisible

After World War II, the relative importance of NTFPs in international trade declined, as exports of tropical timber increased and advances in "inorganic, and especially petroleum-based, chemistry led to the replacement of forest products such as gums, resins, fibers, and medicines by cheaper synthetic alternatives", incentivised in part by disruption of supplies during the war (Alexiades and Shanley 2004). The decline of NTFPs in international trade was paralleled by their disappearance from the international forest policy agenda. For example, the summary of the first World Forestry Congress in 1926 made several references to "forest products other than wood" such as barks, resins, saps, and leaves, but by the seventh World Forestry Congress, the summary made just brief reference to "the social potential of the rather neglected section of minor forest products". According to Padovani (1995), the FAO stopped collecting and publishing data on NTFPs in 1971. A major report on *Tropical Forest Resources* produced by FAO and UNEP in 1982 focused almost exclusively on timber and fuelwood. As described in Box 2.1, this reflected global concern about a "fuelwood crisis", which temporarily drew international attention to fuelwood supply, in the same way that international

Box 2.1 Evolving Perspectives on Fuelwood

Priscilla Cooke St Clair, Dept of Economics, Pacific Lutheran University, Tacoma, WA 98447, stclaipa@plu.edu

Over the last 40 years the prevailing view on fuelwood has fluctuated dramatically. In the 1970s, as rising fossil fuel prices focused attention on energy, it was widely noted that fuelwood was the predominant household fuel for most of the developing world. When initial estimates of future fuelwood supply (based on forest growth) and future demand (based on population growth) indicated a growing gap between supply and demand, massive deforestation and declining welfare for fuelwood-dependent households were envisioned. This became referred to as the "the other energy crisis" or the "fuelwood gap". In response to this perceived crisis, forestry programs to increase fuelwood supply and improved stove programs to encourage efficient fuelwood use increased dramatically (Cooke et al. 2008).

By the mid 1980s, however, it became evident that many fuelwood oriented programs were not meeting expectations. Additionally, new household-level research indicated that fuelwood generally came from easily regenerating twigs and woody scrub, that scarcity of fuelwood could be driven by labour shortages even when forest resources were abundant, and that households responded rationally to economic scarcity of fuelwood by both conserving fuelwood and switching to substitutes (Arnold et al. 2006). The view that prevailed in the 1990s was that fuelwood use was not a major cause of deforestation, and that most households did not see fuelwood scarcity as a big problem. Fuelwood-related programs were sharply cut back.

In the 2000s attention began to turn again to fuelwood, which is still the predominant fuel for rural households in much of Africa and South Asia, even though its use has declined in some areas due to urbanisation and income growth. It is now recognised that in some circumstances fuelwood scarcity can have very adverse consequences for household welfare (Arnold et al. 2003; Cooke et al.

(continued)

2008). In addition, attention has been drawn to fuelwood due to its connection to climate change. As a renewable, nonfossil fuel, fuelwood has links both to energy policy and carbon sequestration programs. It remains to be seen how this will play out in policy, and how policy will impact fuelwood users.

attention would later be temporarily focused on commercialisation of nonwood forest products (NWFPs).

Throughout this period, researchers continued to generate case studies of specific NTFPs, including their ecology, harvest, processing, and trade (Robbins and Matthews 1974). This literature is well represented in the journal *Economic Botany* that was launched in 1947. However, as noted by Tewari and Campbell (1995), "botanists and anthropologists usually confined their interest to descriptions of the variety and local uses of long lists of species, without discussing management options or economic value". Likewise, some Tropical Forestry Action Plans (TFAPs) made note of specific NTFPs, but Flint (1990) concluded that "even where nonwood products are considered [in TFAPs], they tend to be viewed in isolation, and the social and economic effects of, for example, increased logging or conservation on nonwood livelihoods are rarely considered".

To the extent that NTFPs were considered as a class of products or activities, they were likely to be seen through the lens of "the tragedy of the commons" (Hardin 1968). This also characterised the approach of many governments throughout the tropics, who claimed forest areas that they perceived as open access and underutilised, in order to exploit the timber resources or "develop" the land (Lynch and Talbott 1995). Partly as a result of these policies, ongoing extraction of certain NTFPs in some places was undermined by degradation of the resource base (de Beer and McDermott 1996). However, as a general category, NTFPs remained central to the livelihoods of rural peoples, both utilised directly and actively traded in local and regional markets (Fig. 2.1).

2.2.3 Rediscovery

Between the 1987 publication of the Brundtland Report and the 1992 UN Conference on Conservation and Development, there was an explosion of interest in NTFPs. The scientific groundwork for this was laid by studies demonstrating the importance of NTFPs to rural peoples throughout the tropics, including India (Jodha 1986), the Amazon (Padoch 1988; Anderson and Jardim 1989), and Indonesia (Peluso 1983; Caldecott 1988a, b). New labels for this category of goods, including "non-timber forest products", were introduced to the literature (Jacobs 1984). The International Tropical Timber Organisation (ITTO) commissioned a study on the multiple-use of tropical forests that would later be published as the influential book *Not by Timber Alone* (Panayotou and Ashton 1992). Terms that would become integral to the discussion were given prominence (if not coined)

Fig. 2.1 House and fishing tools made of bamboo and rattan, Northern Lao PDR (photo: Claudio Delang)

by Myers (1988) who wrote about “nonwood products”; Hecht et al. (1988) who described NTFPs as a “subsidy from nature”; and the Brundtland Report itself that made “sustainable development” into a widely accepted goal. This was driven home by the widely publicised economic analysis of Peters et al. (1989).

This new interest in NTFPs dovetailed with several other trends in the same time period. First, common property regimes (CPRs) and community-based natural resource management (CBNRM) were garnering new interest and respect from researchers and international donors (Bromley and Cernea 1989; Ostrom 1990; Poffenberger 1990; Menzies 2004). Second, there was increasing recognition of the potential value of traditional ecological knowledge (TEK) for understanding and managing ecosystems (Berkes 1993). Third, environmental organisations, such as the World Wildlife Fund under its Wildlands and Human Needs Program, were seeking to integrate development into their conservation programs for protected areas. Fourth, the international press gave increasing attention to the issue of tropical deforestation, partly motivated by the massive forest fires in Indonesia in 1986 and the Brazilian Amazon in 1987.

In this period, the timber industry was widely blamed for a “deforestation crisis”, and there were many calls for boycotts of tropical timber imports. The concept of NTFPs as an alternative means of earning a livelihood from the forest was brought to prominence by the rubber-tappers’ movement in the Brazilian state of Acre. With support from anthropologists Mary Allegretti and Steve Schwartzman, the movement’s leader, Chico Mendes, gained national and international attention, founding the National Council of Rubber Tappers in 1985 and visiting Washington DC to testify to the US Congress in 1987. Mendes’ assassination in 1988 made international headlines and became a rallying point for advocates of the resource access rights of traditional forest-dependent people, specifically for collection of NTFPs.

In this setting, two 1989 publications crystallised a new paradigm for "sustainable development" of tropical forests. De Beer and McDermott published the first edition of their book on the economic value of NTFPs in Asia in 1989. This study is widely credited with establishing "non-timber forest products" as a category. The authors argued that "the key point of distinction between these materials and timber is that the latter is managed on an industrial scale by and for interests located well outside forest boundaries. While certain non-timber forest products may ultimately become inputs in to large-scale urban-based industries, all of them share the characteristic that they are extracted using simple technologies by rural people living in or near forests" (De Beer and McDermott 1996: 24). In the same year, Peters et al. (1989) published a two-page commentary in *Nature*, which was reported on the front page of newspapers such as the *Washington Post* and has since been cited over 600 times (according to Google Scholar, or 275 times according to Web of Science, by the end of 2010). They estimated the total market value of all products that could be harvested from a hectare of forest in the Peruvian Amazon and presented this as a compelling economic argument for an alternative to timber logging and deforestation.

These publications coincided with and contributed to great interest in NTFPs from tropical forests. In December of 1989, the National Wildlife Foundation convened a conference in the US on "Extractive Economies in Tropical Forests: A Course for Action", which led to an edited volume by Nepstad and Schwartzman (1992). This was soon followed by other international conferences, such as "Rainforest Harvest: Sustainable Strategies for Saving the Tropical Forests?" convened in the UK in 1990 by Friends of the Earth (Counsell and Rice 1992) and "The Sustainable Harvest and Marketing of Rain Forest Products" convened in Panama in 1991 by Conservation International and the Asociación Nacional para la Conservación de la Naturaliza (Plotkin and Famolare 1992). The prefaces to these conference proceedings lay out the case for NTFPs: "The ancient practice of extracting economically valuable, non-timber forest products (NTFPs) leaving the forests structurally and functionally intact, has emerged as a possible means of reconciling the conflicting roles of tropical forests" (Nepstad and Schwartzman 1992); "Unlike wood, non-timber forest products (fruits, fibers, medicines, and so forth) can often be harvested without any damage to the ecosystem" (Plotkin and Famolare 1992); "Governments, scientists and environmentalists now generally regard 'extractive management' of tropical forests as a realistic and economically feasible alternative to conventional logging and clearance" (Counsell and Rice 1992).

These conferences reflected a strong geographic focus of both research and political action on South America and Southeast Asia (cf. Neumann and Hirsch 2000: 11). The promised "win–win" of marketing NTFPs seemed most compelling in these regions, where both biodiversity and traditional peoples were threatened by rapid loss of vast rainforests. Attention focused on the Brazilian Amazon in particular, due in part to the symbolic rallying point of Chico Mendes' assassination, to early political success with the declaration of extractive reserves, and to high-profile marketing campaigns by the Body Shop and Cultural Survival Enterprises. Perhaps because of the recent boom in exports (de Beer and McDermott

1996), rattan from Kalimantan and other parts of Southeast Asia also figured prominently in the discussions.

This rediscovery of NTFPs did not make as big a splash in Africa. Neither the popular nor the scientific press significantly increased coverage of African NTFPs in this period, with a few exceptions such as the well-known ODI study on NTFPs in Ghana (Falconer and Koppel 1990). Perhaps the most obvious explanation is that this time period was marked by numerous other crises in Africa, including famine, civil war, and HIV/AIDS, all of which pushed forestry and the environment down the priority list. Another contributing factor is that many of the best-known products in this region are from dry forests or anthropogenic landscapes, collected by people as part of complex livelihood portfolios, as reflected in the Hidden Harvest Project (Guijit et al. 1995; Campbell 1996) which was in contrast to the people identified primarily as forest extractors in the perceived "pristine" rainforests of the Amazon and Kalimantan. A third explanation is that two of the best-known NTFPs in this region, bushmeat and fuelwood, are not as obviously appealing as nonwood plant products from an environmental sustainability perspective. Initiatives to recreate a local stake in sustainable management of wildlife resources, such as CAMPFIRE in Zimbabwe, drew on many of the same concepts as NTFPs (TEK, CBNRM, CPRs), but generally proceeded on a parallel track.

2.3 Reactions to the Renewed Profile of NTFPs

The explosion of interest in NTFPs provoked varied reactions that played out during the 1990s. First and best known, nongovernmental organisations and multi-lateral agencies moved quickly to establish programs to support the commercialisation of NTFPs. Second, there were early skeptics of both the sustainability and the development potential of NTFPs. Third, research organisations took up the challenge of assessing the potential and the necessary conditions for promoting "productive conservation" of forests via NTFP markets.

2.3.1 Optimism

In the early 1990s, efforts to develop the harvesting, processing, and international marketing of NTFPs were pursued with almost Pollyanna enthusiasm by nongovernmental organisations, donors, governments, and multi-lateral agencies, all animated by the potential for a win–win strategy to conserve forests while improving local welfare. This enthusiasm was also reflected in the creation of new programs focused on NTFPs, such as the FAO's "Promotion and Development of Nonwood Forest Products" established under the Forest Products Division in 1991 and similar programs at international environmental NGOs including Conservation International, Rainforest Alliance, and Friends of the Earth (Hidalgo 1992; Ehringhaus 2006).

These efforts were justified by a number of oft-repeated stylised facts about NTFPs. First was the assumption of negligible environmental impact of NTFP collection (e.g., Godoy and Feaw 1989; Sayer 1991). Second, as pointed out by De Beer and McDermott (1996), there was already a significant international trade in NTFPs (e.g., 150 products with total estimated value of USD 1.1 billion according to FAO 1997). Third, the total value of NTFPs in national economies was widely believed to be vastly undercounted in official statistics (Haripriya 2001; Puustjarvi et al. 2005; Hecht 2007), suggesting enormous potential to develop and bring these products into the formal economy. Fourth, there was optimism that new products and new uses of NTFPs would be developed (Unasylva 190–191). Combined, these beliefs suggested that there was a vast untapped opportunity to bring more NTFPs onto international markets. The international marketing strategy was further justified by concerns that local markets for unprocessed NTFPs are easily flooded and therefore offer low and unstable prices and limited incentive for sustainable use (Pendleton 1992; Padoch and Pinedo-Vasquez 1996).

In these early efforts, there continued to be a strong focus on the Amazon. As described by Coomes and Barham (1997), international NGOs entered the region "*en masse* to work with local communities in implementing a wide range of initiatives that promise, by supporting forest people's traditional life and livelihood practices, to conserve the rain forest and promote locally-led development efforts. ... Once considered the antithesis of forest preservation, forest product extraction by traditional groups has come to represent a major focus of hope and action for groups working in Amazonia". One of the best-known efforts was Cultural Survival Enterprises (CSE), which sought to develop new products and new markets for NTFPs. In the 2 years after its launch in 1989, CSE convinced private businesses to place initial orders for 25 different NTFPs. However, they quickly ran into supply problems and sharp criticism over "rainforest crunch", their most heavily promoted product from the Amazon (Gray 1990; Hanson 1992; Corry 1993).

While rainforest crunch was the most widely publicised failure, many of the early NTFP commercialisation initiatives struggled to make good on this seemingly obvious sustainable development strategy. This was partly due to a lack of understanding of existing commercialisation systems, and the challenges inherent to bringing change to those systems, such as lack of local experience with marketing, market instability, and the difficulty of building institutional relationships (1998 review of DFID projects, cited in Hughes and Flintan 2001). But projects also failed to take into account the broader context, including the heterogeneous and evolving livelihood strategies employed by local people, the multifaceted services provided by traders and middle-men, and the mix of local and external deforestation threats (Hughes and Flintan 2001; Coomes and Barham 1997). Some organisations responded with longer term and more multidimensional interventions (e.g., members and grantees of Biodiversity Conservation Network and NTFP Exchange Program). But regardless, these initial challenges both lent credence to the early skeptics and helped set the tenor of major research programs on NTFPs.

2.3.2 *Skepticism*

Early criticism of NTFPs as a basis for sustainable development flowed broadly in two veins. First, a number of anthropologists argued that linking forest people (and particularly indigenous people) to international markets had never done any good and was fraught with danger for those peoples. They further argued that the siren call of the "rainforest harvest" was drawing attention away from the real deforestation threats (Gray 1990; Hanson 1992; Corry 1993; Dove 1994). The second line of argument was that the economic basis of extractivism (Homma 1992) and of extractive reserves in particular (Browder 1992; Salafsky et al. 1993) was flawed. Based on his 1988 dissertation, Homma emphasised the historical regularity of market busts due to overexploitation, domestication, or substitution of NTFPs. Browder (1992) pointed out the fragility and narrow grounds for the apparent convergence of interests between the residents of extractive reserves and distant environmentalists. Two substrands of this critical literature emphasised the lack of data on the environmental sustainability of NTFP harvest (e.g., Redford 1992 on hunting by extractive populations) and the methodological flaws in the Peters et al. (1989) study, the strong assumptions of which continue to attract comment and criticism (Browder 1992; Cavendish 2000; Pyhälä et al. 2006).

2.3.3 *Research*

The quickly polarised and polemic debate over NTFPs proved fertile ground for researchers, as reflected in both the gray literature (e.g., global expert consultations on NWFPs organised by FAO in Tanzania in 1993, Thailand in 1994, and Indonesia in 1995) and scientific publications. Prior to 1994, only 33 publications in the Science Citation Index mentioned non-timber or NWFPs in the title or keywords, but for the 5 years from 1994 to 1998, the Index lists 111 publications on these topics (see Box 12.1). This was partly a change in labeling, with the terms NTFP or NWFP now attached to studies of particular products or production systems. These case studies employed diverse methods to examine diverse issues (Townson 1995). However, there were also efforts to systematically identify key research questions, recommend consistent methods, and synthesise knowledge.

Several of these efforts were grounded in economics. For example, in 1992, the Smithsonian Institution and the Harvard Institute for International Development convened a workshop that proposed a series of hypotheses to guide research on the role of NTFPs in local economies, published in a widely cited special issue of *Economic Botany* (volume 47). This workshop also gave greater prominence to

NTFPs in South Asia. In the same time period, the Hidden Harvest Project highlighted the role of NTFPs in agricultural systems and rural livelihoods in Africa, with particular attention given to subsistence foods (Scoones et al. 1992) and to the combination of participatory and nonmarket valuation methods to quantify local values of NTFPs (Campbell and Luckert 2002). Both of these efforts encouraged researchers to bring economic methods, including the household production framework and nonmarket valuation methods, to bear on NTFPs (as later reviewed by Wollenberg and Nawir 1998; Tewari 2000; Sills et al. 2003; Vedeld et al. 2004).

The impact of commercialisation on sustainability also emerged as a key research theme. Peters' (1994) "ecological primer" encouraged research on the ecological implications and management guidelines for NTFP harvest, as later reviewed by Wong (2000) and Ticktin (2004). This research found that sometimes commercial extraction of NTFPs was in fact not sustainable (e.g., Peres et al. 2003 on Brazil nuts). In 1995 and 1996, CIFOR hosted several workshops (in Zimbabwe, Spain, and Indonesia) that recommended focusing future research on the impacts of commercialisation on smallholder NTFP use, through systematic reviews of the literature and reporting of case study results (Ruiz Perez and Byron 1999). This led to a literature review by Neumann and Hirsch (2000), comparative case study research (Belcher and Ruiz Perez 2001; Ruiz Perez et al. 2004; Belcher et al. 2005), and examination of the potential role of certification (Shanley et al. 2005). The TROPENBOS Foundation pursued a similar line of research, focusing on identifying patterns and testing key hypotheses regarding commercialisation and sustainability of production (Ros-Tonen et al. 1995). DFID funded a third comparative research project, launched in 2000, to identify characteristics associated with successful NTFP commercialisation in Mexico and Bolivia (Marshall et al. 2006).

In the 15 years since this research agenda started to take shape, the literature has vastly expanded: the Science Citation Index lists 200 publications on non-timber or nonwood products from 1999 to 2003, 335 publications from 2004 to 2008, and 80 in 2009 alone. While the results of this research have always been nuanced and varied across products and sites, the predominant spin on the interpretation and discussion of those results has evolved over time.

2.4 Pessimism

By the turn of the century, enthusiasm over NTFPs had been deflated, as the complexity and constraints on increased commercialisation became more evident. As described by Ros-Tonen (2003), "the picture at the start of the new century is one in which optimism regarding the potential of NTFP extraction as a combined strategy for conservation of natural forests and poverty alleviation has waned, to be replaced with a more cautious approach or even forthright pessimism".

This reality check coincided with and was mutually reinforced by other trends in conservation and development, most notably disenchantment with attempts at integration of these two goals. Integrated Conservation and Development programs (ICDPs) were increasingly criticised for failing to achieve either conservation or development outcomes (Wells et al. 1998; Hughes and Flintan 2001). There was a "resurgent protectionist argument" in favour of strict protected areas (Schwartzman et al. 2001; Wilshusen et al. 2002), and "green consumerism" was starting to lose ground to direct conservation payments, or payments for ecosystem services (Hardner and Rice 2002).

At the same time, there was a renewed focus on poverty alleviation (Arnold 2001; Maxwell 2001), with the Millennium Development Goals (adopted by UN member states in 2001) setting a new analytical framework for governments, donors, and researchers, including in forestry (Wunder 2001). While this could have created an opening for NTFPs as a key resource for the rural poor, there remained insufficient data on NTFPs to meet the standards of "evidence-based policy-making". For example, participants in a workshop on Poverty Reduction Strategy Papers (PRSPs) in Africa identified both lack of statistics on NTFPs and poor dissemination and advocacy by the forestry sector as reasons that NTFPs are generally not considered in those plans (Paumgarten 2009). Further, there was growing suspicion that forests might contribute to rather than alleviate poverty, as evidenced by spatial convergence of tropical forests with areas of chronic poverty (Sunderlin et al. 2007) and the history of boom–bust economic cycles in NTFPs. International forestry assistance in general was becoming highly politicised, and perceived as "a no-win zone for donors" (Street 2006, ETFRN NEWS 47/48, Economist 13 March 2003). This retreat from conservation was reinforced by the ascendance of other policy and aid agendas, including terrorism and public health.

In this context, it was easy to put a negative spin on the often-mixed outcomes of NTFP implementation projects and complex findings of NTFP research, focusing on the failure of NTFP commercialisation to "lift" people out of poverty. This new pessimism framed the academic and policy discussion, as illustrated in an introduction to a special issue of the *International Forestry Review* (IFR) that criticised earlier "exaggerated claims of economic potential [that] tended to overlook the great diversity of products referred to, in terms of biological characteristics, and social and economic value, whilst simultaneously ascribing unreasonably lofty and altruistic goals to some of the world's poorest people". The quintessential myth-busting refrain (cf. Spilsbury and Nasi 2006) became that "NTFPs are not a silver bullet". This assessment was underpinned by a new set of stylised facts, replacing those of the previous decade and labeling NTFPs as inferior, substitutable, and unmanageable.

While claims that NTFPs are inferior are based on a variety of product characteristics (e.g., perishability, seasonality, etc.), the economic meaning of the term is that, all else equal, when incomes rise, demand falls. This holds true for some products such as wild foods that are not very palatable and natural materials that are not very durable. This can be reinforced by phytosanitary policies not adapted to and therefore imposing excessive costs on the NTFP trade. There is more evidence that the *share* of income spent on NTFPs falls as incomes increase. For example, Cavendish (2000) found income elasticities between 0.3 and 0.5, and concluded that "dependence on and use of NTFPs is linked to poverty and to market failure rather than to household choice; the current prevalence of NTFP use by rural households is a result of their low incomes rather than the attraction of NTFPs themselves". However, this needs to be tempered with cultural preferences (see Chaps. 5 and 6).

Many important forest products (e.g., *Hevea brasiliensis* and *Paullinia cupana*) have been substituted by either cultivated crops or synthetic products such as plastic buttons instead of vegetable ivory (*Phytelephas macrocarpa*) and industrially produced repellents in lieu of plant-based pesticides such as barbasco (*Lonchocarpus nicou*). This historical evidence buttresses the second stylised fact that NTFPs are inherently substitutable. In practice, specific products such as Brazil nuts, are more likely to be substitutable than general categories of products such as "nuts" or even "rainforest products".

The third stylised fact is that tropical forests are not and cannot be actively managed for NTFPs. This assumption is embedded in definitions of NTFPs as products obtained from the "wild", implying with no management of their regeneration and production. In the prototypical diverse tropical forest, in which individual species occur at low density, lack of management means that the marginal costs of collecting rise rapidly, resulting in low returns to labour. Depending on the plant part harvested, lack of management may also result in overexploitation, diminished vigour of populations, and economic exhaustion of the resource (Cunningham 2001; Ticktin 2004; Marshall et al. 2006). Prospects for technical solutions are constrained by incomplete scientific understanding of the complex ways that harvesting interacts with the species' life cycles (see Chap. 7).

Different combinations of these three stylised facts underlie several heuristic models of NTFPs that strongly influenced the literature, policy discussions, and funding decisions starting in the late 1990s. Authors often alluded to these models without explicitly laying out the underlying assumptions. Perhaps the best known is the "boom–bust cycle" posited by Homma (1992) as an organising framework for the economic history of the Amazon. Homma (1992) argued that commercial extraction of the most valuable forest products in Amazonia follows a cyclical pattern, which is characterised by an initial stage of expansion, sometimes followed by a stabilisation phase, ultimately leading to a bust when the forest product is replaced by either synthetic substitutes or cultivation of the same or similar species. This model rests solidly on the assumptions that NTFPs are substitutable and unmanageable.

The boom–bust pattern inevitably imposes transition costs, as people who have invested in and become dependent on a particular product during the boom have to adjust to the bust. From a long-term perspective, the benefits of the boom may balance the costs of the bust, but these costs and benefits are unlikely to be equitably distributed. As Dove (1994) points out, when a resource gains value, elites who previously had no interest (or traditional tenure rights) in the product quickly take over its extraction, processing, and trade. These may be "local 'elites' with more capital to invest, better connections, and better skills, or . . . competitors from other areas" (Belcher and Schreckenberg 2007). Regardless of who benefits, the proceeds of the boom may be invested in activities with higher immediate returns but less long-term sustainability (e.g., Escobal and Aldana 2003). Thus, a corollary of the boom–bust model is that the boom undermines local livelihoods.

NTFPs that do not enter international markets are often conceived of as "famine foods" that are inferior and substitutable, and by corollary, not worth managing. Early references to this concept emphasised the critical safety net function of NTFPs (e.g., Falconer and Koppel 1990; Koppert et al. 1993), but in the literature with a more pessimistic slant, it becomes almost synonymous with inferior and substitutable products (Delang 2006). While Pierce and Emery (2005) argued that the use of NTFPs as famine goods remains common during times of crisis throughout the world, more typically the label of famine foods is used to simultaneously recognise the livelihood importance while dismissing the policy importance on NTFPs. For example, Ogden (1996) asserts that "the collection, processing and preparation of such foods is time consuming and they are therefore being progressively abandoned with increasing commercialisation and degradation of forest resources". Thus, in this model, famine foods are considered a stop-gap until markets and public policies can provide better alternatives.

A third model is of NTFPs as poverty traps: inferior goods with low prices that do not compensate for their high collection costs but cannot be managed to reduce those costs (Sheil and Wunder 2002). Belcher and Schreckenberg (2007) classify NTFP activities as poverty traps in cases "where decreasing prices nevertheless result in the need to increase harvesting to maintain a minimum income level". Delacote (2009), referring to what is usually called a "common pool resource", argues that "a poverty trap situation occurs when too much labour is allocated to common property resource (CPR) extraction. In this case, return to labour decreases by a simple tragedy of the commons effect, and the CP resource cannot properly insure the households anymore. Some households thus need to migrate, and the remaining households need to allocate all their labour to the CP resource which can only provide them with their minimum requirement. Thus they are trapped into poverty".

As with most generalisations, the concepts of boom–bust cycles, famine foods, and poverty traps accurately characterise some but certainly not all NTFPs. For example, of the 61 cases in CIFOR's comparative study of commercialisation, only 12% followed a boom–bust pattern (with contracted or unstable market). In a 2006 study of 10 products from 18 marginalised communities in Bolivia and Mexico, none of the NTFP activities were characterised as poverty traps (Schreckenberg et al. 2006). However, the pessimists' heuristic models did serve as an effective antidote to earlier unrealistic expectations that NTFP commercialisation would automatically reconcile development and conservation objectives. Just as important, they encouraged researchers to broaden their sights beyond the highly visible and appealing NTFPs with potential international markets, to a new research agenda that aimed to uncover the actual (as opposed to potential) role of NTFPs in livelihoods.

2.5 Emerging Middle Ground

While the international community swung from optimism to pessimism about the potential to alleviate poverty and incentivise conservation through international markets for NTFPs, forest-dependent peoples continued to use and manage their forests in diverse ways to fulfill diverse functions in their livelihood systems. Between internationally traded NTFPs (which are considered at risk of boom–bust) and famine foods (which households consume only under duress), there is vast middle ground of NTFPs with demand grounded in cultural traditions, traded in local and regional markets, making up a diverse basket of products that insure and enhance quality of life, and managed in subtle ways across a spectrum of forest types. Over the past decade, these existing functions of NTFPs have come into greater focus in the scientific literature.

This transition was evident in the 2003 "International Conference on Rural Livelihoods, Forests and Biodiversity", which included numerous papers examining how rural livelihoods depend on NTFPs (especially in Latin America), forest plantations and agroforestry systems (especially in Asia), and biodiversity (especially in Africa). Many of the papers re-stated the new conventional wisdom that commercialisation of NTFPs had been oversold as a one-size-fits-all solution. However, the conference themes also included the safety-net role of forests. And in his introduction to the special issue of *World Development* (volume 33, issue 9) resulting from the conference, Sunderlin (2005) noted "the complex ways in which forest resources help meet the needs of marginalised people. They can be crucial for mitigating or avoiding poverty, a fact not easily grasped by analysts who only focus on ways of lifting people out of poverty permanently" (Chap. 3).

Another challenge with understanding the multiple functions of forests is the great heterogeneity across products and settings, which has become increasingly apparent with expanded research on NTFPs in Africa (e.g., Shackleton et al. 2007, 2008) and in temperate (e.g., Emery et al. 2006) and boreal zones (e.g., Boxall et al. 2003). In this section, we identify four cross-cutting themes that are emerging in this recent literature (a) centrality of culture and tradition, (b) local and regional markets, (c) value of diversity in and of itself, and (d) continuum of often invisible forest management.

2.5.1 *Culture*

The cultural importance of forests extends well beyond their widely recognised role in indigenous and tribal customs. The literature documents the centrality of NTFPs in rural institutions and social networks across diverse settings. For example, in South and West Africa, marula fruit (*Sclerocarya birrea* subsp. *caffra*) and cola nuts (*Cola* spp.), respectively, help to maintain an important ethic of reciprocity, cultural norms, and social benefits that are central to rural livelihoods (Obeng and

Brown 2004; Shackleton and Shackleton 2006). Similarly, in the Appalachian mountain range of North America, the ritual of digging ginseng roots each spring season constitutes a social institution (Hufford 1997; Pierce 2002). Culture continues to shape use of NTFPs among people with a historical as well as a current connection to forested regions (Cocks 2006; Chaps. 5 and 6). People may simultaneously want to escape a forest-dependent existence, which for some can be isolated and deprived, and at the same time, yearn to maintain some connection with that existence (Pretty et al. 2009). One manifestation of this is the new trade routes that have been created as people who migrate to urban centers and around the globe take their culinary, craft, and healing traditions with them (Clark and Sunderland 2004; Stoian 2005; Padoch et al. 2008).

The market for NTFPs with cultural significance often places a premium on "wild harvested" products, meaning that they cannot be easily substituted by cultivated or synthetic products. This parallels the interest in "ethical consumption", fair trade, and ecological certification, which also reflects concern with the production process and not just the final product. However, without a strong cultural connection, the international market for certified "green" or "fairly traded" food and health care products derived from NTFPs can be "extremely fickle and trend-driven" (Laird and Guillen 2002). Rai and Uhl (2004) provides a good example with the boom of the 'uppage' (*Garcinia gummi-gatta*) market when it was promoted as a weight-loss supplement and the bust of that market when scientific tests showed it to be "ineffective" (Belcher and Schreckenberg 2007). In response to this fragility of international markets, attention is shifting to local and regional markets.

2.5.2 Local Markets

Substantial but largely unquantified local markets for forest goods exist throughout the world (Wiersum and Ros-Tonen 2005; Shackleton et al. 2007, 2008). In Evans' (1996) terms, these markets absorb both wild staples, which "are ingredients of everyday meals which are integral parts of cultural foodways or food patterns" and wild luxuries, which "are rare, valuable or otherwise prestigious items of food from the wild". Box 2.2 describes açaí, a product that represents both of these categories (Fig. 2.2). The potential public health benefits of continued consumption of a diversity of NTFPs and other wild-harvested and traditional products are increasingly apparent as urban populations undergo the nutrition transition (Johns and Sthapit 2004; Johns and Maundu 2006). From the producer perspective, the absorptive capacity and prices paid in these markets are typically lower than in international markets, but barriers to entry, costs, and the risk of "boom–bust" are also lower, partly due to the cultural significance and familiarity of NTFPs. However, as Shanley et al. (2002) point out, even these markets can be very difficult for some rural producers to access.

During the last two decades, case studies have yielded insights regarding the conditions under which commercialisation of NTFPs is most likely to be successful. Lack of resource access, market information and basic infrastructure, as well as weak political representation often hinder small producers seeking to market forest goods. Viable trade is more likely under conditions including adaptable resource management practices; transparency along the value chain; organisation among producers; and inclusion of women, although key entrepreneurs also

Box 2.2 Açaí: Fruit of the Poor Becomes Fruit of the Prosperous

Nathan Vogt, ACT, Indiana University, 179 rua de Obidos, Edificio Montreal, Apto 802, Cidade Velha, Belém, Brasil 66020-440, ndvogt@gmail.com

Açaí fruit (*Euterpe oleracea*) has been harvested from the Amazon forest and consumed for subsistence since the pre-colonial era. Throughout the colonial era, *caboclo* populations (miscegenation of Indigenous, African, and Portuguese peoples) continued to harvest and consume açaí as a subsistence crop, utilising indigenous management techniques in its production. However, accounts from the nineteenth century naturalists exploring Amazonia reveal that açaí was not consumed by urban European elites, was considered a food for the poor lower classes, and thus was not managed for commercial purposes (Bates 1863; Wallace 1853).

When riverine *caboclo* families began establishing urban residences in large numbers in the 1960s, they brought and maintained their habits and cultural preference for açaí fruit, creating a demand which has continued to grow strongly over the past 30 years. In addition to cultural continuity, açaí fruit has provided affordable nourishment and a caloric staple particularly important to low-income urban residents.

In recent years, as açaí became more widely available in urban stalls and restaurants, it came onto the radar of food companies, which saw its potential as an energy and health drink in national and international markets, where products deemed both socially and environmentally responsible, have become fashionable, and can command astounding prices. For instance, the value of açaí fruit pulp resulting from the harvest of 1 ha of managed forest at the farmer's gate (i.e., fruit in nature) ranges from around USD 1000 to USD 1200. The price for the same amount (in equivalent processed pulp) increases 20- to 50-fold (depending on the end product) by the time it reaches consumers in southern Brazil and 70-fold or more (depending on the end product) for international consumers (Brondizio 2008). For instance, pills and vitamin supplements claiming health and anti-aging benefits of açaí cost USD 50 per container.

Commercial companies and the government, both new to the açaí business, perpetuate the idea that they direct the management and intensification of açaí that is carried out by caboclos in native forests. In reality, the production systems being developed by industries are built upon the traditional knowledge and generations of management practice of riverine caboclos, such as the

(continued)

agronomically sophisticated "Roçados de Varzea" (floodplain gardens). The management efforts of caboclos over the past three decades deserve the credit for açaí production reaching its impressive current levels.

This story of the estuarine caboclos' initiative, utilising traditional ecological knowledge, is an excellent example of the potential to reconcile conservation and development goals through commercialisation of an NTFP. However, producers continue to be depicted as mere suppliers of raw material and thus suffer from the stigma of extractivism and the invisibility of their sophisticated forest management systems (Brondizio and Siqueira 1997). Although producers have benefitted from market expansion, they have been unable to participate in new sectors of the economy associated with the commercialisation of fruit stock, its transformation, and its value aggregation along the chain. The increasing demand for açaí has resulted in growing competition for production areas from corporations seeking to control supply. Current trends indicate that açaí may become sourced less from native forests directed and managed by caboclos, and more from industry-dominated monocultural systems (Brondizio 2008).

Fig. 2.2 Harvesting açai (Euterpe oleracea) near Belém in the eastern Brazilian Amazon (photo: Marcos R. Tito)

often play a vital role (Schreckenberg et al. 2006; te Velde et al. 2006). In eight South and Southeast Asian countries, lessons learned from scores of decade-long, on-the-ground projects include the need for realistic, long-term time frames; continuity of commitment; hard work; and mutual respect. "Effective marketing of forest goods and sustainable livelihood development is a long and tedious process, requiring a step by step approach" (Arquiza 2008).

Recent literature also sheds light on the role of NTFP production and trade in urban livelihoods. Both the optimists and the pessimists saw traditional traders as barriers to commercialisation projects that sought to improve livelihoods and promote forest conservation. However, traders clearly play essential roles and take on significant risks in NTFP markets, and in many cases, they are themselves an important category of relatively poor people who make their livelihoods from NTFPs (Ndoye et al. 1998; Shackleton and Shackleton 2004; Ingram 2009). In the Amazon in particular, recent literature points to the growing importance of peri-urban populations, often recently migrated from the interior, in the harvest and distribution systems for NTFPs (Stoian 2005; Padoch et al. 2008; Parry et al. 2010). While there is consensus that collection of NTFPs by urban and peri-urban residents is a growing phenomenon, the implications of this for conservation and forest management are still debated.

2.5.3 Diversity

While specific marketable NTFPs offer a means to earn cash income and continue cultural traditions, the great diversity of NTFPs, in and of itself, is increasingly recognised for its contribution to rural livelihoods. The value of this diversity manifests itself as natural insurance, smoothing of labour demand and incomes, a well-stocked and affordable natural pharmacy, and a diversified and nutritious diet (especially for children), all from a source unlikely to be captured and monopolised by elites.

The safety net or "natural insurance" function of forests in developing countries has been widely noted (Neumann and Hirsch 2000; Pattanayak and Sills 2001; Shackleton and Shackleton 2004; Marshall et al. 2006). Any single NTFP can be subject to supply or demand swings, just as with any other component of rural livelihoods. But NTFPs as a group can serve as a more secure fall-back option because of the diversity of species and plant parts that can be collected for consumption or sale throughout the year. While this potential role of NTFPs is now well accepted in the literature, its strength and applicability in different settings is poorly understood (Paumgarten 2005). Research has identified the type of shock (Takasaki et al. 2004), the available alternatives (Fu et al. 2009), human capital (Fisher et al. 2010), and forest policy (McSweeney 2005) as influencing household reliance on NTFPs as a safety net.

Two other ways in which the diversity of forest products contributes to rural well-being are by supporting health care and nutrition (Fig. 2.3). Many people living on

Fig. 2.3 Wide variety of medicinal plants for sale in the Brazilian Amazon (photo: Marcos R. Tito)

forest margins neither have access to nor can afford to patronise well-stocked modern pharmacies, but use modern medicines for only some ailments and rely on the natural pharmacy of medicinal plants for others (Shanley and Luz 2003; Colfer et al. 2006). It is the diversity of plants freely available that makes this so valuable to local people. Likewise, the diversity of wild foods found in forests can greatly improve the nutritional quality of diets, perhaps particularly for children who snack on fruits, nuts, insects, and other wild foods on their way to school and chores (McGarry and Shackleton 2009). Evans (1996) argues that, "children especially have difficulty in eating enough food to satisfy total calorie requirements unless there is some liquid sauce or stew to accompany these carbohydrate staples. The variety of colour, smell and texture that intrinsically wild food can provide is wide and its role in providing essential vitamins, minerals, trace-elements, proteins and fats is supported by both biochemical analysis and anthropological fieldwork". Many authors have called for more careful study of macro- and micronutrient dense wild species in order to add them to nutrient databases and incorporate them into nutrition policies (e.g., Grivetti and Ogle 2000; Johns and Eyzaguirre 2006).

The array of nutritional contributions and culinary diversity which NTFPs offer is also gaining recognition and appreciation in the developed world. For example, in North America and Europe, advocates for the consumption of locally harvested

Fig. 2.4 Processed NTFPs for sale in Canada: wild lingonberry jam (photo: Sheona Shackleton)

food (i.e., community supported agricultural systems, Slow Food, macrobiotic diets and farmers' markets) are introducing diverse agricultural and forest products to local diets and economies (Fig. 2.4). There is growing concern that diets are dominated by a negligible number of foodstuffs, limited in micronutrients, trace elements, and overall nutritional value. In this context, the wide-ranging dietary value of wild fruits, forest greens, game meat, nuts, and fungi is gaining renewed cultural, culinary, and socioeconomic appreciation (Emery and Pierce 2005; Emery et al. 2006).

2.5.4 Managed Forest

Pure extractivism and monoculture are now recognised as just the extreme endpoints of an entire gradient of management for NTFPs in forested landscapes that are cultural products, the result of centuries of manipulation and management (Dove et al. 2005; Pretty et al. 2009, Chap. 7). Certainly, some well-known products such as tea and rubber in Indonesia, guarana (*Paullinia cupana*) and cashew in Brazil, and coffee and African plum (*Dacryodes edulis*) from Central Africa have transitioned from extraction to monoculture. However, local people also manage existing forests (e.g., through enrichment plantings) and cultivate new forests in a "hortus" model that replicates forest patterns and retains the complexity of the natural ecosystem (Michon 2005). This maintains the valuable diversity of NTFPs and incorporates their production into social networks that regulate access and control (Ros-Tonen and Wiersum 2005). In CIFOR's comparative assessment of 61 cases of commercial NTFPs, many of the best income-earning opportunities were based on intermediate levels of management (Ruiz Perez et al. 2004).

Examples of products which are managed within forest and agro-forest ecosystems include rattan in Kalimantan (de Jong et al. 2003), açaí in Brazil (Box 2.2), and maple sugaring in North America (Hinrichs 1998).

Most NTFPs only gradually and partially transition from extractivism to domestication. Especially in cases where land tenure is clear, demand can catalyse innovation among farmers that favours sustainable practices. For example, diversified forest gardens hosting fruits, latex-producing trees, and rattan are central features of communities and help define territories and land claims in Indonesia (de Jong et al. 2003). In case studies from Mexico and Bolivia, small-scale domestication was widespread, occurring in 35% of the communities facing resource depletion, and only 11% of the cases had no signs of resource management plans or domestication in place (Schreckenberg et al. 2006). Research in South Africa finds that in the initial phases of domestication, farmers draw on the broad genetic base available in the wild, resulting in improved germplasm (Leaky et al. 2004) and trees with higher yields, fruit size, and desired fruit traits (Shackleton et al. 2003).

Based on mounting evidence of traditional, successful but often invisible management for NTFPs, researchers have called for a) treating diversified forest cultivation and management as an alternative rather than just a transition towards more modern or intensive production systems (Wiersum 2003; Michon 2005); b) a participatory approach to research on NTFP management and domestication that could enhance its role in smallholder livelihood systems (Akinnifesi et al. 2008); c) greater recognition of the value of secondary forest, fallows, and other managed ecosystems (Ambrose-Oji 2003); and d) a rethinking the dichotomy between timber and non-timber products (Padoch and Pinedo-Vasquez 1996) (Chap. 8).

As Michon (2005) argues, "dissociating timber from NTFP in scientific forest research, in international discussions on forest management and in development projects indirectly contributes to reinforce policies that deprive local communities from the large benefits of timber management". In the past decade, research has increasingly focused on the potential and trade-offs involved in managing or harvesting both timber and non-timber products from forests (Shanley and Rosa 2004; Herrero-Jáuregui et al. 2009; Menton et al. 2009; Chap. 8). Thus, the wild harvested and internationally marketed NTFPs that generated so much excitement at the beginning of the 1990s have now been placed back in the context of a range of forest products (from medicinal plants to timber), a range of forest management intensities (from pure extraction to intensive forest management), and a range of markets (from international to local).

2.6 Conclusion

NTFPs have become firmly established in the academic research domain, with the Web of Science reporting 50–75 new publications on this topic every year since 2003. Current research considers the full range of livelihood functions provided by NTFPs. This book reflects much of the recent thinking on the "middle ground" of

NTFPs, giving due consideration to the role of culture and tradition, local and regional markets, the way the livelihoods are supported by the sheer diversity of NTFPs, and the forest management practices of local people.

NTFPs have also earned a place in international forest policy discussions, e.g., as a recognised subtheme in the 13th World Forestry Congress in Buenos Aires in 2009. However, policy, donor funding, and implementation projects are often one step behind research (Chap. 11). In the case of NTFPs, this means a continued focus on commercialisation, sometimes discussed as a viable win–win for sustainable development, and other times disparaged as a pipe dream that has misdirected efforts and resources. It is crucial that the policy discussions catch up with current research, for example, understanding the conditions under which forests function as safety nets that prevent the poorest from falling deeper into poverty in the face of shocks such as those expected from climate change. Indeed, the diverse roles of forests in local livelihoods, and correspondingly diverse ways in which local people manage the forests, will be important for adapting to climate change, and therefore should be central considerations in plans to mitigate climate change through forest conservation and management.

References

Akinnifesi FK, Leakey RRB, Ajayi OC, Sileshi G, Tchoundjeu Z, Matakala P, Kwesiga F (2008) Indigenous fruit trees in the tropics: domestication, utilisation and commercialisation. ICRAF/CAB, Wallingford, UK

Alexiades MN, Shanley P (2004) Productos Forestales, Medios de Subsistencia y Conservación. In: Alexiades MN, Shanley P (eds) Forestales, Medios de Subsistencia y Conservación: Estudios de Caso sobre Sistemas de Manejo de Productos Forestales No Maderables. Volumen 3 – America Latina. Center for International Forestry Research, Bogor, Indonesia

Ambrose-Oji B (2003) The contribution of NTFPs to the livelihoods of the 'forest poor': evidence from the tropical forest zone of south-west Cameroon. Int Forest Rev 5:106–117

Anderson AB, Jardim MAG (1989) Costs and benefits of floodplain forest management by rural inhabitants in the Amazon estuary: a case study of açaí palm production. In: Browder JO (ed) Fragile lands in Latin America: the search for sustainable uses. Westview, Boulder, CO

Arnold JEM (2001) Forestry, poverty and aid. CIFOR Occasional Paper No. 33. Center for International Forestry Research, Bogor, Indonesia

Arnold M, Kohlin G, Persson R, Shepperd G (2003) Fuelwood revisited: what has changed in the last decade? CIFOR Occasional Paper No. 39. Center for International Forestry Research, Bogor, Indonesia

Arnold M, Kohlin G, Persson R (2006) Woodfuels, livelihoods and policy interventions: changing perspectives. World Dev 34:596–611

Arquiza, Yasmin (ed) (2008) From seeds to beads. Tales, tips and tools for building a community-based NTFP enterprise. Quezon City, the Philippines, Non-Timber Forest Products Exchange Programme

Bates HW (1863) The naturalist on the Amazons: a record of adventures, habits of animals, sketches of Brazilian and Indian life, and aspects of nature under the equator during eleven years of travel. George Routledge, London

Belcher B, Ruiz Perez M (2001) An international comparison of cases of forest product development: overview, description and data requirements. CIFOR Working Paper No. 23, Center for International Forestry Research, Bogor, Indonesia

Belcher B, Schreckenberg K (2007) Commercialisation of non-timber forest products: a reality check. Dev Policy Rev 25:16–23

Belcher B, Ruız Perez M, Achdiawan R (2005) Global patterns and trends in the use and management of commercial NTFPs: implications for livelihoods and conservation. World Dev 33:1435–1452

Berkes F (1993) Traditional ecological knowledge in perspective. In: Inglis J (ed) Traditional ecological knowledge: concepts and cases. International Development Research Centre, Ottawa, ON, Canada, pp 1–9

Boxall PC, Murray G, Unterschultz JR (2003) Non-timber forest products from the Canadian boreal forest: an exploration of aboriginal opportunities. J For Econ 9:75–96

Bromley DW, Cernea MM (1989) The management of common property natural resources: some conceptual and operational fallacies. World Bank, Washington, DC

Brondizio ES (2008) The Amazonian caboclo and the açaí palm: forest farmers in the global market. New York Botanical Garden, New York, NY

Brondizio ES, Siqueira AD (1997) From extractivists to forest farmers: changing concepts of Caboclo agroforestry in the Amazon estuary. Res Econ Anthropol 18:233–279

Browder JO (1992) The limits of extractivism: tropical forest strategies beyond extractivism. Bioscience 42:174–182

Caldecott JO (1988a) Hunting and wildlife management in Sarawak. World Conservation Union, Gland

Caldecott JO (1988b) Rattan in Sarawak: at the crossroads. Rattan Inf Cent 7(1–2):11

Campbell BC (ed) (1996) The Miombo in transition: woodlands and welfare in Africa. Center for International Forestry Research, Bogor, Indonesia

Campbell B, Luckert MK (2002) Uncovering the hidden harvest: valuation methods for woodland and forest resources. Earthscan, London, UK

Cavendish W (2000) Empirical regularities in the poverty environment relationship of rural households: evidence from Zimbabwe. World Dev 28:1979–2003

Clark LE, Sunderland TCH (2004) The state of the art of non-timber forest products in West Africa. USAID Technical Paper no. 122

Cocks M (2006) Biocultural diversity: moving beyond the realm of 'indigenous' and 'local' people. Hum Ecol 34:185–200

Colfer CJP, Sheil D, Kaimowitz D, Kishi M (2006) Forests and human health in the tropics: some important connections. Unasylva 224(57):3–10

Cooke P, Kohlin G, Hyde W (2008) Fuelwood, forests and community management: evidence from household studies. Environ Dev Econ 13:103–135

Coomes OT, Barham BL (1997) Rain forest extraction and conservation in Amazonia. Geogr J 163:180–188

Corry S (1993) The rainforest harvest: who reaps the benefit? Ecologist 1993:148–153

Counsell S, Rice T (eds) (1992) The rainforest harvest: sustainable strategies for saving the tropical forests? In: Proceedings of an international conference held at the Royal Geographical Society, London, 17–18 May 1990. Friends of the Earth, London

Cunningham AB (2001) Applied ethnobotany: people, wild plant use and conservation. Earthscan, London

de Beer JH, McDermott MJ (1996) The economic value of non-timber forest products in Southeast Asia, 2nd Revised edn. IUCN, Amsterdam

de Jong W, Tuck-Po L, Ken-ichi A (2003) The political ecology of tropical forests in Southeast Asia. Kyoto University and TransPacifica Press, Kyoto

Delacote P (2009) Commons as insurance: safety nets or poverty traps? Environ Dev Econ 14:305–322

Delang CO (2006) Not just minor forest products: the economic rationale for the consumption of wild food plants by subsistence farmers. Ecol Econ 59:64–73

Donkin RA (2003) Between east and west: the Moluccas and the traffic in spices up to the arrival of Europeans. American Philosophical Society, Philadelphia, PA

Dove MR (1994) Marketing the rainforest: 'green' panacea or red herring? Asia-Pacific Issues No. 13. East-West Center, Hawaii

Dove MR, Sajise EP, Doolittle A (2005) Conserving nature in culture: case studies from Southeast Asia. Yale University Southeast Asia Studies, New Haven

Ehringhaus C (2006) Post-victory dilemmas: land use, development, and social movement in Amazonian extractive reserves. PhD Dissertation, Yale University, New Haven, CT

Emery MR, Pierce AR (2005) Interrupting the telos: locating subsistence in contemporary US forests. Environ Plan 37:981–993

Emery MR, Martin S, Dyke A (2006) Wild harvests from Scottish Woodlands Social, cultural and economic values of contemporary non-timber forest products. Forestry Commission, Edinburgh, Scotland

Escobal J, Aldana U (2003) Are non-timber forest products the antidote to rainforest degradation? Brazil nut extraction in Madre de Dios, Peru. World Dev 31:1873–1887

Evans MI (1996) A contextual classification of 'intrinsically wild' food species. Biopolicy J 1: Paper 3

Falconer J, Koppel CRS (1990) The major significance of minor forest products. The local use and value of forests in the West African humid forest zone. Forest, Trees and People, Community Forestry note No. 6, FAO, Rome

Fisher M, Chaudhury M, McCusker B (2010) Do forests help rural households adapt to climate variability? Evidence from Southern Malawi. World Dev. doi:10.1016/j.worlddev.2010.03.005DOI:dx.doi.org

Flint M (1990) The tropical forestry action plan (TFAP) and non-wood forest products. In: Counsell S, Rice T (eds) The rainforest harvest: sustainable strategies for saving the tropical forests? Friends of the Earth Trust, London

Fu Y, Chen J, Guo H, Chen A, Cui J, Hu H (2009) The role of non-timber forest products during agroecosystem shift in Xishuangbanna, southwestern China. For Policy Econ 11:18–25

Godoy R, Feaw TC (1989) The profitability of smallholder rattan cultivation in central Borneo. Hum Ecol 17:347–363

Gray A (1990) Indigenous peoples and the marketing of the rain forest. Ecologist 20:223–227

Grivetti E, Ogle BM (2000) Value of traditional foods in meeting macro- and micronutrient needs: the wild plant connection. Nutr Res Rev 13:31–46

Guijit I, Hinchcliffe F, Melnek M, Bishop J, Eaton D, Pimbert M, Pretty J, Scoones I (1995) Hidden harvest: the value of wild resources in agricultural systems – a project summary. IIED, London

Hanson JH (1992) Extractive economies in a historical perspective: gum Arabic in West Africa. Adv Econ Bot 9:107–114

Hardin G (1968) The tragedy of the commons. Science 162:1243–1248

Hardner J, Rice R (2002) Rethinking green consumerism. Sci Am 286:88–95

Haripriya GS (2001) Integrated environmental and economic accounting: an application to the forest resources in India. Environ Res Econ 19:73–95

Hecht JC (2007) National environmental accounting: a practical introduction. Int Rev Environ Res Econ 1:03–66

Hecht SB, Anderson AB, May P (1988) The subsidy from nature: shifting cultivation, successional palm forests, and rural development. Hum Organ 47:25–35

Herrero-Jáuregui C, García-Fernández C, Sist P, Casado MA (2009) Conflict of use for multi-purpose tree species in the state of Para, eastern Amazonia, Brazil. Biodivers Conserv 18:1019–1044

Hidalgo RC (1992) The tagua initiative in Ecuador: a community approach to tropical rain forest conservation and development. In: Plotkin M, Falamore L (eds) Sustainable harvest and marketing of rainforest products. Island, Washington, DC, pp 263–273

Hinrichs CC (1998) Sideline and lifeline: The cultural economy of maple syrup production. Rural Sociology 63:507–532

Homma AKO (1992) The dynamics of extraction in Amazonia: a historical perspective. In: Nepstad DC, Chwartzman S (eds) Non-timber products from tropical forests: evaluation of a conservation and development. New York Botanical Garden, New York, pp 23–31

Hufford M (1997) American ginseng and the culture of the commons. Orion 16:11–14
Hughes R, Flintan F (2001) Integrating conservation and development experience. A review and bibliography of the ICDP literature. International Institute for Environment and Development, London
Ingram V (2009) The hidden costs and values of NTFP exploitation in the Congo Basin. Paper read at world forestry congress XIII, Buenos Aires, Argentina
Jacobs M (1984) The study of non-timber forest products. Environmentalist 4:77–79
Jodha NS (1986) Common property resources and rural poor in dry regions of India. Econ Polit Wkly 21(27):1169–1181
Johns T, Eyzaguirre PB (2006) Linking biodiversity, diet and health in policy and practice. Proc Nutr Soc 65:182–189
Johns T, Maundu P (2006) Forest biodiversity, nutrition and population health in market-oriented food systems. Unasylva 224(57):34–40
Johns T, Sthapit BR (2004) Biocultural diversity in the sustainability of developing country good systems. Food Nutr Bull 25(2):143–155
Koppert GJA, Dounias E, Froment A, Pasquet P (1993) Food consumption in the forest populations of the southern coastal area of Cameroon. In: Hladik CM, Pagezy H, Linares OF, Hladik A, Semple A, Hadley M (eds) Tropical forest, people and food. Biocultural interactions and applications to development, Man and biosphere series. Unesco-Parthenon, Paris, pp 295–310
Laird S, Guillen A (2002) Marketing issues. In: Shanley P, Pierce A, Laird S, Guillen A (eds) Tapping the green market: certification and management of non-timber forest products. Earthscan, London, pp 322–331
Leaky RRB, Tchoundjeu A, Smith RI, Munro RC, Foundoun JM, Kengue J, Anegbeh PO, Atangana AR, Waruhiu AN, Asaah E, Usoro C, Ukafor V (2004) Evidence that subsistence farmers have domesticated indigenous fruits (*Dacryodes edulis* and *Irvingia gabonensis*) in Cameroon and Nigeria. Agrofor Syst 60:101–111
Lynch OJ, Talbott K (1995) Balancing acts: community-based forest management and national law in Asia and the Pacific. World Resources Institute, Washington, DC
Marshall E, Schreckenberg K, Newton AC (2006) Commercialisation of non-timber forest products: factors influencing success. UNEP/World Conservation Monitoring Center, Cambridge, UK
Maxwell S (2001) Is there a new poverty agenda? Dev Policy Rev 19:143–149
McGarry DK, Shackleton CM (2009) Children navigating rural poverty: rural children's use of wild resources to counteract food insecurity in the Eastern Cape province, South Africa. J Child Povert 15:19–37
McSweeney K (2005) Natural insurance, forest access, and compound misfortune: forest resources in smallholder coping strategies before and after hurricane Mitch, Northeastern Honduras. World Dev 33:1453–1471
Menton M, Merry FD, Lawrence A, Brown N (2009) Company-community logging contracts in Amazonian settlements: impacts on livelihoods and NTFP harvests. Ecol Soc 14(1):39
Menzies NK (2004) Communities and their partners: governance and community-based forest management. Conserv Soc 2:449–456
Michon G (2005) Domesticating forests: how farmers manage forest resources. CIFOR/ICRAF, Bogor, Indonesia
Myers N (1988) Tropical forests: much more than stocks of wood. J Trop Ecol 4:209–221
Ndoye O, Pérez MR, Eyebe A (1998) The markets of non timber forest products in the humid forest zone of Cameroon. Rural Development Forestry Network Paper 22c
Nepstad DC, Schwartzman S (ed) (1992) Non-timber products from tropical forests: evaluation of a conservation and development strategy. Advances in economic botany 9. The New York Botanical Garden, New York
Neumann RP, Hirsch E (2000) Commercialisation of non-timber forest products: review and analysis of research. Center for International Forestry Research/FAO, Bogor, Indonesia

Obeng T, Brown N (2004) Kola Nuts (Cola Acuminata and Cola Nitida). In: Clark LE, Sunderland TCH (ed) The key non-timber forest products of Central Africa: state of knowledge. USAID Technical Paper 122, pp 87–112

Ogden CL (1996) Considering nutrition in national forestry programmes. FAO, Rome

Ostrom E (1990) Governing the commons: the evolution of institutions for collective action. Cambridge University Press, Cambridge

Padoch C (1988) People of the floodplain and forest. In: Denslow JS, Padoch C (eds) People of the tropical rain forest. University of California Press, Berkeley, CA

Padoch C, Pinedo-Vasquez M (1996) Smallholder forest management: looking beyond non-timber forest products. In: Perez A (ed) Current issues in non-timber forest products research. Center for International Forestry Research, Bogor, Indonesia

Padoch C, Brondizio E, Costa S, Pinedo-Vasquez M, Sears RR, Siqueira A (2008) Urban forest and rural cities: multi-sited households, consumption patterns, and forest resources in Amazonia. Ecol Soc 13. Available from: http://www.ecologyandsociety.org/vol13/iss2/art2/

Padovani F (1995) Statistical information on non-wood forest products. International expert consultation on non-wood forest products. FAO, Rome. Available from: http://www.fao.org/docrep/v7540e/V7540e31.htm

Panayotou T, Ashton P (1992) Not by timber alone: the case for multiple use management of tropical forests. Island, Covelo, CA

Parry L, Peres CA, Day B, Amaral S (2010) Rural–urban migration brings conservation threats and opportunities to Amazonian watersheds. Conserv Lett 3:251–259

Pattanayak SK, Sills EO (2001) Do tropical forests provide natural insurance? The microeconomics of non-timber forest product collection in the Brazilian Amazon. Land Econ 77:595–612

Paumgarten F (2005) The role of non-timber forest products as safety-nets: a review of evidence with a focus on South Africa. GeoJournal 64:189–197

Paumgarten F (2009) What do PRSPs mean for forestry and rural livelihoods? Presentation given at Diversitas 2nd open science conference, Cape Town, South Africa, 13–16 October 2009

Peluso NL (1983) Networking in the commons: a tragedy for rattan? Indonesia 35:95–108

Peluso NL (1992) The rattan trade in East Kalimantan, Indonesia. Adv Econ Bot 9:115–128

Pendleton LH (1992) Trouble in paradise: practical obstacles to non-timber forestry in Latin America. In: Plotkin M, Famolare L (eds) Sustainable harvest and marketing of rain forest products. Island, Washington, DC, pp 252–262

Peres CA, Baider C, Zuidema PA, Wadt LHO, Kainer KA, Gomes-Silva DAP, Salomao RP, Simoes LL, Franciosi ERN, Valverde FC, Gribel R, Shepard GH Jr, Kanashiro M, Coventry YuDW, Watkinson AR, Freckleton RP (2003) Demographic threats to the sustainability of Brazil nut exploitation. Science 302:2112–2114

Peters CM (1994) Sustainable harvest of non-timber plant resources in tropical moist forest: an ecological primer. Corporate, Landover, MD

Peters CA, Gentry A, Mendelsohn R (1989) Valuation of an Amazonian rain forest. Nature 339:655–656

Pierce AR (2002) American ginseng (Panax quinquefolius). In: Shanley P, Pierce A, Laird S (eds) Tapping the green market: certification and management of non-timber forest products. Earthscan, London, pp 172–182

Pierce AR, Emery MR (2005) The use of forests in times of crisis: ecological literacy as a safety Net. For Trees Liveli 15:249–252

Plotkin M, Famolare L (1992) Sustainable harvest and marketing of rain forest products. Island, Washington, DC

Poffenberger M (1990) Keepers of the forest: land management alternatives in Southeast Asia. Kumarian, West Hartford, CT

Pretty J, Adams B, Berkes F, de Athayde SF, Dudley N, Hunn E, Maffi L, Milton K, Rapport D, Robbins P, Sterling E, Stolton S, Tsing A, Vintinner E, Pilgrim S (2009) The intersections of biological diversity and cultural diversity: towards integration. Conserv Soc 7:100–112

Puustjarvi E, Mickels-Kokwe G, Chakanga M (2005) The contribution of the forest sector to the national economy and poverty reduction in Zambia. SAVCOR, Forestry Department, Zambia and Ministry for Foreign Affairs of Finland, Lusaka, Zambia

Pyhälä A, Brown K, Adger WN (2006) Implications of livelihood dependence on non-timber products in Peruvian Amazonia. Ecosystems 9:1328–1341

Rai N, Uhl CF (2004) Forest product use, conservation and livelihoods: the case of *Uppage* fruit harvest in the Western Ghats, India. Conserv Soc 2:289–313

Redford KH (1992) The empty forest. Bioscience 42:412–422

Robbins SRJ, Matthews WSA (1974) Minor forest products: their total value is of a major order. Unasylva 26:7–14

Ros-Tonen M, Dijkman W, Lammerts van Bueren E (1995) Commercial and sustainable extraction of non-timber forest products: towards a policy and management oriented research strategy. Wageningen: The Tropenbos Foundation

Ros-Tonen MAF, Wiersum KF (2003) The importance of non-timber forest products for forest-based rural livelihoods: an evolving research agenda. Proceedings (CD) of the International Conference on Rural Livelihoods, Forests and Biodiversity. 19–23 May 2003, Bonn, Germany. Centre for International Forestry Research (CIFOR), Bogor

Ros-Tonen MA, Wiersum KF (2005) The scope for improving rural livelihoods through non-timber forest products: an evolving research agenda. For Trees Liveli 15:129–148

Ruiz Perez M, Byron N (1999) A methodology to analyse divergent case studies of non-timber forest products and their development potential. For Sci 45:1–14

Ruiz Perez M, Belcher B, Achdiawan R, Alexiades M, Aubertin C, Caballero J, Campbell B, Clement C, Cunningham A, Fantini A, de Foresta H, Garcıa Fernandez C, Gautam K, Hersch Martınez P, de Jong W, Kusters K, Kutty M, Lopez C, Fu M, Martı´nez Alfaro MA, Nair TKR, Ndoye O, Ocampo R, Rai N, Ricker M, Schreckenberg K, Shackleton S, Shanley P, Sunderland T, Youn Y (2004) Markets drive the specialisation strategies of forest peoples. Ecol Soc 9:4. Available from: http://www.ecologyandsociety.org/vol9/iss2/art4

Salafsky N, Dugelby BL, Terborgh JW (1993) Can extractive reserves save the rainforest? An ecological and socioeconomic comparison of non-timber forest products extraction systems in Petén, Guatemala and West Kalimantan, Indonesia. Conserv Biol 7:39–52

Sayer J (1991) Conservation and protection of tropical rain forests: the perspective of the World Conservation Union. Unasylva 166:40–45

Schreckenberg K, Rushton J, te Velde DW (2006) NTFP value chains: what happens between production and consumption? In: Marshall E, Schreckenberg K, Newton AC (eds) Commercialisation of non-timber forest products: factors influencing success. UNEP, Cambridge, UK

Schwartzman S, Nepstad D, Moreira A (2001) Arguing tropical forest conservation: people versus parks. Conserv Biol 14:1370–1374

Scoones I, Melnyk M, Pretty JN (1992) The hidden harvest: wild foods and agricultural systems. A literature review and annotated bibliography. IIED, London

Shackleton CM, Shackleton SE (2004) The importance of non-timber forest products in rural livelihood security and as safety nets: a review of evidence from South African. S Afr J Sci 100:658–664

Shackleton CM, Shackleton SE (2006) Household wealth status and natural resource use in the Kat River Valley, South Africa. Ecol Econ 57:306–317

Shackleton CM, Botha J, Emanuel PL (2003) Productivity and abundance of *Sclerocarya birrea* subsp. *caffra* in and around rural settlements and protected areas of the Bushbuckridge Lowveld, South Africa. For Trees Liveli 13:217–232

Shackleton CM, Shackleton SE, Buiten E, Bird N (2007) The importance of dry woodlands and forests in rural livelihoods and poverty alleviation in South Africa. For Policy Econ 9:558–577

Shackleton SE, Campbell B, Lotz-Sisitka H, Shackleton CM (2008) Links between the local trade in natural products, livelihoods and poverty alleviation in a semi-arid region of South Africa. World Development 36(3):505–526

Shanley P, Pierce A, Laird S (2005) Beyond Timber: certification of non-timber forest products. Forest Trends/CIFOR

Shanley P, Luz L (2003) The impacts of forest degradation on medicinal plant use and implications for health care in Eastern Amazonia. Bioscience 53:573–584

Shanley P, Rosa NA (2004) Eroding knowledge: an ethnobotanical inventory in Eastern Amazonia's logging frontier. Econ Bot 58:135–160

Shanley P, Luz L, Swingland I (2002) The faint promise of a distant market: a survey of Belem's trade in non-timber forest products. Biodivers Conserv 11:615–636

Sheil D, Wunder S (2002) The value of tropical forest to local communities: complications, caveats and cautions. Conserv Ecol 6(2):9

Sills E, Lele S, Holmes T, Pattanayak S (2003) Role of non-timber forest products in the rural household economy. In: Sills E, Abt K (eds) Forests in a market economy. Kluwer, The Netherlands, pp 259–281

Spilsbury MJ, Nasi R (2006) The interface of policy research and the policy development process: challenges posed to the forestry community. For Policy Econ 8:193–205

Stoian D (2005) Making the best of two worlds: rural and peri-urban livelihood options sustained by non-timber forest products from the Bolivian Amazon. World Dev 33:1473–1490

Street W (2006) Forestry and poverty reduction: can we lead the way? ETFRN News 47/48. Available from: http://www.etfrn.org/etfrn/newsletter/pdf/etfrnnews4748.pdf

Sunderlin W (2005) Introduction. World Development 33(9):1379–1381

Sunderlin WD, Dewi S, Puntodewo A (2007) Poverty and forests: multi-country analysis of spatial association and proposed policy solutions. CIFOR Occasional Paper No. 47. Center for International Forestry Research, Bogor, Indonesia. Available online at: http://www.cifor.cgiar.org/Publications/Papers/

Takasaki Y, Barham BL, Coomes OT (2004) Risk coping strategies in tropical forests: floods, illnesses, and resource extraction. Environ Dev Econ 9:203–224

te Velde DW, Rushton J, Schreckenberg K, Marshall E, Edouard F, Newton A, Arancibia E (2006) Entrepreneurship in value chains of non-timber forest products. For Policy Econ 8:725–741

Tewari DD (2000) Valuation of non-timber forest products (NTFPs): models, problems, and issues. J Sust For 11:47–68

Tewari DD, Campbell JY (1995) Developing and sustaining non-timber forest products: some policy issues and concerns with special reference to India. J Sust For 3:53–79

Ticktin T (2004) The ecological implications of harvesting non-timber forest products. J Appl Ecol 41:11–21

Townson IM (1995) Forest products and household incomes: a review and annotated bibliography. OFI Tropical Forestry Papers 31. Center for International Forestry Research/Forestry Research Institute, Bogor, Indonesia/Oxford

Turner NJ, Loewen DC (1998) The original 'free trade': exchange of botanical products and associated plant knowledge in Northwestern North America. Anthropologica 40:49–70

Vedeld P, Angelsen A, Sjaastad E, Berg GK (2004) Counting on the environment: forest incomes and the rural poor. Environmental Economic Series Paper No. 98. The World Bank, Washington DC

Wallace AR (1853) A narrative of travels on the Amazon and Rio Negro, with an account of the native tribes, and observations on the climate, geology, and natural history of the Amazon valley. Reeve, London

Weinstein B (1983) The Amazon rubber boom. Stanford University Press, Stanford, CA

Wells M, Guggenheim S, Khan A, Wardojo W, Jepson P (1998) Investing in biodiversity. A review of Indonesia's integrated conservation and development projects. World Bank, East Asia Region

Wiersum KF (2003) Use and conservation of biodiversity in East African forested landscapes. In: Zuidema PA (ed) Tropical forests in multi-functional landscapes. Seminar proceedings, Prince Bernhard Centre, Utrecht University, Utrecht, The Netherlands

Wiersum KF, Ros-Tonen MAF (2005) The role of forests in poverty alleviation: dealing with multiple millennium development goals. North-South Policy Brief, Wageningen UR

Wilshusen PR, Brechin SR, Fortwangler CL, West PC (2002) Reinventing a square wheel: critique of a resurgent "protection paradigm" in international biodiversity conservation. Soc Nat Resour 15:17–40

Wollenberg E, Nawir AS (1998) Estimating the incomes of people who depend on forests. In: Wollenberg E, Ingles A (eds) Income from the forest: methods for the development and conservation of forest products for local communities. Center for International Forestry Research, Bogor, Indonesia

Wolters OW (1967) Early Indonesian commerce: a study of the origins of Śrīvijaya. Cornell University Press, New York, NY

Wong JLG (2000) The biometrics of non timber forest product resource assessment: a review of current methodology. Unpublished manuscript. Available from: http://www.etfrn.org/etfrn/workshop/ntfp/

Wunder S (2001) Poverty alleviation and tropical forests – what scope for synergies? World Dev 29:1817–1833

Part II
Multiple Roles and Values of Non-timber Forest Products

Chapter 3
From Subsistence to Safety Nets and Cash Income: Exploring the Diverse Values of Non-timber Forest Products for Livelihoods and Poverty Alleviation

Sheona Shackleton, Claudio O. Delang, and Arild Angelsen

Abstract Millions of rural and urban dwellers across the world make use of a wide diversity of forest products to fulfill several livelihood requirements, from direct household provisioning to cash income, cultural needs, and as a fall back in times of emergency or a means to income diversification. All these roles are significant, and often NTFPs perform multiple functions simultaneously. Valuing NTFPs therefore requires a holistic perspective that considers these products in relation to multiple livelihood strategies, and within particular contextual settings. The context within which people operate has major implications for the perceived importance and value of NTFPs. Chapter 3 deals with these issues, drawing on data from several countries to illustrate the benefits NTFPs bring to different types of households and how such forest products perform an important function in reducing vulnerability and ameliorating poverty.

S. Shackleton (✉)
Department of Environmental Science, Rhodes University, Grahamstown, South Africa
and
Center for International Forestry Research (CIFOR), Bogor, Indonesia
e-mail: s.shackleton@ru.ac.za

C.O. Delang
Department of Geography and Resource Management, The Chinese University of Hong Kong, Shatin, N.T., Hong Kong
e-mail: cdelang@cuhk.edu.hk

A. Angelsen
Center for International Forestry Research (CIFOR), Bogor, Indonesia
and
Department of Economics and Resource Management, Norwegian University of Life Sciences (UMB), 1430 Ås, Norway
e-mail: arild.angelsen@umb.no

S. Shackleton et al. (eds.), *Non-Timber Forest Products in the Global Context*, Tropical Forestry 7, DOI 10.1007/978-3-642-17983-9_3,

3.1 Introduction

Forest-dwellers across the globe have a long history of dependence on a wide array of forest products for household sustenance and sale, with some 1.4–1.6 billion people worldwide estimated to make use of at least some non-timber forest products (NTFPs) (FAO 2001, Table 3.1). These products are collected from natural forests, woodlands, and other lands (such as fallows, agro-forests, secondary forests and fields) surrounding dweller's villages and homesteads, and may include wild foods, forage, medicinal plants, construction materials, fuelwood, raw materials for handicrafts (rattans, vines, bamboo, grasses, reeds, and other fibres), and other products such as resins and honey (Chap. 1; Figs. 3.1, 3.2, 3.3 and 3.4). While subsistence gathering for direct household provisioning tends to be the most widespread use of NTFPs (including in developed countries), forest products are also often sold, in raw or processed form, in informal local and regional markets. In other instances, NTFPs may form the basis of small-scale commercial enterprises, with some commodities reaching high-value international markets (Chap. 4).

Table 3.1 Estimates of number of forest users in different categories worldwide (Scherr et al. 2003)

Category of user	Estimated population of users
Forest dwellers who depend on natural forests for their livelihoods (hunting, gathering, shifting cultivation)	60 million
Rural people who live in or at the margin of forests and who rely on these as a safety net or for supplemental income	350 million
Small-holder farmers who grow farm trees or manage remnant forests for subsistence and income	0.5–1.0 billion
Artisans or employees in formal or informal forest-based enterprises (often urban based)	45 million
Estimated total	1.0–1.4 billion

Fig. 3.1 Rural collectors bringing their harvest of *Sclerocarya birrea* fruits (marula) to sell at the central depot for Amarula Cream liqueur, Limpopo Province, South Africa (photo: Myles Mander)

Fig. 3.2 Hand brooms made from *Athrixia phyllicoides* for sale in a rural month end traveling market, Bushbuckridge, South Africa (photo: Sheona Shackleton)

Fig. 3.3 Animal woodcarving from local softwood sold along tourist routes outside Kruger National Park, South Africa (photo: Sheona Shackleton)

Fig. 3.4 A teenage girl selling marula beer (made from *Sclerocarya birrea* fruits) at the roadside to passersby, Bushbuckridge, South Africa (photo: Sheona Shackleton)

The value of these NTFP-based activities to households is significant, and they undoubtedly contribute appreciably to the livelihood security and welfare of primarily poor, rural people (Bryon and Arnold 1999). For example, recent work to place an economic value on both the auto-consumption (subsistence or in-kind use) and sale of NTFPs at household level has shown this to be worth several hundreds of dollars per annum per household (Vedeld et al. 2004, Table 3.2). Moreover, the income share (percentage contribution to total income) of these products can be as much as one third to one half of total household income, with an average contribution of about one fifth (Paumgarten and Shackleton 2009). In many regions, the largest proportion of this value can be attributed to fuelwood consumption, followed by wild foods and construction materials and medicines (Vedeld et al. 2004; Paumgarten and Shackleton 2009; Babulo et al. 2009). Disaggregation of these findings by household wealth status suggests that NTFPs are most significant for poorer households, although this does not mean they are not used and appreciated by wealthier groups (Shackleton and Shackleton 2006; Rueff et al. 2009; Paumgarten and Shackleton 2009). Cavendish (2000), for instance, working in the savannas of

Table 3.2 Value of NTFPs to rural households derived from several studies

Region/country	Value (USD per household per year)[a]	Percentage contribution to total household income	Source
South Africa, Mogano, Limpopo Province	1,130	–	Shackleton et al. (2002b)
South Africa, Mametja, Limpopo Province	620	–	Twine et al. (2003)
Zimbabwe	578	–	Campbell et al. (1997)
South Africa, Bushbuckridge, Limpopo Province	572	19.4	Dovie (2004)
South Africa, Ha-Gondo, Limpopo Province	565	–	Shackleton et al. (2002b)
South Africa, Kwajobe, KwaZulu-Natal	469	–	Shackleton et al. (2002b)
Zimbabwe	436	28.4	Cavendish (2000)
Eritrea, Dighe, Gash-Barka Administrative Zone	386	–	Araia (2005)
Botswana	335	20.1	Zitzmann (2000) in Chipeta and Kowero (2004)
Zimbabwe	320	–	FAO (1999)
Thailand	288[b]		Delang (2006)
South Africa, Kat River, E. Cape	241	–	Shackleton et al. (2002a)
South Africa, Fish River, E. Cape	160	–	Cocks and Wiersum (2003)
Zimbabwe, Chivi	99	15.0	Campbell et al. (2002)

[a]Values are not directly comparable as different studies have varying criteria regarding what to include/exclude from the analysis. Local currencies have been converted to USD at the exchange rate for the year fieldwork was completed
[b]Direct household consumption only

Zimbabwe found that "environmental income" (including forage for livestock production) formed some 40% of total income for the poorest households, relative to 29% for more well-off households. On the other hand, Ambrose-Oji's (2003) work in Cameroon suggests that it is the middle income groups who benefit most from the use and sale of NTFPs. It is these initial studies that prompted the Center for International Forestry Research (CIFOR) to initiate the PEN (Poverty and Environment Network) study in which some 40 PhD students worldwide set-out to estimate the contribution of forest income to household income streams by using a standardised survey instrument. Preliminary results demonstrate similar ranges of income share to those in Table 3.2 (http://www.pen.cifor.org).

While clearly there is a strong rural dependence on forests and their products, some NTFPs (such as fuelwood, charcoal, bushmeat, medicinal plants, construction timber, traditional brooms) are also in demand in towns (Chap. 6), either because rural dwellers have migrated and still want to consume traditional forest products, or because town dwellers believe that these products are better or cheaper than mass manufactured alternatives. Urban households thus benefit from the availability of a more affordable source of essential goods, as well as access to products of traditional significance (Chap. 5). Furthermore, many poor urban men and women are artisans, processors, and end traders of high demand NTFPs such as medicinal plants, indigenous foods, charcoal, building materials, furniture, and crafts (Stoian 2005; Shackleton et al. 2007b; Fig. 3.4).

Estimates of the global demand for NTFPs and the significance of these products in securing different aspects of household livelihoods are revealing. Table 3.1 illustrates numbers of NTFP users, according to the use and importance of NTFPs for their livelihoods and survival. For about 60 million people, NTFPs are essential, while a further 350 million use NTFPs in times of crisis, such as a harvest failure or ill health that prevents them from working on farms. Between five hundred million and one billion people manage remnant forests for subsistence or sale of NTFPs, and a further 45 million people are artisans or employees transforming NTFPs into marketable products, often in urban centers.

Against this background of dependency on forest products and opportunities for trade and cash income, this chapter explores the role that forest and other ecosystem products play in reducing vulnerability, securing livelihoods, and increasing incomes among NTFP users. In particular, we aim to answer the following questions: (a) Under what circumstances are the consumption and trade of NTFPs a rational economic choice, and when is it an activity that people engage in because of lack of alternatives? (b) Who benefits most from NTFP use and sale? (c) How do NTFPs assist in securing livelihoods and reducing vulnerability? and (d) How do NTFPs contribute to poverty alleviation and help poor people accumulate assets and move out of poverty? We do this in three sections. The first section describes the role of NTFPs in the subsistence of rural communities, including their importance in meeting basic needs and saving household expenditure. The second examines why households turn to trading in NTFPs, while the third section describes the role that the consumption and trade of NTFPs has in poverty alleviation. We illustrate how NTFPs are decisive in providing poor people with a means to cope with crises and how they can help build household and community resilience. We also discuss where and how NTFPs can contribute to small enterprise development, providing a "stepping stone" (Marshall et al. 2006) for people to escape poverty.

3.2 Household Subsistence Use: NTFPs in Meeting Everyday Needs

3.2.1 NTFPs in Household Provisioning

The collection of NTFPs for subsistence use is prevalent in rural communities across the world (Table 3.3 provides figures from South Africa). This dependence can partly be traced back to geographic constraints. Many NTFP users live, almost by definition, far from market towns. This distance, compounded by poor road networks, makes transport costly, hindering participation in the market economy, with the result that households may have little cash at their disposal and limited opportunity to purchase their daily necessities. In such situations, forests provide essentials that others buy in markets. Wild foods are effectively free (other than the opportunity costs of labour), and if a forest dweller had to choose between gathering food from the forest or spending more time and limited cash resources going to town to purchase commercially grown alternatives, it is rational to choose the first option. Medicinal plants offer free self-medication, while rattan, bamboo, wood, vines, and grasses are free raw materials from which people can make baskets, mats, fences, roofs, walls and agricultural implements, and wood provides a free or cheap energy source. Such daily subsistence use of NTFPs allows households to save their meager cash resources for goods and services that cannot be obtained locally (Shackleton and Shackleton 2004). As a result, although the collection of NTFPs for household use is an activity that is generally available to all households, and in which a majority engage, it is likely to be more important for and more

Table 3.3 Prevalence of use (mean ± SE) of NTFPs from South African savannas (n = 14 villages; 30–60 households per village) (Shackleton and Shackleton 2004)

NTFP	Percentage of users	NTFP	Percentage of users
Wild spinaches	95.6 ± 1.3	Bushmeat	51.6 ± 8.4
Fuelwood	95.5 ± 1.9	Wild honey	50.5 ± 10.6
Wooden utensils	95.1 ± 1.9	Medicinal plants	49.4 ± 7.5
Grass brushes	90.7 ± 4.6	Wood for housing poles	49.0 ± 8.1
Wild fruits	88.2 ± 4.0	Thatch grass	48.8 ± 9.0
Twig brushes	87.1 ± 5.1	Wild mushrooms	25.2 ± 9.2
Wood for fencing	62.0 ± 5.5	Reeds for construction	14.6 ± 6.5
Weaving materials	55.4 ± 9.6	Wood for furniture	6.7 ± 1.7
Edible insects	53.5 ± 9.5	Seeds for decorations	3.2 ± 1.8

widely exploited by poorer groups with limited land and other assets, minimal education and skills, and few other income sources (Cavendish 2000; Paumgarten and Shackleton 2009).

In terms of subsistence use, NTFPs are critical for health, nutrition, shelter, and energy. While the most important NTFPs for rural communities may vary and are often context specific, some products that traditionally played a key role have become less important and others have remained central. Among the former are NTFPs that have been replaced by more "modern" and "efficient" goods (plastic household utensils, furniture), and among the latter are those that are not easily replaceable, or only at high cost (fuelwood, wild food plants, fencing and construction material).

3.2.2 NTFPs for Energy

Fuelwood and charcoal are used by the majority of rural, and in some cases urban, households across the developing world (e.g., Babulo et al. 2009). For example, in South Africa (one of the more developed African nations) over 80% of rural households still use fuelwood to some extent (Williams and Shackleton 2002). Nearly all of this, some 10 million tons annually, is supplied from indigenous forests and savannas and has a gross national value of approximately USD 0.40 billion (R3 billion) annually, or just under USD 182 (R2,000) per using household per year (some 23% of the minimum wage). In Nepal, Shrestha (1998) showed that over 13 million tons of fuelwood is consumed annually, with the residential sector accounting for over 91% of use. It has been estimated that more than 2.4 billion people in Nepal rely directly on traditional biomass fuels for their cooking and heating (IEA 2002). The use of fuelwood has been implicated in deforestation, especially in arid areas (Heltberg et al. 2000) and at high altitudes, such as Nepal (Stevens 2003), but also in tropical countries such as Vietnam (Linde-Rahr 2003). However, the picture is seldom simple, and this conclusion has been challenged conceptually and practically in many situations (Benjaminsen 1997; Nagothu 2001; Box 2.1, Chap. 2). A shortage of fuelwood can result in changed cooking patterns with potentially ill effects on household nutrition and health (Brouwer et al. 1997) (Table 3.4).

Table 3.4 FAO projection of fuelwood consumption (in million cubic meters) to 2030 in developing regions (Arnold and Persson 2003)

Year	1970	1980	1990	2000	2010	2020	2030
South Asia	234.5	286.6	336.4	359.9	372.5	361.5	338.6
South East Asia	294.6	263.1	221.7	178.0	139.1	107.5	81.3
East Asia	293.4	311.4	282.5	224.3	186.3	155.4	127.1
Africa	261.1	305.1	364.6	440.0	485.7	526.0	544.8
South America	88.6	92.0	96.4	100.2	107.1	114.9	122.0

3.2.3 NTFPs for Food

Forest and woodlands offer rural dwellers a wide variety of foods, and contribute to food security and nutrition directly and indirectly by providing fruits, seeds, leaves, bulbs, mushrooms, honey, beverages (Fig. 3.4), bushmeat and fish, forage for domestic animals, and tools and equipment needed for farming, hunting, and fishing. This food security dimension of forests is important; for example in the Gash-Barka administrative zone of Eritrea, local people rated the provision of wild foods as the most important ecosystem service provided by riverine forests (Araia 2005). Wild foods are commonly eaten because they are nutritious and rich in vitamins and supplement cultivated staples (Grosskinsky 2000), and because isolation from markets precludes people from buying food. Delang (2006) has shown that a group of forest dwellers in Thailand would need ten times more time to work for cash and buy food in the market than they need to gather "similar" food in the forest.

The diversity of wild foods used can be considerable. For example, forest-based shifting cultivators in Laos obtain a range of species from fallow areas, which change as the fallow ages (Delang 2007). An 11-year old fallow offers cultivators 126 different usable plant taxa, of which 55 are food plants and 12 animal feed. In comparison a 1-year old fallow provides 33 taxa, of which 16 are used for food, while a 3-year old fallow offers 13 taxa, of which seven are used for food. Plans by the government to reduce the period of fallow to 3 years will thus have negative consequences on people's nutrition, as the number of edible plants available in the fallows would decline.

While wild foods are commonly a regular part of the diet, their consumption may be extended to additional species or become more frequent during droughts, floods, or other lean times, or they may substitute for purchased products during cash flow crises (Kaimowitz 2003). In the dry forest regions wild foods are important in supplementing people's diets, and may assume greater significance in the dry season or dry years when they substitute for failed crops (Addis et al. 2005). The so-called "famine foods" of the Sahel region are wild foods obtained from drought resistant dry forest species that may only be consumed in years of severe drought, but are vital for food security during these times (Guinand and Lemessa 2001; Chap. 2). In this way these NTFPs help insure against food insecurity; something that may assume magnified significance under the threats of climate change.

Wild foods are often consumed most frequently by women and children, although this does not necessary apply in all countries (Box 3.1, Table 3.5). Wild food resources also tend to be most important for vulnerable households. For example, data from South Africa show that 62% of 850 children surveyed were supplementing their diets with wild foods; and for 30% over half their diet was formed by these resources (McGarry and Shackleton 2009a). Furthermore, highly vulnerable children (i.e., those in households with high HIV/AIDS proxies) were found to consume more forest foods in their diet than those from less vulnerable households, with hunting of small animals providing an essential source of protein (McGarry and Shackleton 2009b, Table 3.5). Similarly, Hunter et al. (2007) found

Box 3.1 Gender, Wild Foods, and Food Security: An Example of the Hazda in Tanzania

Sheona Shackleton, Department of Environmental Science, Rhodes University, Grahamstown, 6140, South Africa, s.shackleton@ru.ac.za

A study by Murray et al. (2001) on the diets of the Hazda in Tanzania showed that Hazda men eat more meat and honey than women, while women depend more on plant foods, in particular baobab seeds. From a cultural perspective, men have the "better deal": meat and honey are high status foods. But, from a nutritional perspective, men and women's diets are of similar value: baobab seeds are a good source of protein, fat, and energy, equivalent to that of honey. Furthermore, from an access and seasonality perspective, women's diets are favoured. Although baobab seeds and fruit are not the dominant wild foods eaten by Hazda women, they can be consistently collected throughout the year from a variety of locations. These range from the baobab trees themselves, to foraging for discarded seeds in baboon dung piles during nonproductive periods. When these factors were weighed-up, the authors concluded that the baobab provides the most important and nutritionally reliable food in women's diets.

Table 3.5 Frequency (total count of animals) of wild animals in high vulnerability (HV) and low vulnerability (LV) children's diets over a 2-week period, as well as the percentage of children hunting each taxonomic group at Coffee Bay and Mabehana, South Africa

Taxa	Frequency of wild protein in the diet			Percentage of children hunting		
	LV ($n = 24$)	HV ($n = 25$)	p	LV ($n = 24$)	HV ($n = 25$)	p
Mammals	39	133	0.04*	33	60	0.06
Birds	89	195	0.02*	25	64	0.01*
Reptiles	8	13	0.33*	4.2	12	0.03*
Insects	3	13	0.05	4.2	24	0.05
Coastal resources	13	9	0.37	91	96	0.53

*$p < 0.05$ (McGarry and Shackleton 2009a, b)

that households affected by a recent adult mortality (usually from AIDS) often turn to harvesting wild foods and other NTFPs to cope.

3.2.4 NTFPs for Medicine

Medicinal plants have been important in human healthcare throughout history, and continue to play a key role amongst forest communities. For rural populations, modern medicine and healthcare services are often difficult to access and unaffordable, and thus beyond most people's reach (Maundu et al. 2005). Traditional medicines, on the other hand, are locally available and free or of low cost. For example, users in Chitwan (Nepal) save 80% of costs by using herbal treatments provided by the *Guraus* (local faith healers) rather than commercial alternatives bought in pharmacies (KC 2003). Although in some regions medicinal plants are gradually being replaced by commercial pharmaceuticals, use is still sizeable in many countries. For example, 90% of rural and 40% of urban Nigerians use medicinal plant products (Osemeobo and Ujor 1999). In Ethiopia, over 85% of the rural population, plus an increasing number of people in urban centers, use medicinal plants as their primary source of healthcare (Deffar 1998). Moreover, there is evidence to indicate that some plants used for cultural and ritual medicinal purposes are not substitutable and thus continue to be significant for all wealth groups in both rural and urban populations (Cunningham 1997; Cocks and Dold 2006).

3.2.5 NTFPs in Construction and Craft

In terms of construction material, NTFPs play an important, although in some areas declining, role in building and fencing. In tropical Southeast Asia and South Asia, bamboo is one of the main materials for construction (Pant 2007). Easily gathered and transported to the village, it can be used for the support poles, walls, or roof of a house and is the preferred material for the farmland hut, which is occupied for 5 months every year between rice sowing and harvesting. Bamboo is sometimes also used to build the first house for a newly married couple, before the necessary construction wood can be cut, accumulated, and brought from the forest (Pant 2007). Thatch grass as a construction material remains important in many areas, even though there is a trend toward alternative roofing materials (Hawkes 1992). Among some ethnic groups (e.g., the Xhosa in South Africa), cultural norms dictate that at least one dwelling in the homestead must be thatched (Timmermans 2004). Palm leaves are also used in house construction for both walls (plaited sheets) and roofing in Asia and Africa. In Eritrea, the average traditional house requires some ten "camel loads" of palm leaves a year for routine maintenance (Araia 2005).

Historically, natural fibres from forest plants provided the raw material for a wide range of utilitarian goods such as clothing, ropes, basketry, fishing nets, brooms, mats, and construction materials. Today, a number of alternative products have replaced functions typically provided by fibre products. For instance, bought fabrics have replaced home produced natural fabrics. Cheap or free plastic bags and plastic or metal storage bins now substitute for baskets (de Vletter 2001; Kgathi et al. 2005). Nylon ropes have replaced hand woven bark rope. However, fibre use still remains an option for those too poor to afford alternatives or in isolated communities. For example, communities living close to riverine forests in western Eritrea use 21 different household items made from dom palm, with the direct-use value of these being the highest of all NTFPs surveyed, at USD 80 per household per year (Araia 2005). Woven mats, ropes, and basketry products are now often traded both within local and regional markets for cultural and traditional purposes and in nontraditional and tourist markets for their decorative and novelty value (Pereira et al. 2006, Box 3.2).

Box 3.2 Mat Weaving Empowers Women in Rural Villages of South-Central Bangladesh and Secures their Livelihoods

Sharif Ahmed Mukul, Department of Forestry and Environmental Science, Shahjalal University of Science and Technology, Sylhet 3114, Bangladesh, sharif_a_mukul@yahoo.com

Mat weaving using *Typha elephantina* (locally named *hogla*) has long been a traditional pastime of rural women in agrarian Bangladesh. In villages in the south central part of the country this practice has become a popular income generating activity, particularly for rural women who have not had adequate access to education or training. For decades the low lying and unsuitable agricultural land in the area has been used for the cultivation of *hogla* at minimal cost and effort. A survey conducted during 2007–2008 in the tri-weekly market place (locally called *hat*) of three neighbouring villages in Noakhali district revealed that women from about 256 families in the three villages were involved in *hogla* mat weaving.

These women spend their time between their regular household activities weaving *hogla* mats. On average, each woman sold ten standard sized (1.8 × 2.4 m) mats every week. Their average estimated income was Tk. 700 (USD 10) per week. The cost to purchase dried *hogla* leaves was Tk. 240 (USD 3.5), leaving a profit of Tk. 460 (USD 6.5) per week. This makes mat weaving the second most prominent source of family income after agriculture. Overall some Tk. 315,000 (USD 4,532) worth of *hogla* mats and Tk. 95,000 (USD 1,365) worth of raw *hogla* leaves were sold each week in the three surveyed markets. Although there is a permanent market and source of *hogla* leaves, the major trade in leaves takes place during September–November. The prices of both *hogla* leaves and mats falls by (up to 25%) during this period and women are required to weave more mats to maintain their returns. Some women also store and purchase adequate *hogla* leaves at a comparatively low price during this time for use during the rest of the year.

The entire production and supply chain of *hogla* mats is dominated by women – they weave mats, take them to nearby markets to sell them themselves or to male traders, and return home after purchasing the raw materials for further weaving work, along with other necessary goods for their family. Elderly family members and young school-going children (after school) also often provide assistance in this activity. For some elderly, widowed, divorced, and vulnerable women in the area, mat weaving is their only source of income. Generally women invest their earnings from this business in

(continued)

(a) supporting family nutrition, education, and health checkups for their children; (b) paying off regular installments for small loans; and (c) saving for use during emergencies and the agricultural off-farm period. In fact, these basic contributions have made some apparently simple and conservative women of the area important and active players in their families and communities. Their ability to earn independent income empowers them and provides them with the confidence to voice their opinions more loudly when taking family decisions.

Several recent studies have shown that the use of traditional brooms is extensive (Cocks and Dold 2004; Gyan and Shackleton 2005; Shackleton and Campbell 2007, Chap. 2). Twig brooms are used for sweeping outdoors, while grass or reed brooms are used to clean inside the house. In Bushbuckridge in northeast South Africa, all households and most schools and local businesses use these brooms, resulting in an annual turnover of some 360,000 brooms (Shackleton and Campbell 2007). Brooms are also of customary and spiritual significance and are given as wedding gifts in Xhosa culture. They are believed to help prevent lightning strikes and bring good luck (Cocks and Dold 2004). Brooms made from palm fibres or midribs are also common in many regions. In Dighe administrative subzone in Eritrea, the average household owns eight dom palm brooms (Araia 2005).

The above discussion leaves little doubt that forests and woodlands continue to be a key source of products needed for everyday life for many communities around the world. Indeed, in most instances it makes economic sense to continue to use NTFPs. Without access to these products, households would be worse off and would be forced to use scarce cash resources to purchase basic essentials such as energy and food. Subsistence use of NTFPs thus contributes to reducing vulnerability by ensuring all households have access to at least some of their basic needs at no or little cost. This function can be thought of as a "daily safety net" (Shackleton and Shackleton 2004).

3.3 Trading and Commercialisation: NTFPs for Income Generation

3.3.1 NTFPs for Cash: A Growing Business

Apart from subsistence consumption, the sale of NTFPs offers an important means for rural and urban individuals and households to generate cash income (Shackleton et al. 2007a, b). For example, in sub-Saharan Africa alone it is estimated that several million people earn their primary cash income from the sale of forest products (Kaimowitz 2003). In the forest zone of southern Ghana about 20% of the economically active population earns income from selling forest products, while in the Brazilian Amazon about 1.5 million people derive part of their earnings from NTFPs (Ruiz Pérez and Byron 1999). In South Africa, some 3–14% of rural households within the savanna biome [roughly 230,000 households (Statistics South Africa 2003)] are trading at least one NTFP, albeit often on an irregular basis (Shackleton and Shackleton 2004).

This extensive trade in NTFPs appears to be growing worldwide (Neumann and Hirsch 2000; Campbell et al. 2002; Ruiz Pérez et al. 2004, Table 3.6). Growth is being driven at the local level by a greater need for cash income as households become more integrated into the market economy and have higher expectations, and by economic hardship and increased vulnerability due to, among other factors, unemployment, withdrawal of agricultural subsidies, and HIV/AIDS (Monela et al. 1999; Campbell

Table 3.6 Examples from around the world of the extent, size, and value of the trade in different NTFPs (also see Chap. 4)

Medicinal plants	Bangladesh – some 12,000 tons of dried medicinal plants worth around USD 4.5 million are sold annually from rural areas (SEDF/IC 2003). Southern Africa – the trade in medicinal plants is valued at USD 75–150 million per annum with some 35,000–70,000 tons of plant material traded each year (Mander and le Breton 2006)
Baskets	Botswana – commercial buying started in early 1970 in Ngamiland District. In that first year USD 500 worth of baskets was bought from a handful of women, by 1990 this increased to USD 115,000 per year to more than 2,000 women. By 2000 the value of the trade was some USD 350,000 per year
Gums and resins	Ethiopia – the value of gum and resin exports from 2001 to 2003 amounted to USD 2.8 million, 3.3 million, and 4.1 million respectively. Natural gum tapping and collection activities create seasonal employment opportunities for 20,000–30,000 people (SNV 2005)
Woodcarvings	Kenya – the woodcarving industry is worth over USD 20 million annually in export products and employs some 60,000–80,000 carvers supporting over 400,000 dependents (Choge 2004)
Honey	Zambia and Tanzania are two dry forest countries exporting the largest volumes of honey. In Zambia in 2005, 219 tons of honey were exported with a value of USD 491,000, while Tanzania exported 466 tons with a value of USD 674,000. Volumes exported have risen by 20–30% since 2001 (ITC 2006)
Oils – Shea butter	Burkina Faso – shea butter provides income to about 300,000–400,000 women (Harsch 2001; Schreckenberg 2004). Imports of shea butter to Europe from Sahelian countries were estimated at USD 13 million in 1999 (Schreckenberg 2004)
Insects	Botswana – the trade in mopane worms was valued at UK£4.42 million in 1995 and employed as many as 10,000 local people (Styles 1994; 1995)
Wood and charcoal	Tanzania – in 2002 some 21.2 million m^3 of wood, equivalent to 625,500 ha of woodland, were used for 43.7 million bags of charcoal with a net annual value of USD 4.8 million (Scurrah-Ehrhart and Blomley 2006). Burkina Faso – the fuelwood commodity chain of the city of Ouagadougou is estimated at USD 10.6 million (Ouedragogo 2006)

et al. 2002; Shackleton et al. 2008). Additionally, NTFP commercialisation is being promoted and facilitated through the creation and stimulation of external markets by agencies concerned with linking rural livelihoods to the conservation of natural ecosystems (Neumann and Hirsch 2000; Arnold and Ruiz Pérez 2001, Chap. 10). These interventions are generally aimed at addressing the dual goals of livelihood enhancement and biodiversity conservation, based on the assumption that the livelihood benefits obtained from selling NTFPs will provide an incentive to conserve the resource base (Arnold and Ruiz Pérez 2001, Chap. 10). Consequently, it is not uncommon to find building materials, fuelwood, charcoal, indigenous food-stuffs, medicines, craft items (from wood, grass, reeds, vines), furniture (e.g., Figs. 3.1, 3.2, 3.3 and 3.4), and other more specialised products such as resins, paper, and perfumes for sale in local, national, and even international markets (de Beer and McDermott 1996; Alexiades and Shanley 2004; Sunderland and Ndoye 2004; Marshall et al. 2006).

For some households, NTFPs may be an obvious source of cash because their production and harvest requires little capital and labor inputs, and people have the knowledge and skills to undertake these activities. These low barriers to entry combined with the fact that, in many instances, access rights are held in common and so tenureship may not be claimed by more powerful individuals and groups (Beck and Nesmith 2001), means that this activity provides an important option for poor and marginalised households (Beck and Nesmith 2001; Fisher 2004; Shackleton et al. 2008). These are typically households who would have difficulty accessing other employment opportunities, or who are less able to cope with or insure against risk than better-off households (Fisher 2004; McSweeny 2004; Takasaki et al. 2004). Women in particular benefit widely from the use and sale of NTFPs (Box 3.2), as do older and less educated people who cannot compete effectively in the job market, and young households with few liquid assets (Schreckenberg and Marshall 2006). For example, the Shea butter trade in Burkina Faso provides income to some 400,000 women (Table 3.6). NTFPs may also be harvested by children, and, at times, might be the only source of cash that they are able to secure for themselves or their households (McGarry and Shackleton 2009a; Fig. 3.4).

3.3.2 Why Trade in NTFPs?

Conceptually, there are four main reasons why people might engage in the trade of NTFPs, with some households trading for all of these reasons. Applying the typology presented below, and thinking about each reason separately, can help us understand the role that the sale of NTFPs has to different categories of people and households, and in different contexts and situations.

1. *Trading NTFPs in response to emergency or misfortune:* One reason people might trade in NTFPs is that these products can provide cash in times of emergency or misfortune. In this case, NTFPs can be considered as a form of natural insurance (Pattanayak and Sills 2001; McSweeny 2004; Takasaki et al. 2004, Box 3.3) available at critical times to bridge income gaps, deal with shocks, or to meet specific needs

such as school fees or the costs of celebrations. Such trade may exist on a regular basis, but increase in times of shock, such as harvest failure. In these situations these products function as safety nets, providing people and households with fallback options (McSweeny 2004, 2005; Takasaki et al. 2004). For example, following the devastation left by Hurricane Mitch in Honduras in 1998, McSweeny (2005) showed that households unable to recoup their landholdings sold forest products to self-insure despite government enforcing a ban on forest product exploitation.

2. *Trading NTFPs for livelihood diversification, risk reduction, and income smoothing*: Another reason for trading NTPFs is that the latter may be part of an income diversification or risk reduction strategy, as households or individuals seek ways to supplement other sources of income or smooth their earnings throughout the year. For example, the NTFP trade often complements agricultural production in many regions of the world (Ros-Tonen and Wiersum 2005). In such cases, the use of NTFPs is complementary to a range of other livelihood activities and income sources. "Income smoothing" is a widely mentioned benefit of NTFP trading, especially at times when on-farm labour is in low demand. For example, in south-eastern Nigeria 35.7% of the rural population collected NTFPs daily, and the income from the sale of these NTFPs accounted for 94% of the total income from minor sources, providing a considerable smoothing effect, especially during hunger periods (Nweze and Igbokwe 2000).

3. *Trading NTFPs as a regular or primary source of income and as a "stepping stone" out of poverty:* Some households may trade NTFPs to gain regular sources of income. In such cases, the trade may be the primary source of income for the household, resulting in high levels of specialisation (Ruiz Pérez et al. 2004; Kusters and Belcher 2004). Such a scenario is most likely for high value-added products, often with external markets (e.g., woodcarving in Bali, Rohadi et al. 2004; Chap. 4). As such, relatively few households find themselves in this position because of the relative scarcity of NTFPs with large external markets and the higher barriers to entry for such products (Marshall et al. 2006; Belcher and Schreckenberg 2007). In these cases, if the value of the product increases significantly, then the danger exists that the trade will be captured by elites, or the NTFPs will be exhausted after a few years (Dove 1993). Often engagement in the NTFP trade as a "stepping stone" is a matter of choice rather than necessity (Marshall et al. 2006; Schreckenberg et al. 2006).

4. *Trading NTFPs due to a lack of alternatives*: The last reason households may trade in NTFPs relates to a lack of other livelihood or cash income choices. It is not uncommon to find that individuals and households turn to the informal economy, in this case the sale of readily accessible products, in the absence of alternative income earning opportunities (e.g., Shackleton et al. 2008). Such a situation could be considered as a necessity diversification response (Marshall et al. 2006). However, unlike the short-term safety net functions described above, this may evolve into a long-term source of livelihood if the conditions that initially drove the individual or household into the trade prevail, or if the trade becomes a preferred option. For example, the majority of hardwood carvers in Bushbuckridge, South Africa, have been in business for 30–50 years, and this is despite having started carving as a temporary fallback option after becoming unemployed (Shackleton 2005). Carvers

mentioned how working from home (rather than being migrant workers), being with their families, as well as making adequate income were the primary reasons they decided to continue with the trade.

It is usually the poorest households that have most incentive to trade NTFPs. Wealthier households often have access to alternative sources of cash that are denied to the poorest because of economic barriers to entry, be it livestock husbandry that needs the initial animals or trade that needs a means of transport. However, it is not necessarily always the poorest members of the community who engage in these activities. Ambrose-Oji (2003) argued that in Cameroon the households that are more actively involved in the collection and trade of NTFPs are middle-income households, probably in a "diversification and risk reduction strategy". At times, the poorest members of the community may collect NTFPs in the forest, while middle-income households market these in local towns, allowing them to control a larger proportion of the economic surplus, thus cementing economic inequalities. In India, Campbell et al. (1995) stated that poor rural NTFP collectors receive as little as 10–40% of the sale price in the nearest NTFP market. In other cases, NTFPs are said to have the potential to reduce economic inequalities, since they allow poor members of the community, with limited investments, to earn cash. Fisher (2004), for example, argues the latter after finding that the NTFP trade in Malawi had an income equalising effect. Similar results were found by Shackleton et al. (2008) for four products traded in the north-east of South Africa (Fig. 3.5). Moreover, the poor often sell locally to the more wealthy, thereby distributing local cash more widely from rich to poor (Shackleton and Shackleton 2006; Paumgarten and Shackleton 2009).

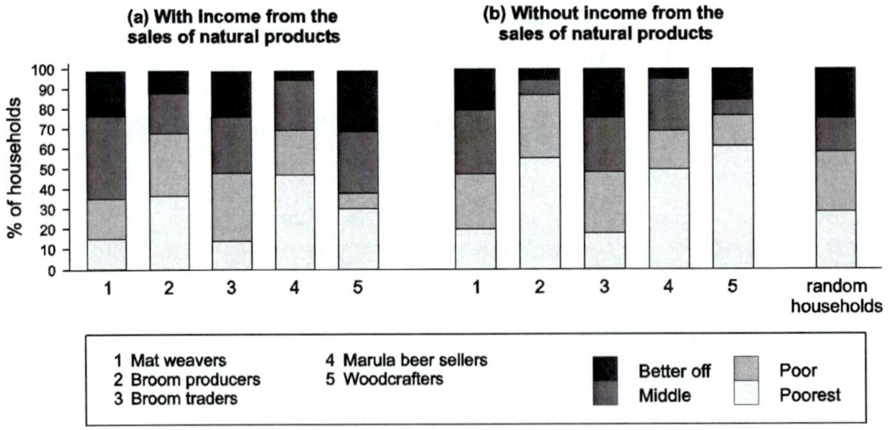

Fig. 3.5 Proportion of households (hh) for each product type that fall within four total annual household cash income classes when: (**a**) the net annual income obtained from the sale of natural products is included and (**b**) when the net annual income obtained from the sale of products is excluded. Classes are based on quartiles for the random household sample, where poorest = ≤ ZAR 3,600; poor = > ZAR 3,600–8,400; middle = > ZAR 8,400–15,090, and better off = > ZAR 15,090 to highest (Shackleton et al. 2008)

Table 3.7 Average and range of incomes generated from small-scale local trade in NTFPs in South Africa (Shackleton et al. 2007a)

Resource	Mean annual income (South African Rand)		Range (ZAR)	Year
Palm brushes	4,272	Net[a]	0–12,000	2002
Marula beer	500 (2 mths)	Net	90–2,230	2001
Fuelwood	338	Gross	12–1,400	1999
Medicinal plants	16,740	Gross	360 to >40,000	1999
Woodroses	2,895	Gross	640–6,000	1997
Softwood carvers	9,840	Net		1997
Medicinal plants	39,480	Gross		1996
Hardwood carvers	3,600	Net	2,580–16,930	1994
Mopane worms	2,500 (1 mth)	Gross		
Aloe gel tappers	12,000	Net		1994
Wild fruits	1,045	Gross	240 to >20,000	1993

[a]Gross income excludes costs while net income represent profits after deducting costs

3.3.3 Varying Benefits for Different Households

Average incomes from the trade in NTFPs, at local or international level, are often marginal, amounting to a few hundred dollars per year (Table 3.7). For this reason it is frequently argued that NTFPs have little role in contributing to the fight against poverty (Wunder 2001). However, the ranges of income for a single product can be considerable, with some households doing well based on high levels of involvement and inherent entrepreneurial ability, while others opt to make the trade a more part-time, supplementary activity (e.g., Marshall et al. 2006; Shackleton and Campbell 2007). In other instances, as previously mentioned, incomes may be unevenly distributed along the market chain, with some actors doing much better than others. Indeed, Marshall et al. (2006) found that many households wanted to find ways to move onto another stage of the value chain to improve their incomes. Thus, while on an average incomes may not appear lucrative, a closer look suggests that it is misleading to make generalisations, as revenues and contributions are highly variable from household to household and product to product (Table 3.7). NTFP incomes cannot be considered in isolation of (a) other livelihood activities; (b) the overall income stream of households; (c) local wage rates; (d) alternative employment options, and (e) the constraints producers and traders face.

3.4 NTFPs in Reducing Vulnerability and Alleviating Poverty

In this section we draw on the above discussions to consider how NTFPs contribute to the global goal of poverty alleviation. We consider poverty alleviation in its broadest sense, which includes notions of (a) reduced vulnerability and greater security; (b) poverty avoidance, prevention or mitigation, i.e., preventing poor people and households from slipping deeper into poverty; and (c) poverty reduction or elimination, i.e., assisting people to escape poverty or raise household incomes above the poverty line (Angelsen and Wunder 2003). These understandings parallel the three sets of policy measures suggested by May et al. (1998) that might be used to alleviate poverty, namely (a) protective measures which are essentially safety net and disaster management measures that provide relief from deprivation; (b) preventive measures that try to avert deprivation, and (c) promotional activities that aim for an improvement of incomes and social consumption.

We have shown above that NTFPs can serve the function of reducing household risk and vulnerability (Box 3.3, Box 3.4), often helping to prevent households, particularly the poorest (Rueff et al. 2009; Shackleton et al. 2008), from sinking lower into poverty during difficult times. NTFPs are accordingly critical in building household resilience, a factor that is likely to become even more important, given the dire predictions around climate change. Thus, the conservation of natural capital, i.e., forest and woodlands, can, according to May et al.'s (1998) typology, prevent and avert deprivation and provide relief from deprivation (see Box 3.4, Babulo et al. 2009). Under these circumstances, NTFPs can be thought of as a safety net and key in mitigating poverty. A safety net is generally regarded as a source of emergency sustenance in times of hardship. Safety nets are mechanisms that mitigate the effects of poverty and other risks on vulnerable households (World Bank 2004). In the poverty literature, the term is often used to refer to public social security transfers such as unemployment benefits, child grants, pensions, subsidies, relief aid, etc., as well as the informal transfer of goods between households (World Bank 2004). In this case, it is the existence of natural capital that can be sustainably drawn in times of need that constitutes the safety net. Biodiversity and forest products, in other words, provide a form of insurance or cushion against risk, often when little else is available (McSweeny 2004).

Box 3.3 Role of NTFPs as a Way to Cope with Shocks in Two Regions of South Africa

Fiona Paumgarten, Center for International Forestry Research, Zambia (PO Box 2105, Johannesburg, 2000, South Africa), fi.paumgarten@gmail.com

A study conducted in two rural villages in South Africa examined the safety net function of NTFPs by determining the households' vulnerability context (i.e., the range of crises, both anticipated and unanticipated, experienced over a 2 year period) and the various coping strategies households rely on. Household wealth was used for comparison. All households reported at least one crisis. In response to these, households relied on a range of coping strategies (Table 3.8). NTFPs were used in response to each crisis offering both consumption- and income-smoothing options.

(continued)

Table 3.8 General coping strategies employed by households in response to anticipated and unanticipated crises (%)

Coping strategy	Total	Wealthy	Poor	X^2	Significance
Kinship	85	80	90	1.9	>0.05
Reduced spending	74	84	64	5.2	<0.05
Changed diet	72	84	60	7.1	<0.05
Saves/budgets	72	88	56	12.7	<0.05
NTFPs	70	68	72	0.2	>0.05
Selling livestock	44	58	30	7.9	<0.05
Stokvels	41	64	18	21.9	<0.05

Table 3.9 NTFPs used as a rural safety net, how the use manifests and the proportion of households using (of households reporting the safety net function)

	Total	Wealthy	Poor	X^2	Significance
Edible herb (fresh)	30.0	29.4	30.6	0.0	>0.05
Edible herbs (dry)	17.1	14.7	19.4	0.3	>0.05
Medicinal plants	40.0	38.2	41.7	0.1	>0.05
Edible fruits	11.4	8.8	13.9	0.4	>0.05
Bushmeat	7.1	0.0	13.9	5.1	<0.05
Fuelwood	25.7	41.8	11.0	8.3	<0.05
Sells fuelwood	10.0	0.0	19.4	7.4	<0.05
Building materials	8.6	2.9	13.9	2.7	>0.05
Sells other NTFPs	8.6	0.0	16.7	6.2	<0.05

With the exception of NTFPs and kinship, all the strategies were relied on by a significantly greater proportion of wealthy households, highlighting that a household's asset base determines the coping strategies at their disposal. For poor households, the use of NTFPs fell within the top three strategies in terms of proportion households.

Of the households that relied on NTFPs in response to crises, 40% used medicinal plants, 30% fresh herbs, 25.7% fuelwood, 17.1% dried herbs, 11.4% fruits, 8.6% construction materials, and 7.1% bushmeat. Ten percent sold fuelwood and 8.6% sold other NTFPs (reed mats, bushmeat, and grass hand-brushes). Although both wealthy and poor households relied on NTFPs, wealth influenced both the products selected and the manifestation of the safety-net function (Table 3.9). For example, no wealthy households sold NTFPs, and a significantly greater proportion of wealthy households relied on fuelwood (as a cost-saving substitute for paraffin).

The safety net function of NTFPs is particularly important to poor households with limited insurance options, although wealthy households also rely on certain products. Findings of the high proportion of households relying on NTFPs as a safety-net recommends further research and suggests that the maintenance of and continued access to communal lands and the resources provided may contribute significantly toward sustaining and improving the welfare of rural households.

Box 3.4 Using Changes in Welfare to Measure Forest Resource Value

Lucrezia Tincani, PhD student, Center for Development, Environment and Policy, SOAS, University of London, 36 Gordon Square, London, WC1H 0PD, UK lucrezia.tincani@soas.ac.uk

Poor rural households typically operate within a complex and risk-prone environment facing a variety of risks on a daily basis. Despite recent studies indicating the critical role of dry forests in mitigating the effects of extreme poverty in sub-Saharan Africa (Campbell et al. 2002; Shackleton and Shackleton 2004), many African governments still fail to make the link between continued degradation of dry forests and increasing poverty. One explanation could be that forest value is predominantly measured with market-value techniques, e.g., indicators such as Direct Use Value (DUV) (the resource quantity consumed and sold, multiplied with their market price). Because DUV is based on market prices, it may not accurately illustrate dimensions which are not captured in an incomplete market, such as opportunity costs.

This study tested this premise, comparing DUV techniques with stated-preference techniques, which uses hypothetical prices stated within an ideal/perfect market which captures all externalities accurately (Tincani 2009). Over 100 households (131) were surveyed within nine villages in rural Burkina Faso in May–July 2008. A hypothetical scenario was used, where the respondent was asked what amount he/she would be Willing To Accept in Compensation (WTAC) for a *marginal* loss (1 ha decrease) in his/her forest resources, over a *limited* time period (one season, lasting 4 months). Throughout this period, a foreigner would rent 1 ha of trees, which would be inaccessible to the respondent throughout the rental-period. The question was phrased to measure the minimum WTAC needed to maintain the same standard of living throughout the rental-period.

Results showed that households with fewer financial, physical (material wealth and livestock), natural, and human assets reported a significantly higher WTAC/ha. These asset-poor households had fewer assets to sell when they needed to buy food in the dry season, and had less labour available to earn money to buy food, therefore suggesting a predisposition to using public assets such as forest resources to maintain their standard of living. As confirmed through focus groups, these asset-poor households relied on the forest for their food, firewood, fodder, and medicinal herbs in the dry season.

Contrarily, DUV/ha was not significantly higher for asset-poor households. WTAC/ha far exceeded DUV/ha for asset-poor households, but both values reached similar levels for asset-rich households with incomes of >700,000 CFA/season. Contrarily to DUV, WTAC values included the opportunity cost of forest resources. This cost was highest for poor households, because these lacked the capital to replace forest resources with a

(continued)

purchased substitute. As household income increased, the opportunity cost declined, and WTAC/ha reached similar values as DUV/ha.

These results show that different methodologies capture different aspects of resource value, with WTAC highlighting opportunity costs, which are particularly relevant to poor households. As these households lack alternative coping strategies, forest products play a vital part in helping poor people cope with hardship. Employing this methodology, which can measure this vital safety-net role, will help to highlight the importance of this function and its important role in increasing households resilience in the face of shocks.

It is apparent from the discussions so far that extensive evidence exists to support the critical role that the NTFPs play in reducing vulnerability, providing useful goods and modest cash income to some of the poorest sectors of society, thus increasing livelihood security and diversification (e.g., Alexiades and Shanley 2004; Kusters and Belcher 2004; Sunderland and Ndoye 2004). However, it is less clear how these products contribute to the third type of poverty alleviation, i.e., poverty reduction or elimination. Can these products assist people to escape poverty, accumulate assets, and improve their standard of living, certainly in any enduring way?

We have discussed how average cash returns are often quite modest for forest products traded in both local and international markets, but we need to see these incomes in the local context and in terms of their wide variability across households (Table 3.4, Shackleton et al. 2007a, b, 2008). There are many individuals and households who are doing well out of NTFP trading, but the numbers are much smaller than those relying on NTFPs for other purposes. Certainly where conditions are conducive and the opportunities for expanded trade have been harnessed and supported, there is evidence to indicate that livelihood and financial benefits can be both raised and extended to much larger numbers of people (Marshall et al. 2006).

However, it is often only in combination with other sources of income that NTFPs may provide a pathway out of poverty. For example, Fisher (2004) and Shackleton et al. (2008) have shown that the sale of forest products together with other sources of income can reduce the proportion of households in the lowest income classes (Fig. 3.5). Similar results were found by Babulo et al. (2009) in Tigray, northern Ethiopia. They showed, using poverty and inequality analyses, that incorporating environmental incomes into household accounts significantly reduced measured poverty and income inequality. Detailed case studies of NTFP commercialisation in Latin America (Marshall et al. 2006) show that the income from NTFP sales may be used to build household physical assets (roofing, fencing, etc.), as well as human capital through providing fees for the schooling and training of children. Thus, while NTFPs seldom make people rich, they are often important in improving quality of life and for ensuring that the next generation has more choices through their contributions to schooling costs. NTFPs can enrich users' lives through, for example, nutritional enhancement and cultural meaning (Kepe 2007), even if such people may be considered "cash poor". Forest dwellers do not necessarily see the use of NTFPs as fitting any of the categories proposed by external commentators, but rather as a way of life that is place based and unique to them (also see Chap. 5).

3.5 Conclusions

While this chapter demonstrates that NTFPs contribute in a variety of ways to the wellbeing and livelihood security of rural and urban households globally, it is apparent that the relationship between people and forest resources is extremely complex, multifaceted, and dynamic. Households are constantly adapting their livelihood strategies to changing circumstances, taking up or dropping their use of and trade in these products in response to a variety of factors (Arnold 2002; Campbell et al. 2002; McSweeny 2004). This complexity undermines and confounds the ability to obtain a comprehensive understanding and generalised picture of the extent to which NTFPs can secure livelihoods, alleviate poverty, and reduce vulnerability now and in the future, and further research is still needed to shed light on this issue (FAO 2003; Angelsen and Wunder 2003, Chap. 1).

Overall, NTFPs tend to be more influential in mitigating or preventing the intensification of poverty than providing pathways out of poverty. This does not, however, mean that they are unimportant and should be dismissed as having little potential in addressing the Millennium Development Goals. Indeed, NTFPs contribute to the welfare of millions of households worldwide, households that would be a lot worse off if they did not have access to these products for daily use, or as a form of insurance in hard times. There is thus a need and obligation for governments worldwide to underpin the safety net offered by forest biodiversity, and to recognise its key importance in subsistence and poverty prevention. At the same time, it is essential to support those people who have turned to NTFP trading as a means to make ends meet in the absence of alternatives and under increasingly harsh economic conditions. The opportunities offered by NTFP commercialisation should be seen not as a silver bullet (as is often the case), but rather as one component of a multisectoral approach for tackling rural and, increasingly, urban poverty. Thus, "NTFP trading alone is not the answer, but nor is arable production, livestock rearing, migrancy, or state welfare grants. It is only through the integration of these livelihood sectors that there will be any lasting positive impact on the welfare of the rural poor" (Shackleton et al. 2008). In terms of the urban poor, it needs to be recognised that NTFPs form a key link in the urban–rural continuum and provide extended opportunity for urban processors and traders, in particular women, as well as an affordable source of goods for consumers.

References

Addis G, Urga K, Dikasso D (2005) Ethnobotanical study of edible wild plants in some selected districts of Ethiopia. Hum Ecol 35:83–118

Alexiades MN, Shanley P (Eds) (2004) Productos forestales, medios de subsistencia y conservación. Estudios de caso solore sisternas de manejo de productos forestales no maderables, vol 3 – America Latino. Centre for International Forestry Research (CIFOR), Bogor

Ambrose-Oji B (2003) The contribution of NTFPs to the livelihoods of the 'forest poor': evidence from the tropical forest zone of south-west Cameroon. Int For Rev 5:106–117

Angelsen A, Wunder S (2003) Exploring the forest-poverty link: key concepts, issues and research implications. CIFOR Occasional Paper No. 40. Centre for International Forestry Research (CIFOR), Bogor

Araia MG (2005) Revealing the forest hidden value: the case study of Eritrea. Masters thesis, University of Stellenbosch, Stellenbosch

Arnold JEM (2002) Clarifying the links between forests and poverty reduction. Int For Rev 4:231–234

Arnold M, Persson R (2003) Reassessing the fuelwood situation in developing countries. Int For Rev 5:379–383

Arnold JEM, Ruiz Pérez M (2001) Can non-timber forest products match tropical forest conservation and development objectives? Ecol Econ 39:437–447

Babulo B, Muys B, Nega F, Tollens E, Nyssen J, Deckers J, Mathijs E (2009) The economic contribution of forest resource use to rural livelihoods in Tigray, Northern Ethiopia. For Pol Econ 11:109–117

Beck T, Nesmith C (2001) Building on poor people's capacities: The case of common property resources in India and West Africa. World Dev 29:119–133

Belcher B, Schreckenberg K (2007) Commercialisation of non-timber forest products: a reality check. Dev Pol Rev 25:355–377

Benjaminsen TA (1997) Is there a fuelwood crisis in rural Mali? GeoJournal 43:163–174

Brouwer ID, Hoorweg JC, van Liere MJ (1997) When households run out of fuel: responses of rural households to decreasing fuelwood availability, Ntcheu district, Malawi. World Dev 25:255–266

Byron N, Arnold JEM (1999) What futures for the people of the tropical forests? World Dev 27:789–805

Campbell JY, Chattopadhyay RN et al (1995) Income generation through joint forest management in India: a case study of the participatory forest management project in Nayagram. Regional Community Forestry Training Centre, West Bengal

Campbell BM, Luckert M, Scoones I (1997) Local level valuation of savanna resources: a case study from Zimbabwe. Econ Bot 51:57–77

Campbell BM, Jeffrey S, Kozanayi W, Luckert M, Mutamba M, Zindi C (2002) Household livelihoods in semi-arid regions: options and constraints. CIFOR, Bogor

Cavendish W (2000) Empirical regularities in the poverty-environment relationship in rural households: evidence from Zimbabwe. World Dev 28:1979–2003

Chipeta ME, Kowero G (2004) Valuation of indigenous forests and woodlands: an international perspective. In: Lawes MJ, Eeley HAC, Shackleton CM, Geach BGS (eds) Indigenous forests and woodlands in South Africa: policy, people and practice. University of KwaZulu-Natal Press, Pietermaritzburg, pp 139–166

Choge SK (2004) The woodcarving industry in Kenya. In: Sunderland T, Ndoye O (eds) Forest products, livelihoods and conservation: case studies of non-timber forest product systems, vol 2, Africa. CIFOR, Bogor, pp 149–168

Cocks ML, Dold AP (2004) A new broom sweeps clean: the economic and cultural value of grass brooms in Eastern Cape Province, South Africa. For Trees Livelih 13:33–42

Cocks ML, Dold AP (2006) Cultural significance of biodiversity: the role of medicinal plants in urban African cultural practices in the Eastern Cape, South Africa. J Ethnobiol 26:60–80

Cocks ML, Wiersum KF (2003) The significance of biodiversity to rural households in Eastern Cape Province of South Africa. For Trees Livelih 13:39–58
Cunningham AB (1997) An Africa-wide overview of medicinal plant harvesting, conservation and health care. Nonwood For Prod 11:116–129
De Beer JH, McDermott M (1996) The economic value of non-timber forest products in Southeast Asia, 2nd revised edn. Netherlands Committee for the World Conservation Union (IUCN), Amsterdam
De Vletter F (2001) Coping with extreme poverty through traditional skills: the case of the xirundzu basket makers of Mozambique. ILO/SAMAT discussion paper no. 18. International Labour Organisation. Southern African Multidisciplinary advisory team (ILO/SAMAT), Harare
Deffar G (1998) Non-wood forest products in Ethiopia. EC-FAO Partnership Programme (1998-2000). Tropical forestry Budget line B7-6201/97-15/VIII/FOR Project GCP/INT/679/EC. http://www.fao.org/docrep/003/X6690E/X6690E00.HTM
Delang CO (2006) Not just minor forest products: the economic rationale for the consumption of wild food plants by subsistence farmers. Ecol Econ 59:64–73
Delang CO (2007) Ecological succession of usable plants in an eleven-year fallow cycle in northern Laos. Ethnobot Res Appl 5:331–350
Dove MR (1993) A revisionist view of tropical deforestation and development. Environ Conserv 20:17–25
Dovie DBK (2004) Economic value of secondary resources in the context of total livelihoods. Box 7.1. In: Lawes MJ, Eeley HAC, Shackleton CM Geach BGS (eds) Indigenous forests and woodlands in South Africa: policy, people and practice. University of KwaZulu-Natal Press, Pietermaritzburg, pp 197–199
FAO (1999) More than wood. Forestry topics report no. 4. FAO, Rome
FAO (2001) How forests can reduce poverty. FAO, Rome
FAO (2003) State of the world's forests 2003. Part II. Selected current issues in the forest sector. FAO, Rome
Fisher M (2004) Household welfare and forest dependence in Southern Malawi. Environ Dev Econ 9:135–154
Grosskinsky B (2000) Nutritional contribution of IWFP. In: Grosskinsky B, Gullick C (eds) Exploring the potential of indigenous wild food plants in southern Sudan. Proceedings of a workshop held in Lokichoggio, Kenya. USAID, Nairobi
Guinand Y, Lemessa D (2001) Wild food plants in Ethiopia: reflections on the role of "wild foods" and "famine foods" in times of drought. United Nations Development Program (UNDP), Emergencies unit for Ethiopia (UNDP-EUE), Rome
Gyan C, Shackleton CM (2005) Abundance and commercialisation of *Phoenix reclinata* in the King Williams town area, South Africa. J Trop For Sci 17:325–336
Harsch E (2001) Making trade work for poor women: villagers in Burkina Faso discover an opening in the global market. Afr Rec 15:6–13
Hawkes RK (1992) Thatching grass: sources, adequacy and potential for local management in the Bulalimamangwe NRMP Area. CASS/MAT Working paper 1-92, Center for Applied Social Sciences, University of Zimbabwe, Harare.
Heltberg R, Arndt TC, Sekhar NU (2000) Fuelwood consumption and forest degradation: a household model for domestic energy substitution in rural India. Land Econ 76:213–232
Hunter LM, Twine W, Patterson L (2007) "Locusts are now our beef" Adult mortality and household dietary use of local environmental resources in rural South Africa. Scan J Public Health 35:165–174
IEA (2002) Energy and Poverty. World Energy Outlook 2002 IEA, Paris
ITC (2006) International Trade Centre statistics. http://langues.p-maps.org/pmaps/index.php?err=sess. Accessed November 2006
Kaimowitz D (2003) Not by bread alone. . .forests and rural livelihoods in sub-Saharan Africa. In: Oksanen T, Pajari B, Tuomasjukka T (eds) Forestry in poverty reduction strategies: capturing the potential. EFI Proceedings No. 47. European Forest Institute, Joensuu, pp 45–64

Kc S (2003) Study of use and marketing potential of medicinal plants in Chitrasen community forest, Chitwan, Nepal. MSc thesis, Pokhara University, Lalitpur
Kepe TV (2007) Beyond numbers: understanding the value of vegetation to rural livelihoods in Africa. Geoforum 39:958–968
Kgathi DL, Mmopelwa G, Mosepele K (2005) Natural resource assessment in the Okavango Delta, Botswana: Case studies of some key resources. Nat Res Forum 29:70–80
Kusters K, Belcher B (eds) (2004) Forest products, livelihoods and conservation: case studies of non-timber forest product systems, vol 1: Asia. CIFOR, Bogor
Linde-Rahr M (2003) Property rights and deforestation: the choice of fuelwood source in rural Vietnam. Land Econ 79:217–234
Lowore J (2006) Miombo woodlands and rural livelihoods in Malawi. CIFOR, Bogor
Mander M, Le Breton GL (2006) Overview of the medicinal plant industry in southern Africa. In: Diederichs N (ed) Commercialising medicinal plants: a southern African guide. Sun, Stellenbosch, pp 1–9
Marshall E, Schreckenberg K, Newton AC (eds) (2006) Commercialisation of non-timber forest products: factors influencing success. Lessons learned from Mexico and Bolivia and policy implications for decision makers. UNEP World Conservation Monitoring Centre, Cambridge
Maundu P, Kariuki P, Eyog-Matig O (2005) Threats to medicinal plant species – an African perspective. In: Proceedings of biodiversity loss and species extinctions: managing risk in a changing world, A global synthesis workshop convened at the IUCN World conservation forum, 18–20 Nov 2004, Bangkok
May J, Budlender D, Mokate R, Rogerson C, Stavrou A, Wilkins N (1998) Poverty and inequality in South Africa – summary report. Report prepared for the Office of the Executive Deputy President and the Inter-Ministerial Committee for Poverty and Inequality. Government of South Africa, Pretoria
McGarry DK, Shackleton CM (2009a) Children navigating rural poverty: rural children's use of wild resources to counteract food insecurity in the Eastern Cape province, South Africa. J Child Pov 15:19–37
McGarry DK, Shackleton CM (2009b) Is HIV/AIDS jeopardising biodiversity? Environ Conserv 36:5–7
McSweeny K (2004) Forest product sales as natural insurance: the effects of household characteristics and the nature of shock in Eastern Honduras. Soc Nat Res 17:39–56
McSweeny K (2005) Natural insurance, forest access and compounded misfortune: Forest resources in smallholder coping strategies before and after hurricane Mitch, Northeastern Honduras. World Dev 33:1453–1471
Monela GC, Kajembe GC, Kaoneka ARS, Kowera G (1999) Household livelihood strategies in the miombo woodlands of Tanzania: emerging trends. Tanz J For Nat Conserv 73:17–33
Murray SS, Schoeninger MJ, Bunn HT, Pickering TR, Marlett JA (2001) Nutritional composition of some wild plant foods and honey used by Hazda foragers in Tanzania. J Food Comp Anal 14:3–13
Nagothu US (2001) Fuelwood and fodder extraction and deforestation: mainstream views in India discussed on the basis of data from the semi-arid region of Rajasthan. Geoforum 32:319–332
Neumann RP, Hirsch E (2000) Commercialisation of non-timber forest products: review and analysis of research. CIFOR, Bogor
Nweze NJ, Igbokwe EM (2000) Non timber forest products in the rural economies of southeastern Nigeria. J Non-timber For Prod 7(3/4):145–155
Osemeobo GJ, Ujor G (1999) Non-wood forest products in Nigeria. EC-FAO Partnership Programme (1998-2000) – Project GCP/INT/679/EC Data Collection and Analysis for Sustainable Forest Management in ACP Countries – Linking National and International Efforts. FAO, Rome
Ouedragogo B (2006) La demande de bois-énergie à Ouagadougou: esquisse d'évaluation de l'impact physique et des échecs des politiques de prix. Online Journal Développement Durable

et Territoires. Varia, mis en ligne le 20 mars 2006. http://developpementdurable.revues.org/document4151.htm

Pant P (2007) Bamboo for life ... for people and red panda: a study on utilitarian value and management of bamboo and its relation with red panda conservation in Maimjhuwa VDC, Ilam, Nepal. MSc Thesis, Wageningen University, Wageningen

Pattanayak SK, Sills EO (2001) Do tropical forests provide natural insurance? The microeconomics of non-timber forest product collection in the Brazilian Amazon. Land Econ 77:595–612

Paumgarten F, Shackleton CM (2009) Wealth differentiation in household use and trade in non-timber forest products in South Africa. Ecol Econ 68:2950–2959

Pereira T, Shackleton CM, Shackleton SE (2006) Opportunities and constraints to trade in reed-based craft products in rural villages in the Eastern Cape, South Africa. Dev S Afr 23:477–496

Rohadi D, Permadi P, Hidayat S (2004) The development of the woodcarving industry and the cultivation of Paraserianthes falcataria in Bali, Indonesia. In: Kusters K, Belcher B (eds) Forest products, livelihoods and conservation. Case studies of non-timber forest product systems, vol 1: Asia. Centre for International Forestry Research (CIFOR), Bogor, pp 265–283

Ros-Tonen MAF, Wiersum KF (2005) The scope for improving rural livelihoods through non-timber forest products: an evolving research agenda. For Trees Livelih 15:129–148

Rueff H, Parizot C, Israel AB, Schwartz M (2009) Dryland afforestation and poverty alleviation: Bedouin and Palestinian non-timber forest collectors in contrasting economic environments. Hum Ecol 36:923–930

Ruiz Pérez M, Byron N (1999) A methodology to analyse divergent cases studies of non-timber forest products and their development potential. For Sci 45:1–14

Ruiz Pérez M, Belcher B, Achdiawan R, Alexiades M, Aubertin C, Caballero J, Campbell B, Clement C, Cunnigham A, Fantini A, de Foresta H, Fernandez CG, Guatam K, Hersch Martinez P, de Jong W, Kusters K, Kutty MG, Lopez C, Fu M, Nair TKR, Ndoye O, Ocampo R, Rai N, Ricker M, Schreckemberg K, Shackleton S, Shanley P, Sunderland T, Youn Y (2004) Markets drive the specialisation strategies of forest peoples. Ecol Soc 9:4. http://www.ecologyandsociety.org/Journal/vol9/iss2/art4

Scherr SJ, White A, Kaimowitz D (2003) *A* new agenda for forest conservation and poverty reduction. Making markets work for low income producers. Forest Trends and Centre for International Forestry Research, Forest Trends, Washington DC

Schreckenberg K (2004) The contribution of shea butter (*Vitellaria paradoxa* C.F. Gaertner) to local livelihoods in Benin. In: Sunderland T, Ndoye O (eds) Forest products, livelihoods and conservation: case studies of non-timber forest product systems, vol 2: Africa. CIFOR, Bogor, pp 149–168

Schreckenberg K, Marshall E (2006) Women and NTFPs. Improving income and status? In: Marshall E, Schreckenberg K, Newton AC (eds) Commercialisation of non-timber forest products: factors influencing success. Lessons learned from Mexico and Bolivia and policy implications for decision makers. UNEP World Conservation Monitoring Centre, Cambridge

Schreckenberg K, Marshall E, Newton A, te Velde DW, Rushton J, Edouard F (2006) Commercialistion of non-timber forest products: what determines success? ODI Forestry Briefing, Number 10, ODI, London

Scurrah-Ehrhart C, Blomley T (2006) Amani butterfly enterprise case study, Tnag region, Tanzania. In: Community-based forest enterprise in tropical forest countries: Scoping study. International Tropical Timber Organisation (ITTO), Forest Trends, and Rights and Resources, Washington DC

SEDF/IC (2003) Medicinal Plants Marketing in Bangladesh. SEDF/Inter-corporation, Swiss Development Foundation

Shackleton SE (2005) The significance of the local trade in natural resource products for livelihoods and poverty alleviation in South Africa. PhD thesis, Rhodes University, Grahamstown

Shackleton SE, Campbell B (2007) The traditional broom trade in Bushbuckridge, South Africa: helping poor women cope with adversity. Econ Bot 61:256–268

Shackleton CM, Shackleton SE (2004) The importance of non-timber forest products in rural livelihood security and as safety nets: a review of evidence from South Africa. S Afr J Sci 100:658–664

Shackleton CM, Shackleton SE (2006) Household wealth status and natural resource use in the Kat River Valley, South Africa. Ecol Econ 57:306–317

Shackleton CM, Shackleton SE, Ntshudu M, Ntzebeza J (2002a) The role and use of savanna non-timber forest products to rural households in the Kat River Valley, South Africa. J Trop For Prod 8:45–65

Shackleton SE, Shackleton CM, Netshiluvhi TR, Geach BS, Balance A, Fairbanks DHK (2002b) Use patterns and value of savanna resources in three rural villages in South Africa. Econ Bot 56(2):130–146

Shackleton CM, Shackleton SE, Buiten E, Bird N (2007a) The importance of dry forests and woodlands in rural livelihoods and poverty alleviation in South Africa. For Pol Econ 9:558–577

Shackleton SE, Shanley P, Ndoye O (2007b) Viable but invisible: recognising local markets for non-timber forest products. Int For Rev 9:697–712

Shackleton SE, Campbell B, Lotz-Sisitka H, Shackleton CM (2008) Links between the local trade in natural products, livelihoods and poverty alleviation in a semi-arid region of South Africa. World Dev 36:505–526

Shrestha DL (1998) Present Status of Wood Energy in Nepal. Woodfuel Trade in Nepal, Dhulikhel. FAO Regional Wood Energy Development Programme in Asia, Rome, pp 15–21

SNV (2005) Value chains identification for intervention. Progress report on identification process. Support to Business Organisations and their Access to Markets (BOAM). Addis Ababa

Stevens S (2003) Tourism and deforestation in the Mt Everest region of Nepal. Geogr J 169:255–277

Stoian D (2005) Making the best of two worlds: Rural and peri-urban livelihood options sustained by non-timber forest products from the Bolivian Amazon. World Dev 33:1473–1490

Styles CV (1994) There's big value in mopane worms. Farmer's Weekly, 22 July, 20–22

Styles CV (1995) The elephant and the worm. BBC Wildl 13:22–24

Sunderland T, Ndoye O (eds) (2004) Forest products, livelihoods and conservation: case studies of non-timber forest product systems, vol 2: Africa. CIFOR, Bogor

Takasaki Y, Barham BL, Coomes OT (2004) Risk coping strategies in tropical forests: floods, illnesses, and resource extraction. Environ Dev Econ 9:203–224

Timmermans HG (2004) Rural livelihoods at Dwesa/Cwebe: poverty, development and natural resource use on the Wild Coast, South Africa. MSc thesis, Rhodes University, Grahamstown

Tincani L (2009) The application of stated-preference techniques to forest resource valuation. Presented at the conference on "Potential roles of local fruit trees for poverty alleviation in Sub-Saharan Africa" organised by the Institute for Environment and Agricultural Research (INERA) in Ouagadougou, Feb 2009. Full paper accessible from http://soas.academia.edu/LucreziaTincani

Twine WC, Moshe D, Netshiluvhi T, Siphugu V (2003) Consumption and direct-use values of savanna bio-resources used by rural households in Mametja, a semi-arid area of Limpopo province, South Africa. S Afr J Sci 99:467–473

Vedeld P, Angelsen A, Sjaastad E, Kobugabe Berg G (2004) Counting on the environment: forest incomes and the rural poor. Environment Department Papers, Paper No. 98. Environmental Economics Series. The World Bank Group, Washington DC

Williams A, Shackleton CM (2002) Fuelwood use in South Africa: where to in the 21st century? S Afr For J 196:1–7

World Bank (2004) Social safety nets. World Bank social protection. http://www1.worldbank.org/sp/safetynets/

Wunder S (2001) Poverty alleviation and tropical forests – what scope for synergies? World Dev 29:1817–1833

Chapter 4
Non-timber Products and Markets: Lessons for Export-Oriented Enterprise Development from Africa

Anthony B. Cunningham

Abstract Micro-enterprise development is widely recognised as important in many developing countries, although support to the NTFP sector is only relatively recent. Much recent literature has questioned the benefits to local livelihoods, suggesting they are relatively small or not widely apparent. This chapter considers a number of examples of export market NTFPs across a wide range of products, which demonstrate that the value of these markets is enormous and each involves thousands of people. These export NTFP products have many characteristics in common. Particularly important is the need to understand supply chains, consumers, middlemen, prices, roles and market potential. Successful enterprises have concentrated on a few species that are potential winners. Common weaknesses of enterprises based on harvests by small-scale producers are their inability to get the large volumes to meet market demand, and lack of attention to quality, quantity and production on time. The characteristics of winning products in successful enterprises are considered.

4.1 Introduction

This chapter synthesises lessons learnt from indigenous and naturalised plant species enterprises by addressing three questions. First, what is missing in the comparative analyses of micro- and small enterprises based on indigenous or naturalised non-timber forest products (NTFPs)? Second, what are the characteristics of "winning products" and how are these likely to alleviate poverty? And third, what lessons can successful enterprises teach policy-makers, the private sector and small-scale farmers? To answer these questions, published literature is

A.B. Cunningham
Centre for International Forestry Research, Bogor, Indonesia
and
School of Plant Biology, University of Western Australia, 35 Stirling Highway, Crawley, WA 6009, Australia
e-mail: tonyc05@bigpond.net.au

S. Shackleton et al. (eds.), *Non-Timber Forest Products in the Global Context*, Tropical Forestry 7, DOI 10.1007/978-3-642-17983-9_4,

reviewed in Sect. 4.2, drawing on experiences from successful enterprises. Thereafter follows a summary of how this chapter differs from previous analyses of trade in NTFPs and why the focus tends towards export oriented micro-enterprises. Section 4.3 sets out the characteristics of winning products, illustrating lessons drawn from successful enterprises, highlighting some from case studies in Africa, Asia, Australia and Latin America. Section 4.4 identifies unresolved issues and new opportunities that need research, followed by Sect. 4.5, which concludes the chapter.

4.2 The "Reality Check" and Beyond

Micro-enterprise development is widely recognised as important in many developing countries, where self-employment helps create economic opportunities for low-income households with limited employment options or earning power (Woller 2004). While most assessments of medium, small and micro-enterprises have dealt with the manufacturing, mining, agricultural or timber sectors, it is only recently that comparative studies have been carried out on the NTFP sector. These studies of NTFPs commercially traded in Asia, Africa or Latin America have all cautioned against undue optimism (Ruiz-Pérez et al. 2004; Kusters et al. 2006; Marshall et al. 2006), with Belcher and Schreckenberg (2007) calling for a reality check on what these micro-enterprises can, or cannot, deliver. Similarly, Wunder (2001) has suggested that forms of land-use other than sustainable harvest of NTFP's from tropical forests may provide a better route out of poverty than forest based enterprises. While this is likely on soils with high arable potential, land-use options in regions with low arable potential are more complex, with indigenous plants playing a crucial social safety net role and as a source of income (Cunningham 1985; Shackleton et al. 2007, 2008).

Although the comparative studies of commercial NTFPs examined a wide range of cases, relatively few were from Africa. Ruiz-Pérez et al. (2004) examined 61 cases but only 17 of the 61 cases (i.e., 28%) were African examples. And Kusters et al. (2006) compared a sub-set of 55 of these same cases. In addition, most African cases were products for domestic markets rather than for export. Eighteen separate Latin American cases were analysed by Marshall et al. (2006). In his recent book, Collier (2007) points out the extent to which the African manufacturing sector has stagnated due to protectionism and low productivity. More recently, competition from India and China stresses how important it is for African economies to break into export markets (Collier 2007). It is useful, therefore, to examine the exports of NTFPs that have been missed in previous comparative analyses to tease out wider lessons for other products.

Unlike previous comparative studies that mainly focused on informal production of NTFPs for local and national markets, this chapter also considers two very different export sectors for the practical and policy lessons they may hold:

- Cases where export business is linked to micro-enterprises, often through producer associations as part of Private Sector Development (PSD) or "venture socialism". Here private enterprise deliberately assists small-scale producers through profit-making businesses. For example, Phytotrade Africa (started in 2001), a trade association across southern Africa, had over 10,000 producers and an income of USD 2.5 million per year by 2005. Generating 40% of its annual recurrent costs, it is expected to have 40,000–50,000 producers by 2010 (IFAD 2005).
- The export of narcotic plants. Although illegal, these are highly lucrative but poorly studied enterprises, such as sale of khat (*Catha edulis*) and marijuana (*Cannabis sativa*).

In the case of PSD, during most of the nineteenth century, natural product export was achieved with grading and basic processing done in developing countries and extensive value-adding and re-exporting done in Europe or North America. Since the 1990s, this PSD has been changing on two fronts. Firstly, support for regional Trade Hubs and processing facilities, such as the West African Trade Hub (WATH) based in Accra, Ghana. Secondly, the role of China and India as growing consumer markets for several important natural product market niches such as natural cosmetics, flavours and fragrances. These particular niche markets are described below.

4.2.1 Natural and Organic Cosmetics

Global sales of natural and organic cosmetics were worth almost USD 7 billion in 2007, and are expected to reach USD10 billion by 2010 (NPI centre 2007). This is driven by demand from Europe and North America, but India and China are the important emerging markets. Since buying Sanoflore and The Body Shop in 2006, L'Oreal is working with producers of cosmetic oils in a range of developing countries. Also at the high end of the market, "Origins Natural Resources", part of Esteé Lauder, has launched a certified organic cosmetic called "Origins Organics". In addition, European supermarkets are marketing natural and organic cosmetics under private labels. New Product Development (NPD) has become a key feature of the natural cosmetics market. Certification is an important component of this market, with Ecocert and UK Soil Association working closely with cosmetic manufacturers, who are innovative in developing new product formulations, including natural preservatives, surfactants and colourants in a shift away from synthetic chemicals. There is already a significant export trade in African plant products for this market, such as seed oils from marula (*Sclerocarya birrea* subsp. *caffra*), Kalahari melons (*Citrullus lanatus*) and two sour-plum species (*Ximenia caffra* and *X. americana*) (Box 4.1).

Box 4.1 A Tale of Cosmetic Oils from Six African Plant Species

Anthony Cunningham, Centre for International Forestry Research, Bogor, Indonesia and School of Plant Biology, University of Western Australia, 35 Stirling Highway, Crawley, 6009, Australia, tonyc05@bigpond.net.au

The African cosmetic oil-seeds market is characterised by several commercial successes across a wide range of countries and differing policy circumstances, from Morocco in north Africa to Namibia in southern Africa. These successful enterprises hold wider lessons for micro- and small enterprise development based on NTFPs. The oil seeds involved are from different plant species in a range of life-forms (three tree species, two shrub species and one creeping melon): *Argania spinosa* and *V. paradoxa* (Sapotaceae), *C. lanatus* (Curcurbitaceae), *S. birrea* (Anacardiaceae), *X. americana* and *X. caffra* [Ximeniaceae (formerly in the Olacaceae)]. Argan (*Argania*) oil and shea butter (*Vitellaria*) exports from Africa have a long history, with huge quantities exported (1,000–2,000 tonnes of argan oil/year, 160,000 tonnes/year of shea butter). In contrast, the export trade in *Citrullus, Sclerocarya* and most recently *Ximenia* seed oils is less than a decade old and export quantities are much lower, with 29 tonnes of marula (*Sclerocarya*) oil exported in 2007 and even less *Citrullus* and *Ximenia* oil. Two oils, argan oil and shea butter, are from slow growing tree species in the Sapotaceae, *A. spinosa* and *V. paradoxa*. Both are from monotypic genera, with both tree species dominating the vegetation where they occur. *A. spinosa* characterises the Mediterranean *Acacia–Argania* Dry Woodland and Succulent Thickets of northern Africa, mainly in Morocco; being one of the few high diversity vegetation types remaining, it is considered Critical/Endangered by WWF as local people have given up their traditional management practices, with consequent intensive use of the *Argania* woodlands (WWF 2001). *V. paradoxa* parklands are even more extensive, occurring across west Africa, with a different sub-species (with different oil chemistry) extending into east Africa. Seeds of *V. paradoxa* have two main components, stearin (used in chocolates and margarine) and olein (used in cosmetics) (Holtzman 2004), and an FOB ("Free on Board" or domestic) value of about USD 37.5–45 million per year (Lovett 2004). Despite differences in taxonomy, trade history, life-form and geographic location, all these oil-seed sources share common characteristics. All are multiple-use species with a long history of human use; all occur in cultural landscapes that have been transformed (and continue to change) due to clearing for agriculture, browse by livestock, fire or a combination of these; and in all cases, women are the primary harvesters and processors of all of these oil seeds. They also face a range of significant challenges, particularly for the slowest growing species, argan (*A. spinosa)* and shea (*V. paradoxa*). Both *A. spinosa* and *V. paradoxa* are long-lived tree species, respectively living to 400 years (Githens and Wood 1943) and 300 years (Burkill 2005). Although

(continued)

argan trees start fruiting at 3–5 years (Nerd et al. 1998; Morton and Voss 1987), they take 60 years to reach peak fruiting, with growth rates dependent on water, nutrients and mycorrhizal associations (Nouaim et al. 2001). Shea trees take even longer, with fruit production peaking at 50–100 years old (Burkill 2005). *S. birrea* is faster growing, with 25% of cultivated trees producing 27 kg/year of fruit when 4 years old under optimal management conditions (Nerd and Mizrahi 1993) and 500 kg of fruit/year at 12 years old (Van Wyk and Gericke 2000) and, unlike argan and shea, are shorter lived softwoods, living to 60–80 years. Slow growth rates and the long wait for any economic returns mean that competition from cultivation is unlikely. The flip-side of this is while *Sclerocarya* densities and distribution have been increased over centuries by deliberate planting (Cunningham and Shackleton 2004), this is not the case with *Argania* and *Vitellaria*, due to a combination of clearing for farmland and browse of young seedlings. In contrast, *C. lanatus* is an annual re-seeder, domesticated by local farmers in north-central Namibia, with a variety of land-races, including selection for oil-seeds (Maggs-Kölling and Christiansen 2003).

4.2.2 Flavours and Fragrances

The world flavour and fragrance market was predicted to reach USD 18.6 billion in 2008. Flavours are used commercially in beverages, foods (confectionery, bakery, savoury and snack foods), pharmaceutical products and mouth-washes. Fragrances are used in perfumes, cosmetics and toiletries, soaps and detergents, household cleaners, air fresheners and aromatherapy. Four companies currently account for around 40% of the global market. Two are Swiss (Givaudan SA and Firmenich International SA), one (Symrise) is based in Germany, while the fourth [International Flavours and Fragrance (IFF)] is based in New York. At this stage, few African countries can compete in the manufacturing of flavours and fragrances, but they can supply quality, value-added products to a market looking for innovative new products and blends. International flavour and fragrance houses are dealing with rising labour costs in Europe, North America and Japan by shifting manufacturing to hubs in India and China. In addition, both China and India are growing markets for flavour and fragrance products, as people there increase their spending on sophisticated personal care products. The Indian flavours and fragrance market is valued at around USD 225 million per year; 55% of this market value is for fragrances found in wash products, including soap, hand washes and shower gels (42%), shampoo (5%), hair oil (3%) and fine fragrances (2%). Success in this market depends on the ability to select key commercial partners (Fig. 4.1a), to use national "natural advantage" in endemic aromatic plants to develop new flavours and fragrances at an early stage, and to function on a globally integrated scale.

Fig. 4.1 (**a**) Aromatic *Commiphora* resin harvest in NW Namibia in a partnership between conservancy members and Aveda, brokered by the NGO, Integrated Rural Development and Nature Conservation. (**b**) Mechnisation can reduce drudgery and save women time (Shea processing, Burkina faso). (**c**) Bulking up centre for marula oil processing, Namibia. (**d**) Grading *Schisandra sphenanthera* fruits, China. (**e**) Tenure: a signboard with harvest guidelines demarcating a *Schisandra* wild harvest area, China. (**f**) Ripe *Schisandra* fruits, an ingredient in herbal teas, medicines and energy drinks. (**g**) *Catha edulis* chemotypes distinguished by petiole colour. (**h**) Wrapping *Catha edulis* in banana leaves to prolong the short shelf-life. (**i**) Trust and text messaging: prior to take-off of a planeload of *Catha edulis* from Nairobi, Kenya to Mogadishu, Somalia

4.2.3 Functional Foods (or Nutriceuticals)

These foods are promoted as having health benefits in addition to their nutritional value. Natural vitamin C from Australian *Terminalia ferdinandiana* fruits is a recent example (Cunningham et al. 2009). The global functional food market is over USD 73 billion/year. Breakfast cereals account for 26% of this market. Energy drinks, often containing guaraná from the South American liana, *Paullinia cupana,* form another 20% of the market (Box 4.2). Use of *Schisandra*, a Traditional Chinese Medicine (TCM) (Box 4.3) (Cunningham and Brinkmann 2009) is also increasing in the energy drink market, with potential for quality control and sustainable sourcing (Fig. 4.1d, e). With ageing populations in Europe and North America becoming more health conscious, functional foods are a growing but sometimes costly market to penetrate. In the UK, the functional food and drink sector grew by 159% between 1999 and 2001 and is now valued at £667 million per year.

Box 4.2 Guaraná: Market Organisation and Battles for Market Share

Anthony Cunningham, Centre for International Forestry Research, Bogor, Indonesia and School of Plant Biology, University of Western Australia, 35 Stirling Highway, Crawley, 6009, Australia, tonyc05@bigpond.net.au

Fruits of guaraná (*Paullinia cupana*) are from a vine traditionally used as a medicine to provide energy (Erickson et al. 1984). Today they form the basis of a major industry in Brazil, centred in the Majes region of the Amazon, with production mainly for the carbonated soft drink industry (1,250 tonnes/year, 44% market share) and production for the extract, syrup, and powder industries (700 tonnes/year, 24.5% market share). The history of the soft-drinks market in Brazil holds valuable lessons for how (and with whom) the market could be developed for other NTFPs. Commercial development of guaraná started in 1958. In 1973, a producer association, the Sociedad Agrícola de Maúes S.A., was created to process guaraná seeds, who later formed the Hacienda Santa Helena, to regenerate land set aside for the cultivation of guaraná. Today, at least 50% of 6,000 farmers in the Majes region produce guaraná. This production supplies 80% of the Brazilian market through organic production. Once the fruit is picked, it is toasted and the seed can be processed into different products. In the past few years, two rival Brazilian companies in the soft drink and beer market in Brazil, Brahma and Antarctica, merged to form the company AmBev, which now controls 70% of the local beverage market (Kuri 2008). A link with Pepsi-Cola was an important part of getting the Brazilian government support for the merger process. AmBev is now the seventh largest soft drink company in the world and plans to internationalise guaraná as the "only true Brazilian soft drink". In 2007, sales of guaraná Antarctica in Portugal rose 305%, reaching 4.5 million litres and guaraná Antarctica is also widely sold in Spain and Puerto Rico. Flavour and ecological rainforest and organic production are part of the appeal.

Box 4.3 *Schisandra* in the Growth of Markets for Traditional Chinese Medicines

Anthony Cunningham, Centre for International Forestry Research, Bogor, Indonesia and School of Plant Biology, University of Western Australia, 35 Stirling Highway, Crawley, 6009, Australia, tonyc05@bigpond.net.au

The growing international popularity of TCM internationally is widespread, not only across Europe and North America, but even in west Africa, in competition with strong African traditions. In Britain, for example, more than 3,000 clinics sell TCM (Vines 2008). At a global scale, China is by far the largest exporter of medicinal and aromatic plants. In 2007, China exported 241,561 tonnes of medicinal plants classified under HS Code Chapter 1211 with a reported value of USD 418 million (mainly to Hong Kong (for re-export), Japan, South Korea, Vietnam, and other Asian countries), and exported 709,781 tonnes of aromatic plants classified within HS Codes 0902 through 0910, mainly to Japan, Morocco, USA, Hong Kong, and Malaysia (Brinckmann 2008). According to a report published by Hong Kong Trade Development Council, the global sales for Chinese medicines grew at 8% a year since 1994. In 2002, the total global sales were USD 23.2 billion (Phillip Securities Research 2003). This may grow even faster with the growing link to formal, industrialised production and export of TCM preparations.

Today, TCM is in a new development phase, with the plan that by the 2010, a modern TCM innovation system would be established along with a series of standards for modern TCM products to create a competitive, modern TCM industry through new technology and standardisation (Fig. 4.2).

(continued)

Fig. 4.2 The Chinese traditional medicine trade is the largest in the world: a new TCM market in Chengdu, Sichuan, China (photo: A.B. Cunningham)

In the mid-1990s, the term "Cinderella trees" was used for useful tree species that had been overlooked and their potential left largely untapped (Leakey and Newton 1994). *Nan wuweizi* (southern schisandra, *Schisandra sphenanthera*, Schisandraceae), is the "Cinderella" cousin to the better-known *bei wuweizi* (northern schisandra, *Schisandra chinensis*). Both are scandent climbers whose fruits are recognised in the Chinese pharmacopoeia. Most use and trade in the past has been in the larger fruits of *S. chinensis*, however, as this species is more popular in China. However, the southern *wuweizi* can play an important role in improving income to local people through sustainable harvest outside of nature reserves in the Upper Yangtze ecoregion (Fig. 4.3).

Fig. 4.3 Grading *Schisandra sphenanthera* – better prices linked to sustainably managed harvest in the "panda landscapes" of the Upper Yangtze, China (photo: A.B. Cunningham)

4.2.4 *Colloidal Gums*

The specifications of gum arabic are defined within the European Union as "the dried exudate from the trunks and branches of *Acacia senegal* or *Acacia seyal* in the family Leguminosae". This limits competition from other gums with hydrocolloidal properties, which are not only found in other *Acacia* species, but also in *Combretum* and *Terminalia* (Combretaceae). Gum arabic is used in the food industry as a stabiliser and is found in soft drink syrups (including Coca-cola), gummy sweets, chewing gums, marshmallows, shoe polish and watercolour paints. In 2002, Africa exported over 54,000 tonnes of gum arabic, with Sudan dominating the world market and exporting to about 30 countries. Ethiopia and Eritrea, contributing 1.6% and 0.6% of world production, respectively, are small-scale producers. While raw material harvesting, processing and grading take place in Africa, most value-adding takes place in importing countries. The same applies to flavours and fragrance products. France is the leading importer of gum arabic, accounting for more than 40% of global imports, and Colloïdes Naturels International in Rouen is the world leader in gum arabic processing.

4.2.5 Alcoholic Beverages Using Indigenous Plants

Across Africa, Asia and Latin America, the addition of NTFPs to alcohol is used for value-adding and marketing alcoholic beverages. Amarula, a cream liqueur produced in South Africa, is a good example of this trend (Box 4.4). The South African Distell Group, a publicly listed company merged from the Distillers Corporation and Stellenbosch Farmers Winery, created South Africa's internationally known consumer product by masterful branding of two African icons – the elephant and the marula (*S. birrea* subsp. *caffra*) fruit. Although income from marula fruit sales to Distell was similar to that earned by women brewing marula beer for local markets, marula fruit traders still earned up to approximately two times more than the average local farm-worker (Wynberg et al. 2003).

Box 4.4 Marula (*Sclerocarya birrea*), African Elephants and Branding: Turning Ideas into Assets to Expand Markets

Anthony Cunningham, Centre for International Forestry Research, Bogor, Indonesia and School of Plant Biology, University of Western Australia, 35 Stirling Highway, Crawley, 6009, Australia, tonyc05@bigpond.net.au

Despite a high diversity of tasty indigenous fruits, few African species have entered the international market (Cunningham and Shackleton 2004). *Sclerocarya caffra* subsp. *birrea* is a notable exception, with commercialisation of both the cosmetic oil and fruit pulp in a range of business models, each with advantages and disadvantages (Wynberg et al. 2003). The best known of these is the product "Amarula", a cream liqueur from marula fruits which is probably the internationally best-known South African consumer brand. This branding success was carefully achieved through a marketing success through the confluence of corporate creativity, myth, African mystique and advertising power of a large company. Clever branding was at the heart of the marketing campaign blending market fascination with African wildlife (elephants) and African culture (uses of marula) to market a quality product (http://www.amarula.co.za). Amarula was launched as a spirit in 1983 and as a cream liqueur in 1989. By 2003, the company was processing about 800 tonnes of pulp or 2,000 tonnes of fruit into Amarula Cream annually (Shackleton et al. 2003). At that time, marula fruits were collected by about 420 households in rural communities in the vicinity of Phalaborwa and processed at a pulp factory and depot run by a subsidiary, the Mirma Product Development

(continued)

Company. This company was set up by Distell with local community representation. A research survey concluded that Mirma contributed over USD 60,000 into the local area, including a yearly donation from Distell of some USD 10,000 (Shackleton et al. 2003). About 60 people are employed directly by the factory, but only four permanent staff. By 2007, Amarula Cream was being exported to over 90 countries, with Brazil, Canada, Germany and the UK as significant importing countries. In 2008, for example, Amarula respectively represented 23%, 25% and 30% of Distell exports to Latin America, Europe and African countries outside South Africa, with a 15% increase in sales volume (Distell Group Limited 2008). Despite the fact that "Amarula" contains a relatively small percentage of *S. birrea* fruit pulp, the success of marketing and scale of production generates benefits to local communities in the Phalaborwa area, South Africa (Shackleton et al. 2009).

4.2.6 Marijuana (**Cannabis sativa**)

Cannabis is the world's largest illicit drug, produced in over 140 countries, with a consumer market of about 160 million people. Global cannabis production in 2005 was 42,000 metric tonnes, compared to global heroin production (472 metric tonnes) and global cocaine production (980 metric tonnes) (UNDOC 2007). Although more cannabis was produced in the Americas (46%) than in Africa (26%) in 2005 (UNDOC 2007), cannabis remains an illegal, yet important source of income for small-scale farmers, particularly in South Africa, Malawi, Zambia, Swaziland, Nigeria, Ghana, the Democratic Republic of the Congo, Tanzania and Morocco. In South Africa, where the cannabis crop is worth approximately USD 900,000/year, cultivation can be an important source of cash income in some rural areas and even more so for urban traders (Kepe 2003). However, Kepe (2003) argues that, legalisation in South Africa may not benefit small-scale producers, as they would be out-competed by large-scale commercial producers. Due to high demand and high cannabis prices in Europe, the late, great journalist Carlos Cardoso suggested legalised production in Mozambique for export to Holland (Fauvet and Mosse 2003). Holland, along with Albania, is currently a major re-exporter of cannabis to the rest of Europe. Although cannabis cultivation is illegal in sub-Saharan Africa, the value of the *khat* (*Catha edulis*) trade to the Ethiopian government bears out Cardoso's reasoning.

4.2.7 Khat (or miraa) from **Catha edulis**

In 1998–1999, the *khat* trade accounted for 13.4% of Ethiopia's export revenue (US Department of Commerce 2000). In Kenya, cross-border trade, including daily flights from Nairobi's Wilson Airport to Mogadishu, is largely unmonitored.

In 1993 the Kenya–Somalia trade (Fig. 4.1h, i) was considered to be worth USD100 million per year (Randall 1993). The trade in Ethiopia was estimated at USD500 million annually (Green 1999). Farmers in Meru district, Kenya, and in Harrarghie, Ethiopia, are the world's major producers, with a selection of chemotypes produced (Fig. 4.1g). Farmers in the Habro district in Ethiopia earn 70% of their income from *khat,* as a maize-*khat* intercropping system is 2.7 times more profitable than maize monocropping (Feylsa and Aune 2003). More recent estimates of *Catha* leaf imports to the UK are six tonnes per week. This export feeds into a smuggling network to the US, where *khat* sells for USD 300–440/kg (Crenshaw and Burke 2004), putting the UK–USA trade at approximately £150 million per annum.

4.3 Characteristics of Winning Products: Lessons from Successful Enterprises

Despite their diversity, African natural products produced for international export markets have many characteristics in common. Particularly important is the need to understand supply chains, consumers, middlemen, prices, roles and market potential. Successful enterprises have concentrated on a few species that are potential winners. Common weaknesses of enterprises based on harvests by small-scale producers is their inability to get the large volumes to meet market demand, through lack of attention to quality, quantity (Lovett et al. 2005) and timely production. The characteristics of winning products in successful enterprises hold useful lessons for natural product development with African producers, which are considered below.

- *Winning natural products are built through maintaining or developing an abundant natural resource base.* With the exception of the *Cannabis* trade (based on a naturalised species from south Asia), all export enterprises have developed on the basis of Africa's natural advantage. Endemism can give a national advantage, for example rooibos (*Aspalathus linearis*) and honey-bush (*Cyclopia*) teas, which are endemic to the Cape region, South Africa. However, most successful enterprises are based on species that are widely distributed. African examples are *A. senegal, A. seyal, Adansonia digitata, S. birrea* subsp. *caffra, Vitellaria paradoxa, Ximenia americana* and *X. caffra*. This advantage is further enhanced in several cases by farmer-based selection of genotypes or chemotypes [such as *S. birrea* subsp. *caffra* (Leakey 2005), *V. paradoxa* (Sanou et al. 2006) and *C. lanatus* (Rodin 1985)]. How long this advantage is maintained depends on production shifting to other places and germplasm exports. With abundant, relatively slow growing tree species, small-scale producers, at least in rural Africa, have a head start. Any long-term advantages are lost, sometimes forever, when local extinction of species or unusual genotypes occurs. This has happened to endemic cycad populations (*Encephalartos cerinus* in South Africa, *E. pterogonus* in Mozambique), which have been plundered by wealthy collectors. Growth of the *Prunus africana* industry, which has an estimated over-the-counter (OTC) value of USD 220 million/year (Cunningham et al. 1997) for example, is threatened by resource-mining rather than resource management, weak tenure and a slow transition to cultivated production in agroforestry systems or plantations (Cunningham 2005).

- *Clear rights to land and tree tenure and use.* Secure tenure is an important component of any strategy that aims to deliver fair and equitable benefits to African farmers from the natural product commercialisation. Commercialisation has a positive effect through increasing incentives to conserve trees, but where tenure is unclear, it can result in conflict (Wynberg et al. 2003), leading to unsustainable use.
- *Local self-sufficiency is not undermined.* Sustainable harvesting cannot be assumed, particularly in situations with commercial harvest pressures and weak land or resource tenure. There are many examples where this assumption has been made, with the result that supplies for local household consumption are jeopardised or local people have ended up walking further and further to get the same resource or paying more for a now scarce resource. Commercial enterprises need to be based on species that are not only resilient to harvest, but where there is a surplus above subsistence need. For this reason, Phytotrade Africa has deliberately focused on enterprises where products can be sustainably harvested, for example harvesting fruits rather than roots or bark. To maintain resource stocks, it can be useful to develop participatory management plans with simple, enforceable rules. Where monitoring at a community level is required, it needs to focus only on key issues, as people have many other things to do.
- *Build on and out from existing markets, information access and strategic choices.* There are many cases across developing countries where development workers, with good intentions and no business acumen, start natural product industries to generate income for poor local people, only to see them fail and shatter local peoples' hopes. Unlike the movie "*Field of Dreams*", the approach of "build it and they will come" or "produce it and markets will buy", generally fail. What is far more effective is to produce for an existing market. Natural resource based enterprises first need to understand the characteristics of the market, then organise producers to get the right product to the appropriate partners in sufficient quantity, on time, at the right price. These successful enterprises take business oriented strategic choices first, often focusing on growth sectors such as natural and organic cosmetics, flavours and fragrances, functional foods, colloidal gums or plant products added to alcoholic beverages.
- *Visionary champions make a difference: insight, innovation and staying power* Visionary champions often play a fundamentally important role. They are found in family owned companies, publicly listed companies, co-operative clusters of small businesses under one name (e.g., Phytotrade Africa) or clan-based businesses (e.g., WaKamba woodcarving enterprises in Kenya). The visionary of the world's leading *Acacia* gum importer and processor, Colloides Naturels International (CNI), which started in 1895, is Charles Dondain, the grandfather of the current family owners. He experimented with water-soluble *Acacia* gums and

realised their industrial value. Today, CNI operates in over 75 countries, remains a family-owned company and retains its reputation for quality products and hydrocolloid innovation. In South Africa, the rooibos tea (*A. linearis*) export industry was started by A.B. Ginsberg in 1901 (Morton 1983) and has grown into a multi-million dollar export industry. Similarly, the Devil's Claw (*Harpagophytum procumbens* and *H. zeyheri*) export trade was started in the 1950s by a G.H. Mehnert. It was based on traditional uses of the tubers which were then sent to the University of Jena for additional study. Later, in 1962, the Namibian company Harpago (Pty) Ltd started exporting tubers in larger quantities to Erwin Hagen Naturheilmittel GmbH in Germany (Wegener 2000). In Kenya the woodcarving industry, now worth USD 20 million/year in export revenue, was started in the 1920s by Mutisya Munge from Wamunyu, Machakos district. More recently, the Marula Oil producers Network started by Cyril Lombard and developed further by Pierre du Plessis at CRIAA-SADC, Namibia, has grown into a rural producer association that also produces Kalahari Melon Seed oil and *Ximenia* oils for the international cosmetics markets.

- *Sustaining a market requires quality in sufficient quantity, on time.* Quality standards are crucial for natural products exports, particularly if they are used in cosmetics or as functional foods. For example, adulteration and poor grading of gum arabic in Nigeria resulted in the loss of its US market share. The share dropped from 100 tonnes in 1997 to zero imports of Grade 1 gum arabic in 1999–2001 (Harrison and Roberge 2002). Like any relationship, once a reputation is lost, in this case for good quality, it is difficult to regain. To assist Nigerian exporters regain their market share, USAID/Nigeria, a US-based consultancy company, the National Association of Gum Arabic Producers, Processors and Exporters of Nigeria (NAGAPPEN) and two major US gum arabic importers (Importers Service Corporation (ISC) and TIC Gums) prepared a training guide for improved quality standards (Harrison and Roberge 2002). To avoid West African Trade Hub (WATH) losing their reputation as exporters of shea butter, Lovett et al. (2005) have similarly prepared an easy to use export guide on international quality standards. This raises awareness amongst West African producers and exporters of what quality requirements are most important to cosmetics manufacturers. Developing and maintaining market share is not just an issue of quality, but also of sufficient quantity getting to the market, on time. Reliability of supply is commonly affected by fluctuating yields from year to year and requires an organisational structure to source products, for example Kalahari Melon Seed (KMS), shea nuts or marula fruits, from across a wide geographic area. In some cases decentralised tree planting in fields or enrichment planting of tenured trees in savanna, woodland or forest systems has an added benefit. It avoids a situation where horticulture becomes a form of elite capture by wealthy farmers with greater access to land and technology, thereby cutting poorer farmers out of the supply chain. Harvest from wild or local managed trees certainly offers opportunities for organic or FairTrade marketing, but harvesting sufficient quantities requires hundreds, or even thousands, of rural farmers to collect and process these products. Today, well-established and effective local institutions are using processing and communications technologies (particularly mobile phones, Fig. 4.1i) to coordinate bulking up of resources, reduce transport costs and improve supply chain capability. In Mali and Burkina Faso, "multi-functional platform" machinery reduces the drudgery of processing shea products (Fig. 4.1b). In Namibia, the Eudafano Women's Cooperative, who produce marula (*S. birrea* subsp. *caffra*) seed oil, coordinates over 5,000 members for collection and oil pressing to get high value cosmetic oil to the European market (Fig. 4.1c). Similarly in North-western Zambia, Forest Fruits Pty Ltd has invested in quality control training along the supply chain, honey certification and is able to coordinate supplies to get them to export markets in Europe.

- *Pricing: incentives to collect, incentive to buy: high price/volume*. Poor African farmers generally are price-takers, with limited bargaining power. For some products, this is necessary for new enterprises to become competitive. However, in several other cases, value-added processing and price negotiations between external brokers and producer associations, enable producers to get better prices and returns well above average local daily wage rates. Good and fair prices also act as an incentive to harvest larger quantities of a product, enabling exporters to meet export shortfalls [in the case of *Vitellaria*, 52% of the total shea harvest, in the major WATH producing countries, is not collected or even used locally (Lovett 2004)]. Higher prices are an incentive for women to collect more *Sclerocarya, Ximenia, Citrullus* and *Vitellaria* fruits or seeds during peak harvests over a wider geographic area. For species that regenerate from seed it is important to leave some seed for natural regeneration (Cunningham 2001), however, tagua palm (*Phytelephas seemannii*) populations have been shown to survive when as much as 85% of fruits are harvested (Bernal 1998), with a figure of 92% estimated for *S. birrea* subsp. *caffra* (Emanuel et al. 2005).
- *Diverse niche markets to reduce competition*. Niche markets give African producers an edge, whereas with mainstream crops, massive subsidies to OECD producers reduce the chances of success for African farmers. Diversification helps grow the overall niche market. Marula oil, for example, is now found in over 140 products produced by The Body Shop. Shea butter, mostly used as a cocoa butter equivalent, can also be used for in range of cosmetic products. Marula, shea butter and *Ximenia* all contain bioactive constituents relevant to skin-care and opportunities for product diversification.
- *The power of strategic partnerships: business, producer associations and universities backed through policy support*. Unlike natural products sold in local markets, local farmers cannot be expected to have detailed knowledge about export markets, quality control or research requirements. For this reason, the research and development (R&D) capabilities of national or regional universities or the R&D facilities in partnership firms, meets an important need. It is also important to meet necessary intellectual property rights (IPR) protocols. A model example of how to avoid decision-making inertia is the Namibian Indigenous Plants Task Team (IPTT), where government departments, NGO's and a business-minded champion from CRIAA-SADC coordinate strategy. This

is based on the "Product Pipeline Approach", which selects and backs potentially winning species for key export markets. The IPTT has backed a series of successes in indigenous cosmetic oils sourced from poor rural farmers, for example oils from *Sclerocarya, Ximenia,* and *C. lanatus* (Kalahari Melon Seed). Phytotrade-Africa, with help from CRIAA-SADC, is an example of a successful strategic partnership formed with European businesses. Starting with small-scale exports of marula oil extracted in an urban area (Katatura), scale and local value-adding have been achieved through upgrading processing facilities close to supplies of *Sclerocarya, Citrullus* and *Ximenia* and establishing good relationships with large buyers in the UK and France. In West Africa, the shea nut trade has been assisted by the West African Trade Hub (WATH), US-AID, expert advice and recommendations by Addaquay (2004) resulting in greater efficiencies and the opening of a major shea butter refining facility (Addaquay 2004).

- *Conflict resolution mechanisms.* Although rural communities are widely spoken about as if they were cohesive, in many cases they are not. Instead, they are divided along the lines of families, clans, wealth and power relations. Natural product commercialisation can widen these rifts, particularly where value-chains are complex and neither costs nor export markets are well understood by producers. Profit-sharing within producer associations is another potential pitfall. In Mexico, for example, less than half (15 of 42) of communities running Community Forest Enterprises (CFE's) distributed profits (Klooster 2000). The decision to distribute all or part of the profits to legal members of CFE's depends on levels of trust, poverty and the probability of investing in other enterprises. Establishing transparent, acceptable and fair means of conflict resolution is therefore a necessary investment.
- *Regional co-operation in order to compete.* With endemic species, there are commercial advantages in maintaining germplasm within the single source country, but for widely distributed species such as *A. digitata, S. birrea* subsp. *caffra* and *V. paradoxa* commercial advantages have resulted through regional cooperation. These advantages start with the ability to harvest commercially viable quantities of fruits or seeds. This is extended to sharing R&D, legal or coordinated, multi-country lobbying for policy change, for example EU recognition that baobab pulp is not a "novel food" (Wilkinson and Hall 2007). The two best known African examples are PhytoTrade Africa and the West African Trade Hub (WATH). PhytoTrade Africa is a trade organisation in southern African. The West African Trade Hub (WATH) covers more shea producing countries, representing an estimated 81% of sub-Saharan Africa production potential and 94% of actual shea collection across all *V. paradoxa* producing countries (Lovett 2004).
- *Upgrading within value-chains.* As Giuliani et al. (2003) point out in their seminal paper: "upgrading within a value chain implies escalating on the value ladder, moving away from activities in which competition is of the 'low road' type and entry barriers are low. However, upgrading also has a sectoral dimension, and may differ depending on the specific features of different groups of industries". The African enterprises examined fit into four types of upgrading identified by Humphrey and Schmitz (2000) and used by Giuliani et al. (2003), namely:

1. *Process upgrading:* where the transforming production process has been re-organised or improved through introduced technology. For example, marula, KMS and *Ximenia* collecting and oil pressing in Namibia or shea butter refining in Ghana.
2. *Product upgrading*: where natural products are developed into diverse and more sophisticated product lines, with higher values per unit volume. For example, marula oil is now used by The Body Shop in more than 140 high value cosmetic and personal care products. In addition to trade-marked products from gum arabic (Spraygum™ and Instantgum™), CNI has developed the Fibregum™ product line as a soluble fibre source for the functional food market and Equacia™, a blend of *Acacia* gum and wheat fibres.
3. *Functional upgrading* refers to cases where new, superior functions are developed in the value chain. Examples are new marketing and packaging for Amarula liqueur in South Africa or organic Devil's claw production in Omaheke area, Namibia. In some cases, traditional labour intensive, low-value added local markets are completely abandoned in favour of higher value added exports, as is the case with *Ximenia* seed oils in Namibia.
4. *Intersectoral upgrading* occurs when new research or technology enables a product to shift from the one sector into a different, new sector. For example, shea butter, usually exported as a cocoa butter substitute, has unsaponifiable and anti-inflammatory components have enabled a shift into high value cosmetics and personal care products (Alander 2004).

- *Strategies to reduce or avoid elite capture.* In their study of locally traded non-timber forest products in South Africa, Shackleton et al. (2008) found little evidence of elite capture, perhaps due to their relatively low value. As Mansuri and Rao (2003) point out, however, elite capture is widespread: "Even in the most egalitarian societies involving the community in choosing, constructing and managing a public good is a process that will almost always be dominated by elites because they tend to be better educated, have fewer opportunity costs on their time, and therefore have the greatest net benefit from participation". In the cases where there is elite capture of natural products market chains, it occurs at various levels and in different forms in any stage of production, transport, processing or manufacture. At the supply end, large-scale commercial production of organic certified rooibos tea (*A. linearis)* is a threat to the wild harvested, FairTrade product from small-scale producers. Illegal or semi-legal natural products are particularly vulnerable to elite capture. In Kenya for example, the lucrative *Catha edulis* trade is controlled by prominent elites. In terms of permits and trade, well connected Cameroonian businessmen with access to transport dominated the *Prunus africana* bark trade once the French company Plantecam Medicam lost its monopoly on harvest, leading to overexploitation of wild stocks. Careful choice of the types and locations of processing technology is also required. Wynberg et al. (2003) point out an example, in the case of *S. birrea* subsp. *caffra* processing in southern Africa, that the introduction of new mechanised technologies, for pressing the oil, allowed men to gain control in a sphere from which they were traditionally absent when pressing was done by hand by women. Choosing the location of processing facilities also influences who benefits. In West Africa, for example, many large shea cooperatives are located in urban centres, disproportionately benefitting from donor support and benefiting urban rather than rural women (Elias et al. 2006). The use of process patents can also restrict high value harvest for trade as a form of elite capture. The process patent for *Terminalia sericea* root bark extract, reportedly worth USD15,000/kg, is held by the Italian company, Indena Spa. The patent restricts independent processing and sale of sericoside to competitors. Several strategies can reduce or avoid elite capture so that more benefits flow to poorer households. Firstly, the choice of enterprises based on fruits from slow growing tree species, such as marula and shea nut, and where the bulk of supplies are in farmed landscapes where farmers have tenure over the trees. Secondly, efficient coordination of producer association members to produce a quality product in sufficient quantity at a competitive price. Thirdly, use of patents in a joint venture between a producer association and private enterprise or use of trade-marks and branding. An example is the marulein patent between Phytotrade and Aldivia. Finally, the establishment of African regional Trade Hubs (see http://www.watradehub.com), processing facilities or enterprise clusters that support natural products value-adding for small or micro-enterprises (McCormick 1999). For example marula oil in Namibia and shea butter in Ghana.

- *Traceability*. Unlike local informal-sector markets, the ability to trace where a product came from is a necessary requirement in most export markets. From January 2005 for example, the EU required that all agricultural products, including shea nuts, are traceable from source (Lovett 2004). New technologies such as bar-coding offer opportunities for training producers to track products to meet these requirements. Although certification is too costly to implement, as is discussed below, chain-of-custody requirements are a useful form of traceability that are an increasingly common feature of rural enterprises such as shea butter production in Burkina Faso, West Africa (Fig. 4.4a, b) and *Schisandra sphenanthera* fruit harvest in China (Fig. 4.4c).
- *Strategic use of labelling, branding, trademarks and certification*. Product branding, trademarks and certification can play an important role in natural product export markets. Examples of *Acacia* gum trademarks used by CNI, have already been given. In addition, CNI also has organic and kosher certification for its products, which are also halaal. In Europe, farmer-owned brands play an important role across a range of products (Hayes and Lence 2004). For branding, trademarks or certification to work anywhere requires two things; (1) a "caring" market prepared to pay a significant premium for these products and (2) access to a wider share of the market due to consumer awareness. Although many consumers care about the quality and price they often are an "uncaring" market. Export markets are increasingly interested in "clean, green" products. If claims of sustainability are made, then it is crucial to ensure that those claims can be substantiated. This requires a system of traceability through "chain of custody" systems.

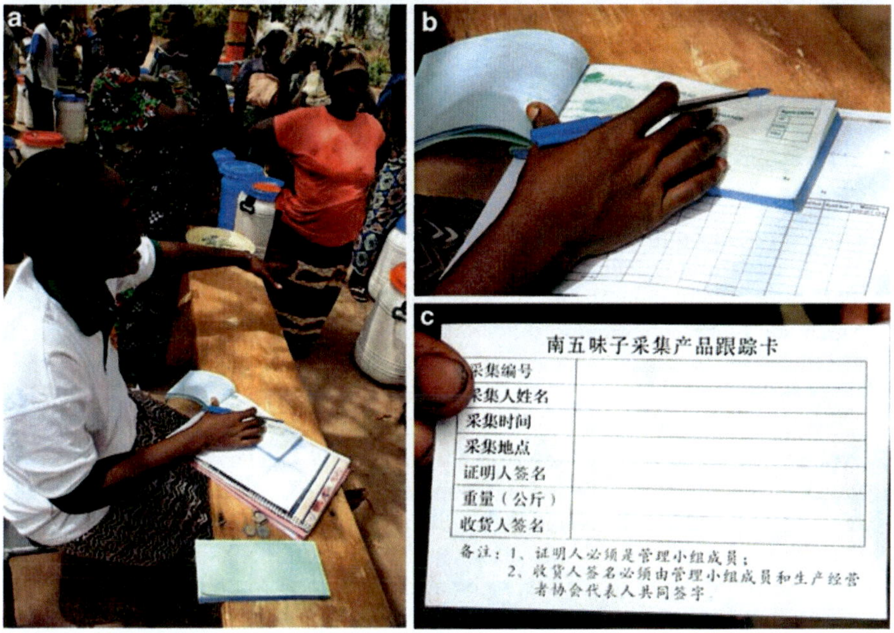

Fig. 4.4 (a-c) Traceability is a key requirement for product quality and safety. (a) Traceability forms being completed while women bringing processed shea butter for sale at a co-operative in Burkina Faso. (b) Detail of form showing copy kept for co-operative records in case batch numbers need to be traced. (c) Traceability form for *Schisandra sphenanthera* fruits, Sichuan, China. (photos: A.B. Cunningham)

- *Effective trade fair participation.* This offers the opportunity to become fully familiar with the necessary knowledge, tools and skills to prepare and co-ordinate professional group/country presentations in international trade fairs in Europe.
- *Donor support to help level the playing field.* Although the Millennium Development Goal on the global partnership for development calls for an open trading system that is rule-based, predictable and non-discriminatory, yet recognises the special needs of the least developed countries in relation to tariff- and quota-free access for their exports, this goal is far from being achieved (IFAD 2004). In 2001, the Organisation for Economic Co-operation and Development (OECD) countries' total public support for agriculture (USD 311 billion), was six times the total amount spent on official development assistance. However, producer support through domestic subsidies, import tariffs and export subsidies was estimated to equal nearly one-third of total farm receipts (IFAD 2003). Aside from mainstream farm crops, private enterprises based on plant products for niche markets in OECD countries like Australia, including joint ventures with major multi-national companies, have received significant subsidies in the form of State research support. Two examples are *Macadamia* nut production and production of opium-poppies for alkaloids. International donor support to level the playing field for small- and micro-enterprises is therefore encouraging, and

has been a key factor in the development of Phytotrade Africa. Phytotrade Africa received a start-up grant of USD 1 million and a subsequent grant of USD 1.5 million from IFAD (2002–2003), plus additional funds from other donors (IFAD 2005). Similarly, US-AID support to Nigerian gum Arabic traders and to African regional Trade Hubs has helped develop commercial networks so that African natural product producers are being taken seriously in the international market.

- *Limited policy bottlenecks*. Although market demand is a key factor influencing growth of natural product exports, export success is directly affected by government policies within producer countries and importing countries such as Europe, Japan and North America. In the US, the African Growth and Opportunity Act (AGOA), for example, provides useful opportunities for African enterprises. Even seemingly small policy changes can make a big difference. Unlike the EU, which allows up to 5% of chocolate to consist of cocoa butter substitutes, thus expanding the market for shea butter trade, the US does not permit non-cocoa vegetable in products labelled as chocolate. The EU, on the other hand, has recently allowed use of baobab (*A. digitata*) pulp. In Australia, fruits of baobabs (*Adansonia gregorii*) are not considered a "novel food" by the Australian Food Standards agency and were given food status in March 2005 (Wilkinson and Hall 2007).

4.4 Future Steps

Over the past decade, extensive research and analysis has developed lessons related to enterprise development, but at vastly different scales. At a global scale, Fagerberg et al. (2007) analysed factors behind the performance of 90 countries, examining why some countries succeed and others fail. They highlight how technology, capacity demand and competitiveness have enabled Uganda to catch up in terms of real GDP growth, while South Africa has lost momentum and Burundi, Cote d'Ivoire and Zambia fell behind during the period 1980–2002 (Fagerberg et al. 2007). Collier (2007) has similarly analysed factors influencing poverty in developing countries.

Small businesses anywhere face many challenges, even in urban areas of developed countries, and failure rates are high. Research in both developed (Bruderl et al. 1992; Stokes and Blackburn 2002) and developing (Rogerson 2004) countries has assessed why firms fail or succeed, enabling policymakers and small business advisors to better serve the small business sector. For example, since 1994, when a democratic government was formed in South Africa, there has been strong policy support for rural enterprises through an incorporated Rural Economic and Enterprise Development (REED) framework that has been integrated into development planning (IDP) or strategic local economic development (LED) plans. Despite this policy support, failure rates are high and support for micro-enterprises has been low compared to small and medium enterprises (Rogerson 2004). In addition, while agriculture gets significant policy support, there is little recognition that rural farmers obtain significant household income through use of indigenous plants or plant-derived products such as honey (Mutamba 2007), edible caterpillars (Ghazoul 2006) or ectomycorrhizal mushrooms (Shackleton and Shackleton 2004).

Identifying and acting on the ingredients of success (recognising this has various interpretations), in small, natural resource-based enterprises will not solve rural unemployment issues, but it can, and does, provide an important source of cash income to many. And if the lessons can be understood, communicated and adopted at both policy and implementation levels, these numbers could be increased and livelihoods made somewhat more secure.

Acknowledgement I am grateful to SIDA for their funding support to CIFOR through the Dry Forest project. I would also like to thank Mathurin Zida in Burkina Faso and my colleagues Xu Qiang, Luo Peng, Yang Jianjun and Lin Ling in the traditional medicines project funded through the EU-China Biodiversity programme. Josef Brinkmann, Pierre du Plessis, Gus le Breton, Cyril Lombard and Peter Lovett are also thanked for enjoyable discussions over several years.

References

Addaquay J (2004) The shea butter value chain: refining in West Africa. WATH Technical Report No. 3, USAID, Washington, DC

Alander J (2004) Shea butter – a multifunctional ingredient for food and cosmetics. Lipid Technol 16(9):202–205

Belcher B, Schreckenberg K (2007) Commercialisation of non-timber forest products: a reality check. Dev Pol Rev 25:355–377

Bernal R (1998) Demography of the vegetable ivory palm *Phytelephas seemannii* in Colombia, and the impact of seed harvesting. J Appl Ecol 35:64–74

Brinckmann JA (2008) China and other East Asian countries. In: Market news service for medicinal plants and extracts, vol 27. International Trade Centre (ITC)/UNCTAD/WTO, Geneva, Switzerland, pp 16–18

Bruderl J, Preisendorfer P, Ziegler R (1992) Survival chances of newly founded business organisations. Am Sociol Rev 57:227–242

Burkill HM (2005) Useful plants of West Africa, vol 5. Royal Botanic Gardens, Kew

Collier P (2007) The bottom billion: why the poorest countries are failing and what can be done about it. Oxford University Press, New York

Crenshaw MJ, Burke T (2004) Focus on illegal drugs: khat – a potential concern for law enforcement. FBI Law Enforcement Journal 2004. http://www.fbi.gov/publications/leb/2004/august04/august04leb.htm#page_11. Accessed 20 Oct 2007

Cunningham AB (1985) The resource value of indigenous plants to rural people in a low agricultural potential area. PhD thesis, Faculty of Science, University of Cape Town

Cunningham AB (2001) Applied ethnobotany: people, wild plant use and conservation. Earthscan, London

Cunningham AB (2005) CITES significant trade review of *Prunus africana*. CITES Management Authority, Geneva, Switzerland

Cunningham AB, Brinkmann J (2009) "Cinderella" *Schisandra*: linking conservation and local livelihoods in the upper Yangtze Ecoregion of China. HerbalGram 85:28–39

Cunningham AB, Shackleton CM (2004) Use of fruits and seeds from indigenous and naturalised plant species. In: Lawes M, Eeley H, Shackleton CM, Geach BG (eds) Indigenous forests and woodlands in South Africa: policy, people and practice. University of KwaZulu-Natal Press, Pietermaritzburg, pp 603–646

Cunningham M, Cunningham AB, Schippmann U (1997) Trade in *Prunus africana* and the implementation of CITES. Bundesamt fur Naturshcutz, Bonn, Germany

Cunningham AB, Garnett S, Gorman J, Courtenay K, Boehme D (2009) Eco-enterprises and *Terminalia ferdinandiana*: "best laid plans" and Australian policy lessons. Econ Bot 63:16–28

Distell Group Limited (2008) 2008 Annual Report. http://www.distell.co.za. Accessed 1 July 2009

Elias M, Bayala J, Dianda M (2006) Impediments and innovations in knowledge sharing: the case of the African shea sector. KM4D J 2(1):52–67. http://www.km4dev.org/journal

Emanuel PL, Shackleton CM, Baxter JS (2005) Modelling the sustainable harvest of *Sclerocarya birrea* subsp. *caffra* fruits in the South African lowveld. For Ecol Manage 214:91–103

Erickson HT, Corrêa MPF, Escobar JR (1984) Guaraná (*Paullinia cupana*) as a commercial crop in Brazilian Amazonia. Econ Bot 38:273–286

Fagerberg J, Srholec M, Knell M (2007) The competitiveness of nations: why some countries prosper while others fall behind. World Dev 35:1595–1620

Fauvet F, Mosse M (2003) Carlos Cardoso: telling the truth in Mozambique. Double Story, Cape Town

Feylsa TH, Aune JB (2003) Khat expansion in the Ethiopian highlands: effects on farming system in the Habro district. Mt Res Dev 23:185–89

Ghazoul J (ed) (2006) Mopane woodlands and the mopane worm: enhancing rural livelihoods and resource sustainability. Forest Research Programme Report R7822, DFID, UK

Githens TS, Wood CE (1943) The food resources of Africa. African Handbooks 3. University of Pennsylvania Press, Philadelphia, PA

Giuliani E, Pietrobelli C, Rabellotti R (2003) Upgrading in global value chains: lessons from Latin American clusters. World Dev 33:549–573

Green RH (1999) Khat and the realities of Somalis: historic, social, household, political and economic. Rev Afr Pol Econ 26:33–50

Harrison G, Roberge D (2002) Best practices, best markets: training the Nigerian gum arabic producer and small trader. USAID/Nigeria Gum Arabic Program/ARD, Washington, DC

Hayes DJ, Lence SH (2004) Farmer-owned brands? Agribusiness 20:269–285

Holtzman J (2004) The shea butter value chain: study synthesis and recommendations for WATH. WATH technical report no.1. USAID, Washington, DC

Humphrey J, Schmitz H (2000) Governance and upgrading: linking industrial cluster and global value chain research. IDS Working Paper, No.120, Institute of Development Studies, University of Sussex, Brighton

IFAD (2003) Promoting market access for the rural poor in order to achieve the millennium development goals. Roundtable Discussion Paper for the Twenty-Fifth Anniversary Session of IFAD's Governing Council. IFAD, Rome

IFAD (2004) Trade and rural development: opportunities and challenges for the rural poor. Governing Council – Twenty-Seventh Session, Rome, 18–19 Feb 2004, IFAD, Rome

IFAD (2005) Report and recommendation of the president to the executive board on proposed grants under the global/regional grants window for agricultural research and training by non-Cgiar-supported international centres. EB 2005/85/R.26/Rev.1, Agenda Item 15(b). International Fund for Agricultural Development, Executive Board, Eighty-Fifth Session, Rome, 6-8 September 2005. http://www.ifad.org. Accessed 12 Dec 2007

Kepe T (2003) *Cannabis sativa* and rural livelihoods in South Africa: politics of cultivation, trade and value in Pondoland. Dev S Afr 20:605–615

Klooster D (2000) Institutional choice, community, and struggle: a case study of forest co-management in Mexico. World Dev 28:1–20

Kuri CMB (2008) The Guaraná industry in Brazil. Int Bus Econ Res J 7:87–98

Kusters K, Achdiawan R, Belcher B, Ruiz Pérez M (2006) Balancing development and conservation? An assessment of livelihood and environmental outcomes of non-timber forest product trade in Asia, Africa, and Latin America. Ecol Soc 11(2):20. http://www.ecologyandsociety.org/vol11/iss2/art20/

Leakey R (2005) Domestication potential of Marula *(Sclerocarya birrea subsp. caffra)* in South Africa and Namibia: 3. Multiple trait selection. Agrofor Sys 64:51–59

Leakey RRB, Newton AC (1994) Domestication of 'Cinderella' species as the start of a woody-plant revolution. In: Leakey RRB, Newton AC (eds) Tropical trees: potential for domestication and the rebuilding of forest resources. HMSO, London, UK, pp 3–5

Lovett PN (2004) The shea butter value chain: production, transformation and marketing in west Africa. WATH Technical Report No. 2, US-AID West Africa Programme

Lovett P, Miller E, Mensah P, Adams V, Kannenberg C (2005) *Shea* butter export guide. West Africa Trade Hub. USAID West Africa Regional Program, September 2005

Maggs-Kölling GL, Christiansen JL (2003) Variability in Namibian landraces of watermelon (*Citrullus lanatus*). *Euphytica* 132:251–258

Mansuri G, Rao V (2003) Community based (and driven) development: a critical review. development research group. The World Bank, Washington

Marshall E, Schreckenberg K, Newton AC (eds) (2006) commercialisation of non-timber forest products: factors influencing success: lessons learned from Mexico and Bolivia and policy implications for decision-makers. UNEP World Conservation Monitoring Centre, Cambridge

McCormick D (1999) African enterprise clusters and industrialisation: theory and reality. World Dev 27:1531–1551

Morton JF (1983) Rooibos tea, Aspalathus linearis, a caffeinless, low-tannin beverage. Econ Bot 37:164–173

Morton JF, Voss GL (1987) The Argan tree *(Argania sideroxylon*, Sapotaceae), a desert source of edible oil. Econ Bot 41:221–233

Mutamba M (2007) Farming or foraging? Aspects of rural livelihoods in Mafulira and Kabompo districts of Zambia. Paper presented at the workshop Policies and Incentives for Miombo Woodland Management, October 30–31, Lusaka, Zambia

Nerd A, Mizrahi Y (1993) Domestication and introduction of marula (*Sclerocarya birrea* subsp. *caffra*) as a new crop for the Negev desert of Israel. In: Janick J, Simon JE (eds) New crops. Wiley, New York, pp 496–499

Nerd A, Irijimovich V, Mizrahi Y (1998) Phenology, breeding system and fruit development of argan *[Argania spinosa*, Sapotaceae] cultivated in Israel. Econ Bot 52:161–167

Nouaim R, Lineres M, Evans J-M, Chaussod R (2001) Mycorrhizal dependency of micro-propagated argan tree (*Argania spinosa*): II. Mineral nutrition. Agrofor Syst 27:67–77

NPI Center (2007) Global natural cosmetic sales approaching USD7 billion. Monday, 17 September 2007. http://www.npicenter.com. Accessed 26 Dec 2007

Phillip Securities Research (2003) Stock Update – Eu Yan Sang International Ltd. Jan 17, pp 1–3

Randall T (1993) Khat abuse fuels Somali conflict, drains economy. J Am Med Assoc 269:12–14

Rodin RJ (1985) The ethnobotany of the Kwanyama Ovambos. Monog System Bot Miss Gard, 9. Allen, Kansas

Rogerson CM (2004) The impact of the South African government's SMME programmes: a ten-year review (1994–2003). Dev S Afr 21:765–784

Ruiz-Pérez M, Belcher B, Achdiawan R, Alexiades M, Aubertin C, Caballero J, Campbell B, Clement C, Cunningham AB, Fantini A, de Foresta H, García Fernández C, Gautam KH, Hersch Martínez P, de Jong W, Kusters K, Kutty MG, López C, Fu M, Martínez Alfaro M A, Nair TR, Ndoye O, Ocampo R, Rai N, Ricker M, Schreckenberg K, Shackleton S, Shanley P, Sunderland T, Youn Y (2004) Markets drive the specialisation strategies of forest peoples. Ecol Soc 9(2):4. http://www.ecologyandsociety.org/vol9/iss2/art4

Sanou H, Picard N, Lovett PN, Dembélé M, Korbo A, Diarisso D, Bouvet J-M (2006) Phenotypic variation of agromorphological traits of the shea tree, *Vitellaria paradoxa* C.F Gaertn, in Mal*i*. Gen Res Crop Evol 53:145–161

Shackleton CM, Shackleton SE (2004) The importance of non-timber forest products in rural livelihood security and as safety nets: a review of evidence from South Africa. S Afr J Sci 100:658–664

Shackleton SE, Wynberg R, Sullivan C, Shackleton CM, Leakey R, Mander M, McHardy T, den Adel S, Botelle A, du Plessis P, O'Reagan D, Laird S (2003) Marula commercialisation for sustainable and equitable livelihoods: synthesis of a southern African case study. Unpublished report, Centre for Ecology and Hydrology, Wallingford, UK, 51 pp. http://www.ceh-wallingford.ac.uk/research/winners/

Shackleton CM, Shackleton SE, Buiten E, Bird N (2007) The importance of dry forests and woodlands in rural livelihoods and poverty alleviation in South Africa. For Pol Econ 9:558–577

Shackleton SE, Campbell B, Lotz-Sisitka H, Shackleton CM (2008) Links between the local trade in natural products, livelihoods and poverty alleviation in a semi-arid region of South Africa. World Dev 36:505–526

Shackleton SE, Shackleton CM, Wynberg R, Sullivan C, Leakey R, Mander M, McHardy T, den Adel S, Botelle A, du Plessis P, Lombard C, Laird S, Cunningham T, O'Regan D (2009) Livelihood trade-offs in the commercialisation of multiple-use NTFPs: lessons from marula (*Sclerocarya birrea* subsp. *caffra*) in southern Africa. In: Uma Shaanker R, Hiremath AJ, Joseph GC, Rai ND (eds). Non-timber forest products: conservation, management and policy in the tropics. ATREE, Bangalore, pp 139–173

Stokes D, Blackburn R (2002) Learning the hard way: the lessons of owner-managers who have closed their businesses. J Small Bus Enterp Dev 9:17–27

UNDOC (2007) World drug report, 2007. UN Office on Drugs and Crime. ISBN 978-92-1-148222-5

US Department of Commerce (2000) Ethiopia: overall export down, khat sales up. http://www.tradeport.org (cited by Feylsa and Aune, 2003) WATH Technical Report No. 2., US-AID West Africa Programme

Van Wyk B-E, Gericke N (2000) People's plants: a guide to the useful plants of southern Africa. Briza, Pretoria, 351 pp

Vines G (2008) Solving Chinese puzzles. Kew Summer 18–21

Wegener T (2000) Devil's claw: from African traditional remedy to modern analgesic and anti-inflammatory. HerbalGram 50:47–54

Wilkinson J, Hall M (2007) Baobab fruit: the upside down tree that could turn around the drinks industry. Soft Drinks International, 26–28 April 2007

Woller G (2004) A review of impact assessment methodologies for micro-enterprises development programmes in evaluating local economic and employment development. OECD, Paris

Wunder S (2001) Poverty alleviation and tropical forests: what scope for synergies? World Dev 29:1817–1833

WWF (2001) Terrestrial ecoregions: Mediterranean *Acacia-Argania* dry woodlands and succulent thickets (PA1212). http://www.worldwildlife.org/wildworld/profiles/terrestrial/pa/pa1212_full.html

Wynberg RP, Laird SA, Shackleton SE, Mander M, Shackleton CM, du Plessis P, den Adel S, Leakey RRB, Botelle A, Lombard C, Sullivan C, Cunningham AB, O'Regan DP (2003) Marula commercialisation for sustainable and equitable livelihoods. For Trees Livelihood 13:203–215

Chapter 5
Cultural Importance of Non-timber Forest Products: Opportunities they Pose for Bio-Cultural Diversity in Dynamic Societies

Michelle Cocks, Citlalli López, and Tony Dold

Abstract There is an increasing awareness that monetary value does not fully represent the complete value and significance of NTFPs. Consequently, there is growing interest in the cultural dimensions of biodiversity and the role that it plays in human well-being. This chapter presents two case studies, one on traditional brooms in South Africa, and the other on *amate* paper in Mexico, to demonstrate the importance of cultural values on driving demand for NTFPs. Because cultural values are so deeply embedded, the demand for culturally valued NTFPs continue across the rural-urban divide, and are maintained even by modernising urban communities. This poses particular challenges, not only for conservation of the NTFPs, but also to sustain cultural diversity in a rapidly changing world.

5.1 Introduction

In the late twentieth century most tropical forest resources were conventionally divided into two main groups: timber resources and non wood or minor forest products, most commonly called non-timber forest products (NTFPs). Chambers and Leach (1987) were among the first to recognise the importance of NTFPs for

M. Cocks (✉)
Institute for Social and Economic Research (ISER), Rhodes University, P.O. Box 94, Grahamstown 6140, Eastern Cape, South Africa
e-mail: m.cocks@ru.ac.za

C. López
Centro de Investigaciones Tropicales, Universidad Veracruzana, Calle Araucarias s/n, Interior de la Ex-hacienda Lucas Martin, Colonia Periodistas, C.P. 91019, Apartado Postal 252, Xalapa, Veracruz, México
e-mail: citlalli_lb@yahoo.com

T. Dold
Botany Department, Selmar Schonland Herbarium (GRA), Rhodes University, P.O. Box 94, Grahamstown 6140, Eastern Cape, South Africa
e-mail: t.dold@ru.ac.za

S. Shackleton et al. (eds.), *Non-Timber Forest Products in the Global Context*,
Tropical Forestry 7, DOI 10.1007/978-3-642-17983-9_5,

their contribution to rural livelihoods in the form of either a source of cash and or as a form of savings and assets. Peters et al. (1989) highlighted the potential monetary significance of NTFPs by providing evidence that they potentially yielded higher net revenues per hectare than tropical timber and that the former could also be harvested with considerably less negative impact on the forest. Since then, much attention was given to the contribution of NTFPs to rural livelihoods through food production and household welfare (Hladik et al. 1993; Wollenberg and Ingles 1998). Consequently, wild harvested plant products were classified as having either a subsistence value or a commercial value (Goebel et al. 2000; Campbell and Luckert 2002).

Most of the available NTFP guidelines and evaluation methods have largely focussed on the ecological and economic aspects of NTFPs (e.g., Peters 1996; Campbell et al. 1997; Wong 2000; Campbell and Luckert 2002; Shackleton et al. 2007; Burgener and Walter 2007). This led to a proliferation of scholars from a range of disciplines attempting to document the various contributions of natural resources to peoples' livelihoods (Kepe 2008).

While most of these studies provided some expression of economic value of NTFPs and greatly improved understanding in this regard, there is an increasing awareness that monetary value alone does not fully represent the complete value and significance of NTFPs (Sheil and Wunder 2002; Cocks and Dold 2004; Cocks et al. 2006; Kepe 2008). For example, Sheil and Wunder (2002) note that forest products are often deeply embedded in the political, institutional, and cultural life of the people who are involved in their collection and consumption, and therefore question whether everything that people "value" in the broader sense can, or should, be "priced" in the narrow quantitative sense. As argued by Alexiades and Shanley (2004) "the multidimensionality of NTFPs is evident in the myriad of processes, actors, and factors that shape their management, processing, and commercialisation, and therefore, forest products often have distinct, long and complex, historical trajectories" (see Box 5.1). To date, however, within the multidisciplinary literature on NTFPs the cultural dimension is less studied and seldom understood (Alexiades and Shanley 2004; Kepe 2008).

Cultural values of natural resources has been given some attention within the disciplines of anthropology, ethnoecology and ethnobotany (Toledo 2002; Alexiades 2003; Cocks 2006b), but for the most part have been ignored in valuation studies of NTFPs. There is however, a growing interest in the cultural dimensions of biodiversity, and the role that it plays in human well-being. Consequently, greater attention is now directed towards the relationship between biodiversity and human diversity. The Declaration of Belem (in Posey 1988) describes this relationship as an "inextricable link" between biological and cultural diversity (Posey 1999). The notion of the inextricable link implies not only that biological and cultural diversity are linked to a wide range of human–nature interactions, but that they are co-evolved, interdependent, and mutually reinforcing. In this context, novel approaches to sustainable natural resource management have been suggested by researchers and conservation bodies. For example, sacred sites are especially valued since these areas often represent the oldest form of habitat protection and still constitute a large and mainly

Box 5.1 Public Access Rights in Sweden: A Demonstration of Cultural Importance of Forest and Their Products

Annika Dahlberg, Dept of Geography and Quaternary Geology, Stockholm University, 10691, Stockholm, Sweden, annika.dahlberg@natgeo.su.se

The use of NTFPs has mainly been described for developing countries, which is not unexpected since people living in these regions often depend heavily on these products for livelihood security. However, NTFPs can also be important to the well-being of many in developed countries, albeit seldom crucial from a livelihood perspective. While the harvesting of NTFPs can bring material benefits in such settings, the cultural and recreational values are generally of higher importance. I will argue that not only do specific NTFPs have a cultural value, but that the right of access to the land and NTFP resources is in itself an indicator of cultural importance.

In the Nordic countries of Sweden, Finland and Norway a custom of public rights of access to the countryside exists that includes the rights to harvest NTFPs. 'Allemansrätten', as this is called, can be described as a code of conduct that secures the right of the public to move freely about the countryside, irrespective of land ownership, provided one does not cause disturbance or damage. This freedom of movement also applies to rivers, lakes and sea, and includes the right to camp and light a fire, and to collect NTFPs of low economic value, such as mushrooms, berries and flowers (Sandell 2006).

In Sweden aspects of this right are traceable back to the Middle Ages. Reasons behind the early development of Allemansrätten are varied, and include historical traditions of common property regimes. The later development of Allemansrätten can be linked to the transformation of social and political identities in Sweden in the first half of the 20th century. National identity became coupled to a growing interest in 'the Swedish landscape and nature' and to the perception of the 'nature loving Swedes'. Rapid urbanization and industrialisation resulted in a growing interest in physical leisure activities, and the time and money needed to engage in outdoor recreation was assured through an overall increase in welfare. The tradition of public right of access facilitated these developments and was in turn strengthened by them. Allemansrätten became regarded as a concept in legislation after the 2nd World War, and incorporated in the Swedish constitution in 1994. However, in spite of this the right of public access is not defined in law, except in legislation concerning where and when it does *not* apply. Allemansrätten can thus be described as the free space between different restrictions related to: 1) economic interests such as timber and crops; 2) people's privacy; 3) conservation interests; and 4) ongoing land use activities. Today, in spite of an increasing range of organised forms of outdoor recreation and access to foreign tourism destinations the right of public access to the countryside

(continued)

holds a very strong position in Sweden (Fredman et al. 2008; Dahlberg et al. 2010).

Both urban and rural dwellers engage in a wide range of activities made possible through the existence of public access. These take many forms, from more ambitious and adventurous hiking, canoeing and skiing trips, to the Sunday afternoon stroll through the nearby forest. In the summer people wander through forests and meadows to pick flowers or berries, such as raspberries and blueberries, which are used to make jam and juice. During late summer and autumn many different species of mushrooms are harvested. Other NTFPs, such as cones, lichen and mosses, are sometimes collected for crafts and decorations for traditional events. The species and amounts harvested vary between years due to natural fluctuations in abundance, between regions due to availability and traditions, and between people due to knowledge, time and interest. Overall, the total number of species utilised is high and for certain popular species the amounts harvested substantial. As described in detail by Dahl (1998), many utilised species have a cultural significance, which is illustrated through the numerous references found in Swedish songs, films, books, and commercial advertisements.

That the public right of access in itself has a cultural dimension can be illustrated and discussed from different perspectives. Dahl (1998) describes how both official documentation as well as private individuals argue in favour of Allemansrätten based on its perceived antiquity (in a Swedish context) and on its uniqueness, i.e., as something uniquely Swedish. Although both these arguments can be contested – especially the latter – their perceived validity illustrates how important Allemansrätten is as a cultural symbol. Colby (1988) argues that the long tradition of various forms of access rights has resulted in a land ethic where an established morale defines rights and determines behaviour. He illustrates this through interviews with Swedish landowners who have noted a difference in the behaviour of Swedish and foreign visitors on their land. Where most Swedes intuitively know, and respect, what is allowed within the public access rights, many foreigners find the concept strange and often misinterpret it as 'everything is allowed'. However, as pointed out by both Colby (1988) and Dahl (1998), although Allemansrätten causes conflicts when misused by tourists, it is also seen and used as a means to attract tourists. Allemansrätten can thus be described as part of Swedish identity. This is not to say that all Swedes know or follow the code of behaviour inherent in Allemansrätten and some complain that urban dwellers do so less than rural people. On the other hand, to many people living in urban areas the links to the countryside are very important, and they may at times value the rights provided by Allemansrätten higher than rural people. To urban people, the right of public access is one means of maintaining cultural and historical links to their rural roots in pre-industrial Sweden.

(continued)

Furthermore, a positive awareness of Allemansrätten is commonly found also among people who very seldom utilise this right – to them it is a cultural symbol and inherent in Swedish identity.

Through Allemansrätten the concepts of nature and culture are linked in intricate ways, both materially and immaterially. The right of public access, in Sweden and elsewhere, has a cultural value that exists independently of the areas and resources that may be accessed. Or as stated by Dahl (1998: 296) "*Being allowed to pick berries, mushrooms and wild flowers stands as a metaphor for 'having access to Nature*". That is, the right to harvest material NTFPs can be seen as a metaphor for the right to access nature, seen as part of Sweden's national (or human) identity.

unrecognised network of sanctuaries around the world (Secretariat of the Convention on Biological Diversity 2004). A number of guidelines, such as those of Oviedo and Maffi (2000) and Dudley et al. (2005), have been produced to inform environmental practitioners of the value of associating traditional cultural practices and spiritual meanings to biodiversity conservation. In 2004, the Secretariat of the Convention on Biological Diversity proposed ways to incorporate cultural, environmental, and social considerations of indigenous and local communities into new or existing environmental impact-assessment procedures. In line with these proposals, other initiatives have aimed to develop cultural indicators to assess indigenous peoples' food and agro-ecological systems, which are also closely related to use of forest resources and NTFPs (Woodley et al. 2008).

It is important to recognise that most indigenous communities no longer represent homogenous linguistic groups, but are increasingly becoming more plural, in response to changing socio-economic and political influences, such as formal and mainly western-shaped education systems, the globalisation of economic relationships, migration, and processes of secularisation (Cocks 2006b; Mathez-Stiefel et al. 2007). To date, the majority of case studies selected to illustrate the link between biodiversity and cultural diversity have largely focussed on "exotic" communities which, by their nature, are remote and isolated (Cocks 2006a). However, the exchange of knowledge, ideas, and products form an intrinsic part of the history of all civilisations across the globe (Alexiades and Shanley 2004), including of increasing exchange along the rural–urban continuum over the last century (Padoch et al. 2008).

Despite this growing exchange along the rural–urban continuum several anthropological studies have shown that people frequently continue to maintain their links with their place of origin while moving to different settings, and as such have become multi-local households or transnational communities. Historical and ongoing processes of internationalisation and transnationalisation demonstrate that social, cultural, and economic actions are not defined only at national levels anymore, but are rather embedded in complex layers of transnational, national, and sub-national institutional frames revealing the emergence of new networks and patterns that cross over multiple boundaries (Appadurai 1996). For example, in recent years, medicinal plant products are becoming increasingly traded through formal markets such as pharmacies, internet-based marketing and mail-order systems, and international markets (Mander and Le Breton 2006). The nature of the international trade in medicinal plants makes it difficult to obtain precise information about its structure and scale but a report given by the International Trade Centre, stated that eight countries belonging to the European Economic Commission in Europe imported 80,738 tonnes of plant material from Africa (Hamilton 1992). It is estimated that there are more than 2,000 herbal medical companies operating in Europe and more than 22,000 in the USA; with Germany representing the largest market in the world for herbal medicines, boasting annual sales of approximately USD 1.2 billion (Hamilton 1992). This trade is estimated to be escalating between 12% and 15% per year in the UK, USA, and Italy (Hamilton 1992), largely because of the greater than ever popularity for natural-based, environmental friendly products.

Other commercially traded NTFP's form part of what is being referred to as the "nostalgic market". For example, Gockowski et al. (2003) describe how the introduction of exotic agricultural crops appears to have had little effect on the sales and consumption of traditional leafy vegetables in Cameroon. Similarly Cocks et al. (in press) have reported that the consumption of wild leafy vegetables remains stable across peri-urban areas and urban centres within the Eastern Cape province of South Africa, as has many wild foods in Europe (Box 5.2). The authors therefore argue that there is a need to study the consumption of food as a bio-cultural phenomenon, and not simply from a nutritional understanding or as a poverty relief strategy, as the preparation and consumption of these dishes is predominantly done by women and children as a social pastime with family and friends (Cocks et al. in press). It has therefore been recommended that in studying communities' food habits, one needs to "invite the curious eyes of historians, geographers, sociologists, and folklorists" into the analysis (Fieldhouse 1998).

On the other hand, local NTFPs have been internationalised by their commercialisation in markets around the world. For example, the *uxi* fruit (*Endopleura uchi*) growing in the states of Pará and Amazonas, once considered the fruit of the poor, is today sold in many cities of Brazil in different presentations such as a flavouring in ice cream (Shanley and Gaia 2004). Another example constitutes the fruits from *shea* trees (*Vitellaria paradoxa*) in southern Sahel and Sudan zones of Africa, traditionally consumed as a nutritious snack, prepared as soap, and as cooking fat. Today, it is sold internationally as a key ingredient in natural cosmetic products (Schreckenberg 2004). A liquor flavoured with the pulp of marula (*Sclerocarya birrea* subsp. *caffra*) from South Africa is exported to dozens of countries worldwide (Shackleton et al. 2009b).

In view of the developments outlined above, this chapter demonstrates, through selected cases (traditional grass brooms to South Africa and *amate* paper in Mexico), the cultural and evolving value of NTFPs. The implications for bio-cultural diversity conservation are raised and discussed.

Box 5.2 Cultural Significance of Berries and Mushrooms in Northern Europe: NTFPs Between Tradition and Acculturation

Yvonne M. Scherrer, University of Basel, Program Sustainability Research, Klingelbergstrasse 50 CH – 4056, Basel, Switzerland, y.scherrer@pantek.org

While the cultural significance of NTFPs in the context of development among mostly rural societies has received increasing attention in the past years, the role of NTFPs in industrialised, developed societies is still rather unexplored. Although the sheer economic need to rely on wild natural resources for subsistence is significantly less accentuated in Europe, their cultural significance is still surprisingly high. Traditions related to gathering and processing NTFPs are either an uncontested part of cultural heritage, are being re-discovered or, as described below, even transferred to regions/countries where a specific NTFP was formerly less or not at all considered.

In Northern European countries such as Finland, Norway, or Sweden, there is a long tradition of gathering berries that is considered to be of nationwide significance. In the first place, blueberry (*Vaccinium myrtillus*), lingonberry (*Vaccinium vitis-idea*), cloudberry (*Rubus chamaemorus*), and raspberry (*Rubus idaeus*) are collected mainly for domestic consumption. The custom of gathering berries more than just collecting them to eat, represents a culturally embedded tradition, and has developed into one of Sweden's major recreational activities. It is considered to be an enjoyable and satisfying activity, providing a diverse range of health and educational benefits. By strengthening links between people and the environment, it is regarded as helping to foster appreciation of the natural world and to sustain a connection with the countries' cultural heritages. Similar results have been reported from Switzerland (Kilchling et al. 2009). Consequently, gathering berries such as blueberries and ligonberries from the woods is not only a popular pastime in Scandinavian countries, but is even supported by the educational system, when school children are at times allowed time off from school specifically for gathering wild berries (Kangas and Markkanen 2001; Kardell 1980).

The cultural connotations and significance of another widely occurring NTFP – wild edible mushrooms – clearly manifests in its spatial distribution. Although trends may vary among countries, in Northern and Central Europe (the Germanic, so-called "mycophobic" populations), mushrooms have traditionally been mistrusted and excluded from the diet. Other countries, especially Eastern European ones and some in Southern Europe (so-called "mycophilic" countries, inhabited by Slavs and Romans) have a strong tradition of collecting and eating wild edible mushrooms and have developed a robust marketing network to satisfy the high domestic demand. The frontier regions between "mycophobic" and "mycophilic" countries display interesting examples of acculturation of NTFPs. In Sweden, where collecting mushrooms is highly popular today, the custom of eating fungi was imported from France, and was initially exclusively adopted by the nobility. Even the years

(continued)

of near-famine in the nineteenth century could not change the adverse attitude of the general public towards mushrooms, despite considerable official effort and propaganda. It was, eventually, the migration to the cities in the twentieth century, and the related awareness of other eating habits that brought the Swedish to consider wild fungi to be worth harvesting and eating (Kardell 1980).

A similar pattern of initial reluctance can be found in Finland. Through France and Sweden, the old Roman mushroom tradition, favouring chanterelles (*Cantharellus cibarius*) and cep/porcini (*Boletus edulis*), came to the educated, mostly Swedish speaking people of southwest Finland. In the eastern part of the country, the ordinary country folk diffidently adopted a second mushroom tradition that came from the east via Karelia favouring acrid milk caps, the *Lactarius* species. However, it was only after the famines in the 1860s and World War II that the attitude of the Fins towards mushrooms started to change fundamentally. Unlike in Sweden, it was not the urban life style that set the impulse to broaden the local diet, but the immigration of 400,000 evacuees after the war from the part of Karelia conquered by the Soviet Union. The Karelians were resettled among farming families all over Finland, and seemed to have been successful in integrating their mushroom preferences into the Finnish diet. Even today, the majority of marketed mushrooms come from Eastern Finland with the *Lactarius tivialis* species still being the favourite choice (Haerkoenen 1998).

5.2 Case Study: Traditional Grass Brooms in an Urban Context

The purpose of the following case study is to demonstrate the reciprocal interaction between urban and rural communities regarding the use and trade of traditional grass brooms in the Eastern Cape province of South Africa. Information pertaining to the urban use of grass brooms was collected from Nelson Mandela Metropole (NMM) (Fig. 5.1). NMM is home to 775,255 people, of whom 56% are black Africans (Statistics South Africa 1996), and represents the largest urban centre in the province. The grass brooms used in NMM are mostly produced in Mxhalanga, a village in the former Ciskei homeland of the Eastern Cape. The homelands are the result of racially based policies implemented by the former apartheid government, and are characterised by poor infrastructure, high population densities, and high poverty levels (Viljoen 1994). A large proportion of the population of Mxhalanga is unemployed, and relies on government welfare payments, or on urban earnings, rather than on subsistence based economies (Cocks and Dold 2004).

Two types of grass brooms are made in Mxhalanga and are sold within the NMM, these being large wooden handled brooms (*umtshayelo wentonga*) and small

Fig. 5.1 Eastern Cape Province with detail of the grass broom production area

grass brooms (*umtshayelo wesandle*) (Fig. 5.2). The producers consider the smaller – *umtshayelo wesandle* – to be of traditional Xhosa origin, a sentiment supported by early historical records (McLaren 1919). The long-handled broom is made in the same way, but a wooden handle is attached. According to Shaw and Van Warmelo (1981) this type of broom was originally copied from a European broom, but has been in existence for more than a century. The grass hand-brooms are made from turpentine grass (*Cymbopogon validus* (Stapf) Stapf ex Burtt Davy), an aromatic tufted perennial reaching 2.4 m tall that is widespread and common throughout the eastern regions of South Africa (Van Oudtshoorn 1992). The Xhosa name for this species is *irwashu*. The grass is steam-bent into the desired shape (Fig. 5.3).

Interviews were conducted with 204 broom buyers while they were purchasing brooms in NMM. Questions asked in the interview included demographic profiles of the broom buyers, who the broom was being purchased for, and the intended use of the broom. Fifty percent ($n = 31$) of all the broom producers in the broader Mxhalanga district were interviewed. Demographic profile and household socio-economic status were recorded together with particulars regarding broom making, such as their production and marketing activities.

5.2.1 Who Buys Grass Brooms and Why?

Two methods of sale are employed in NMM; sellers may sell from door to door in residential areas, or they may set up an informal roadside stall near taxi ranks and

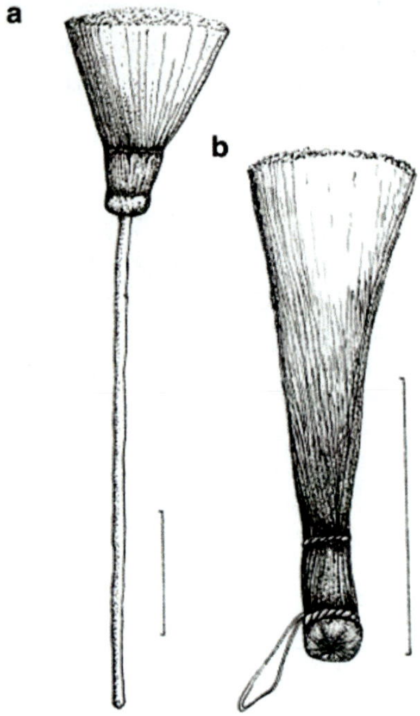

Fig. 5.2 Hand made grass brooms from Mxhalanga: (**a**) *umtshayelo wentonga*, (**b**) *umtshayelo wesandle*. Scale bar 300 mm. Illustrations A.P. Dold

Fig. 5.3 Manufacturing grass brooms from *Cymbopogon* leaves in the Eastern Cape, South Africa (photo: Tony Dold)

bus stations where commuters are targeted as potential customers. No advertisement or discourse other than simply displaying the brooms is employed during the selling process.

Eighty-four percent of grass broom buyers were female, while only 16% were male. Seventy-five percent of the buyers purchased the grass broom for themselves, 19% purchased it for their mothers, and 6% for their wives. The reasons for purchasing a broom varied, with 59% purchasing a broom for cultural purposes. The remaining 41% indicated that they used the grass broom for cleaning purposes only as it was considered to be more effective than a commercially available broom. The three main cultural uses of grass brooms were the following:

- *As a traditional wedding gift*. As many as ten grass hand-brooms, together with sedge mats (*amakhukho*), are presented as a traditional wedding gift from a mother to her daughter a few weeks after the wedding ceremony. The traditional ceremonial presentation of the broom is called *ukutyiswa amasi* (literally to present a gift of sour milk) (Hunter 1936). The broom is symbolic of traditional Xhosa culture and symbolises respect to the ancestral faith in the newlyweds' home, irrespective of religious affiliation, economic status, and geographical location. These brooms are later used for daily sweeping and are replaced when necessary.
- *As a protective talisman against lightning*. A broom purchased for this purpose is not used for cleaning at all, but is hung on the wall next to the main entrance door of the home as a talisman. The presence of a hand-broom in the home is also believed to protect the inhabitants from lightning, often attributed to sorcery (Soga 1931; Hunter 1936).
- *As an implement for the application of traditional protective medicine*. The ritual in which an infusion of various plant medicines is splashed or sprayed on the floor, walls, and roof is called *ukutshiza* (Dold and Cocks 2002). This ritual cleansing and purification ensures the good health and prosperity of the inhabitants (Soga 1931; Hunter 1936; Cocks and Møller 2002). The broom used in this ritual is also not used for cleaning.

The urban survey revealed that almost all the urban buyers were middle-aged to elderly women who had bought brooms for *amasiko* (Xhosa customs). Several socioeconomic factors influenced the use of grass brooms in the study site. The most significant of these was level of education, family origin, followed by economic status and age. The cultural importance of grass brooms decreased with an increase in education level. Economic status (expressed by means of housing type) was significant, showing that people from lower economic groups were more likely to purchase a grass broom for cultural purposes. Older people also tended to buy brooms more for cultural purposes than for cleaning. It was found that the length of the period of urban living of households was not significant (Cocks and Dold 2004).

5.2.2 Broom Makers and Sellers

The grass hand-broom makers are predominantly middle-aged women from disadvantaged backgrounds with little or no formal education who are also often the sole income earners in their households. The skill to make brooms is traditionally learnt from mothers and grandmothers. Material to produce brooms is harvested on a privately owned farm on the Nico Malan Pass in the Seymour district (Fig. 5.1) as the grass is not located near the village of Mxhalanga. This equates to a 260 km return trip for the grass broom producers each season. Shackleton (1990) showed that harvesting of the *C. validus* grass promotes annual growth, and observations in the production area indicated that *C. validus* is a sustainable resource for Mxhalanga broom producers at the current level of harvesting. Both types of broom are produced at Mxhalanga but seldom sold within Mxhalanga; the producers travel extensively to sell them. Sale localities include Alice, Cathcart, East London, King William's Town, NMM, Queenstown, and Stutterheim (Fig. 5.1). Approximately 7,200 large brooms and 1,400 small brooms are produced in Mxhalanga per year.

This case study reveals that traditional Xhosa cultural practices are still significant even in highly urbanised contemporary settings, and the materials needed to perform these are sourced from the rural resource base and accessed via trade links. This has been made possible by the ever increasing linkages between rural and urban areas. Today rural communities which may be geographically isolated are no longer isolated in economic and social terms as cash, goods, and remittances flow from urban to rural areas, and rural goods and culture to urban areas (Wiersum and Shackleton 2005). The grass broom trade also provides an opportunity for disadvantaged rural people to earn an income. Their involvement in the trade is often of necessity and a coping strategy (Shackleton and Campbell 2007; Shackleton et al. 2008). Similar socio-economic circumstances have been recorded amongst broom makers from other regions of the Eastern Cape, including those using palm leaves (Gyan and Shackleton 2005), and those using forest restios as raw materials (Shackleton et al. 2009a). Broom producers in the northern region of the country, in the Bushbuckridge district, similarly represent the poorest households and their engagement in the trade offers them a vital safety net, particularly for those households impacted by HIV/AIDS, for widows, and so-called "granny"-headed households (Shackleton and Campbell 2007).

5.3 Case Study: *Amate* Paper in Mexico

Amate is an indigenous paper made from bark fibres of tropical trees since pre-Hispanic times. Although there is no certainty about the time or place of origin of bark paper, Lenz (1973) and Von Hagen (1945) argue that evidence of the manufacture of bark paper dates from as early as 300 A.D. and this indicates the manufacture and use of bark clothing by Mayan people. The period in which bark paper was most intensively used corresponds to the beginning of the 16th century, when the Aztecs dominated most of Mesoamerican territory. Mesoamerica stretched from central Mexico to northern Honduras, and was populated by related

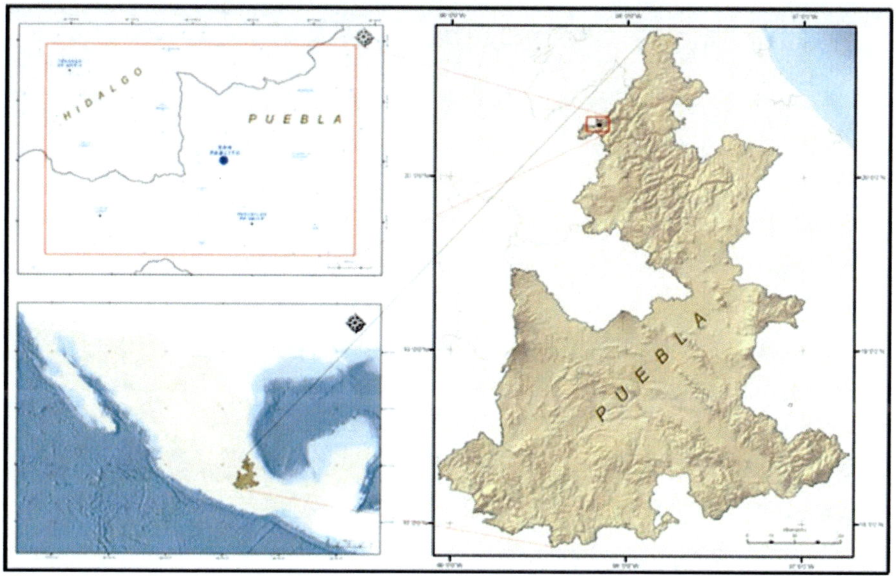

Fig. 5.4 Location of the amate paper producing village, San Pablito in the Sierra Norte de Puebla, Mexico

language groups that achieved a high level of social, political and economic life. Bark paper was bestowed with sacred and commercial values and was used profusely in form of Codexes, offerings and clothing among other uses (Lenz 1973; Seeman 1990).

During the Spanish colonisation it was banned due to its use being linked with native gods, beliefs and political institutions. Despite this prohibition, clandestine production of *amate* persisted among a few indigenous groups in scattered villages along the Huasteca and the Sierra Norte de Puebla. According to Lenz (1973) and Galinier (1987), the inaccessibility of this region and the distance from centres of dominion contributed to the survival of this pre-Hispanic tradition. Among the few indigenous groups who have maintained the production of this paper are the Ñahñu of San Pablito village in the Sierra Norte de Puebla (Fig. 5.4).

The production of *amate* paper as a commercial handicraft started in the 1960s, when tourism increased in Mexico and also as a result of the institutional interest in popular art and handicrafts (Amith 1995). From the start, *amate* paper attracted attention and nowadays it is one of the most versatile and widely distributed Mexican handicrafts in national and international markets (Torres 1982; Amith 1995). A great variety of products are manufactured, including lamps, notebooks, and decorative wallpaper. *Amate* is also produced in plain, coloured, or painted form. It is sold by artisans and intermediaries in streets, open markets, or plazas in most tourist cities, or through a long chain of mediators in handicraft markets, stores, and art galleries (Fig. 5.5).

Fig. 5.5 Sample of an amate paper decorated with cut-out traditional figures

5.3.1 *New Sources of Raw Material*

During fieldwork (1999 and 2009) it has been observed that demand for *amate* is constantly increasing; therefore harvesting pressure on trees for bark production is high. To satisfy this demand, the bark extraction area has expanded, and artisans are constantly experimenting with new trees for use of their bark. In order to investigate which species are used for *amate* production (López 2003), their respective fibre qualities (Quintanar et al. 2004) and bark harvest strategies, a survey was carried out during 1999–2000 and updated during 2009 in San Pablito and at several bark harvest sites. A combination of methods derived from ethnography, ethnobotany and forest sampling techniques were applied.

A total of 15 plant species have been identified, including the traditional species mainly belonging to the *Ficus* genus and new species adopted over the last 30 years belonging different plant families (Moraceae, Ulmaceae, Euphorbiaceae, Urticaceae). According to the field survey results, the new species *Trema micrantha* L. (Blume) is the most intensively used. This is the only tree that can be harvested throughout the year. It is a pioneer plant, has a wide distribution, with high germination rates (Vázquez-Yanes 1998), and is commonly used to provide shade to coffee plants in the plantations of Sierra Norte de Puebla. The use of the bark from *Trema micrantha* constitutes the base of the current production of *amate* for the market (López 2003).

5.3.2 *Ritual Use of* Amate *Paper*

For the Ñahñu of San Pablito, *amate* paper has two values and uses; it is traditionally a ritual product and, more recently, a commercial product. While for commercial paper *Trema micrantha* trees are mainly used, for ritual purposes the traditional species (*Ficus* sp) are preferred. *Amate* paper is traditionally used to make cut-out figures representing fruits and grains, men, women, and children, and also diverse gods, such as the God of the Mountain, or the God of the Earth.

The principle behind this use of paper figures is animistic. The Ñahñu believe that all beings, human and super-human, have a living animating force, which they call *nzahki.* It is the Otomi version of a basic Mesoamerican Indian soul concept. Within this conception trees are a symbol of potency. *Za* -derived from the term *nzahki-* refers to the vital energy situated in the stomach, the centre of the body (López-Austin 1994). Thus, according to Galinier (1990), the bark paper becomes the upholder of this force. People, plants and animals have reciprocal relations and the shamans can influence them. With their word and through the cut-out technique, shamans gain power over the motivations of the being represented by the figure and become a mediator between deities, the sick person and his or her family (Dow 1984; Galinier 1987).

Paper figures are always cut out and used in groups symbolising social restructuring. Galinier (1987) mentions that cures are the process of restoring equilibrium to household groups; they guarantee the return of the cured person to the community and allow the reconciliation of the healing body with ancestors and divinities. *Amate* figures were used for the petition for good crops, cleaning ceremonies, and healing. However as today, most agricultural activities have been abandoned, and artisan work is combined with migration (Censo Programa IMSS 1999), and most rituals are performed to help people to re-adapt after working outside San Pablito (López 2003).

This case is an example to show that regardless of the type of tree used as source for bark, both the ritual and commercial paper are part of the identity of the Ñahñu of San Pablito (López 2003) and how – as Appadurai (1997) and Kopytoff (1997) argue – cultural products will persist whenever the practices surrounding them persist and adapt in a practical and meaningful way.

5.4 Discussion and Conclusion

The examples of grass brooms and *amate* paper demonstrate that the use of wild plant products is not restricted to rural utilitarian use, but constitutes an important element in the performance and conservation of cultural practices and traditions. For example, in the case of the traditional broom, 59% of the buyers in the urban centre bought brooms for cultural purposes. This indicates that many households living in urban centres still adhere to their cultural norms and practices, and rely to some degree, on NTFPs to fulfil these functions. With regard to the commercialisation of the *amate* bark paper, commercial and sacred uses and values co-exist, as all the *amate* paper produced by artisans is sold commercially, but the paper used for rituals acquires a sacred value when shamans confer strength through the technique of cut-outs and their Word (Galinier 1990). These developments can be understood by recognising that culture is not static, but adapts to modern times through the re-articulation of tradition (Canclini 1995). This is made possible on the premise that culture is a selective force which is illustrated through the following example,

> A Cherokee Indian medicine woman who lives in a solidly middle-class suburban community in Washington, DC has worked off and on in administrative jobs within the US Air Force, and has a growing clientele of mostly White Anglo patients with various physical and mental aliments. She heals by invoking spirit forces from the Cherokee pantheon, and serving as a medium for their healing powers, as her grandmother had taught her (Groenfeldt 2003, p. 921).

This example shows that the Cherokee Indian medicine woman's religious worldview is highly traditional, while her social and material cultural context is basically that of a mainstream American. Therefore, there is a deliberate choosing from the cultural assemblage at her disposal (Groenfeldt 2003). This approach to the concept of culture as a selective force has particular merit when trying to explain the phenomena that occur within societies, where lifestyles have been affected and transformed by global processes, and where livelihood strategies of communities have become diversified. As Groenfeldt (2003) observes, peoples' worldviews are maintained to a large extent, but their day-to-day lives are radically transformed. This process has been identified as "truncated innovation", as it is not a retreat into cultural essentialism, but rather one that involves the creation of subtle crisscrossing links between different cultural orientations and experiences that have been mediated by the re-articulation of tradition (Canclini 1995) as demonstrated by the use of *amate* paper by Ñañhu in rituals. For example, the *amate* paper is used ritualistically to re-adapt those who return home after working outside their community as migrant labour to the United States. The extent to which urban people in the case of the broom study still adhere to their cultural practices is complex and appears to be influenced by variables such as family origin, economic status, level of education, and age. However both these case studies present clear examples of the continuities in cultural practices and cultural resilience in spite of increasing commercial changes occurring in each of the locations they are situated within in.

Moreover, one does not have to live geographically close to the natural environment for it to hold spiritual, social, and cultural values for its users (Cocks 2006a). Authors have described how families who have migrated to urban, industrialised, or other contexts in southern Africa return to their ancestral lands to partake in cultural festivities and ceremonies featuring NTFPs (Shackleton et al. 2002; Wiersum and Shackleton 2005; Cocks 2006b). This is made possible by a number of interrelating factors, firstly, the increase in the diversification in rural livelihoods and increased mobility, as well as the incorporation of rural areas into commercial trade networks (Wiersum and Shackleton 2005). Secondly, by the ruralisation of cities as discussed by Padoch et al. (2008) whereby rural lifestyles, attitudes, and occupations continue to persist despite ones' physical relocation to an urban city. This is because many new urban households are multi-sited with families maintaining houses and economic activities in both urban and rural areas (Krüger 1998). This has been facilitated by the improvements in communication and transportation. These linkages ensure multi-functional networks of support and interactions with rural kin (Padoch et al. 2008). It is these very processes, in the case of the broom study, which have encouraged women to sell cultural artefacts in the form of traditional brooms in urban areas.

These changes have, however, resulted in the increased demand for NTFPs (Campbell et al. 2001; Sunderland and Ndoye 2004; Wiersum and Shackleton 2005) as the selling of NTFPs is often one of the primary means for rural households to cope with their economic hardship. The processing and the sale of natural products offers a low barrier to entry into the market and generates cash income (Shackleton and Shackleton 2004; Shackleton et al. 2008). Such strategies could be enhanced by the implementation of local NTFP initiatives which carefully consider the social–cultural aspects within which they operate. Such endeavours could revitalise traditions, reinforcing people and community pride, securing intergenerational transmission of knowledge and skills. For example, marginalised communities within the mountainous area of Guerrero, Mexico, use the leaves of the palm (*Brahea dulcis*) for local and domestic artefacts. New productive projects have been initiated in the area to improve the production of palm handicrafts for external markets. It has been found that the acquisition of knowledge for managing and for producing traditional and new products has assisted the local women in not only valuing their own work more, but that their skills have also become valued by the rest of the community members (Martha Miranda, Grupo Autónomo para la Investigación Ambiental A.C. personal communication 2009). As shown in this chapter, cultural context implies the traditional use, management, perception of natural resources, from extraction and processing techniques to the symbolic aspects of the resources and practices around them. Compared to the evaluation of other NTFPs aspects, such as the ecological, legal-institutional, or market ones, the documentation and interpretation of the cultural context imposes more difficulties and requires considerably more time to fully understand. However, in many cases the cultural context constitutes the main reason of failure or success of the expansion of use of NTFPs.

The degree to which biological diversity is linked to cultural diversity is only beginning to be understood, but, despite these recent advances these complex systems are under threat (Pretty et al. 2009). Despite the emerging acknowledgement of the need for an integrated approach to the conservation of biological and cultural diversity for ensuring sustainable development (for example, the Millennium Development Goals), policy responses to this integrated paradigm have been slow to emerge and be instituted. While this integrated paradigm presents unique challenges, it is considered crucial that more serious attention be paid to the cultural context of NTFPs, to not only ensure that appropriate strategies are adopted to manage NTFPs and ecosystems, but to also encourage the survival of cultural diversity in a rapidly changing world.

References

Alexiades MN (2003) Ethnobotany in the third millennium: expectations and unresolved issues. Delpinoa 45:15–28

Alexiades MN, Shanley P (2004) Forest products, livelihoods and conservation: case studies of non-timber forest product systems. In: Alexiades MN, Shanley P (eds) Forest products, livelihoods and conservation: case studies of non-timber forest product systems, vol 3. Center for International Forestry Research, Bogor, Indonesia, pp 1–22

Amith DJ (1995) La creación de imágenes indígenas: de la pesadilla privada a la protesta pública. The creation of indigenous images: from private nightmares to public protest. In: Amith DJ (ed.) La tradición del amate. Innovación y protesta en el arte mexicano. The amate tradition. Innovation and Dissent in Mexican Art. Mexican Fine Arts Center Museum Chicago and La Casa de las Imágenes, México, pp 41–100

Appadurai A (1996) Modernity at large: cultural dimensions of globalisation. University of Minnesota Press, Minneapolis

Appadurai A (1997) Introduction: commodities and the politics of value. In: Appadurai A (ed) The social life of things. Commodities in cultural perspective. Cambridge University Press, Cambridge, pp 3–63

Burgener M, Walter S (2007) Trade measures-tools to promote the sustainable use of NWFP? An assessment of trade related instruments influencing the international trade in non-wood forest products and associated management and livelihood strategies. FAO Non-Wood Forest Products Working Document No 6

Campbell BM, Luckert MJ (2002) Towards understanding the role of forests in rural livelihoods. In: Campbell BM, Luckert MJ (eds) Uncovering the hidden harvest: valuation methods for woodland and forest resources. People and Plants Conservation Manuals. Earthscan, London, pp 1–17

Campbell BM, Luckert ML, Scoones I (1997) Local-level valuation of savanna resources: a case from Zimbabwe. Econ Bot 51:59–77

Campbell B, Mandondo A, Nemarundwe N, Sithole B, De Jong W, Luckert M, Matose F (2001) Challenges to proponents of common property resource systems: despairing voices from the social forests of Zimbabwe. World Dev 29:589–600

Canclini G (1995) Hybrid cultures: strategies for entering and leaving modernity. University of Minnesota Press, Minneapolis

Censo Programa IMSS (1999) Levantamiento de Universo de Trabajo en la comunidad de San Pablito. IMSS Instituto Mexicano del Seguro Social, México

Chambers R, Leach M (1987) Trees to meet contingencies: savings and security for the rural poor. ODI Social Forestry Network Paper No. 5a, Overseas Development Institute, London

Cocks ML (2006a) Bio-cultural diversity: moving beyond the realm of 'indigenous' and 'local' people. Hum Ecol 34:185–200

Cocks ML (2006b) Wild resources and practices in rural and urban households in South Africa: implications for bio-cultural diversity conservation. PhD. Thesis, Wageningin University, The Netherlands http://library.wur.nl/wda/dissertations/dis4026.pdf

Cocks ML, Dold AP (2004) A new broom sweeps clean: the economic and cultural value of grass brooms in the Eastern Cape Province, South Africa. For Trees Livelihood 14:33–42

Cocks ML, Møller V (2002) Use of indigenous and indigenised medicines to enhance personal well-being: a South African case study. Soc Sci Med 54:387–97

Cocks ML, Bangay L, Wiersum KF, Dold AP (2006) Seeing the wood for the trees: the role of woody resources for the construction of gender specific household cultural artefacts in non-traditional communities in the Eastern Cape, South Africa. Environ Dev Sustain 8:519–533

Cocks ML, Dold AP, Husselman M, Sizani N (in press) New shoots, old roots – the incorporation of alien weeds into traditional food systems. In: Singh RK, Pretty J, Turner NJ, Reyes-Garcia V, Swiderska K (eds) Opportunities from bio-cultural diversity: traditional foods, local knowledge and biodiversity conservation

Colby KT (1988) Public access to private land – Allemansrätten in Sweden. Landsc Urban Plan 15:253–264

Dahl G (1998) Wildflowers, nationalism and the Swedish law of commons. Worldviews Environ Cult Relig 2:281–302

Dahlberg AC, Rohde R, Sandell K (2010) Access, conservation and environmental justice: a comparison of national parks in three countries. Conserv Soc 8(3)

Dold AP, Cocks ML (2002) The trade in medicinal plants in the Eastern Cape Province, South Africa. S Afr J Sci 98:589–597

Dudley N, Higgins-Zogib L, Mansourian S (2005) Beyond belief: linking faiths and protected areas to support biodiversity conservation. WWF/Equilibrium and the Alliance of Religions and Conservation (ARC), UK

Fieldhouse P (1998) Food and nutrition. Stanley Thornes, Cheltenham

Fredman P, Karlsson S-E, Romild U, Sandell K (eds) (2008) Besöka naturen – hemma eller borta? Delresultat från en nationell enkät om friluftsliv och naturturism i Sverige. Rapport No. 3, Forskningsprogrammet, Friluftsliv i Förändring, Östersund

Galinier J (1987) Pueblos de la Sierra Madre. Etnografía de la comunidad Otomí. Colección Clásicos de la Antropología 17. Instituto Nacional Indigenista INI, Centro Francés de Estudios Mexicanos y Centroamericanos CEMCA, México

Galinier J (1990) La mitad del mundo. Cuerpo y Cosmos entre los Rituales Otomies. Universidad Nacional Autónoma de México UNAM, Centro Francés de Estudios Mexicanos y Centroamericanos CEMCA, Instituto Nacional Indigenista INI, México

Gockowski J, Mbazo'o J, Mbah G, Moulende TF (2003) African traditional leafy vegetables and the urban and peri-urban poor. Food Pol 28:221–235

Goebel A, Campbell B, Mukamuri B, Veeman M (2000) People, values, and woodlands: a field report of emergent themes in interdisciplinary research in Zimbabwe. Agric Hum Values 17:385–396

Groenfeldt D (2003) The future of indigenous values: cultural realism in the face of economic development. Futures 35:917–929

Gyan CA, Shackleton CM (2005) Abundance and commercialisation of *Phoenix reclinata* in the King Williamstown area, South Africa. J Trop For Sci 17(3):334–345

Haerkoenen M (1998) Uses of mushrooms by Finns and Karelians. Int J Circumpolar Health 57:40–55

Hamilton A (1992) International trade in medicinal plants: conservation issues and potential roles for botanic gardens. In: Proceedings of the third international botanic gardens conservation congress, Botanic Gardens Conservation International. http://www.bgci.org/congress/congress_rio_1992/hamilton.html

Hladik CM, Hladik A, Linares OF, Pagezy H, Semple A, Hadley M (eds) (1993) Tropical forests, people and food: bio-cultural interactions and applications to development. Man and Biosphere Series No. 13. UNESCO, Paris and Parthenon, New York

Hunter M (1936) Reaction to conquest: effects of contact with Europeans on the Pondo of South Africa. International Institute of African Languages and Cultures, Oxford University Press, London

Kangas K, Markkanen P (2001) Factors affecting participation in wild berry picking by rural and urban dwellers. Silva Fennica 35:487–495

Kardell L (1980) Forest berries and mushrooms: an endangered resource? Ambio 9:241–247

Kepe T (2008) Beyond the numbers: understanding the value of vegetation to rural livelihoods in Africa. Geoforum 39:958–968

Kilchling P, Hansmann R, Seeland K (2009) Demand for non-timber forest products: surveys of urban consumers and sellers in Switzerland. For Pol Econ 11:294–300

Kopytoff I (1997) The cultural biography of things: commoditisation as process. In: Appadurai A (ed) The social life of things. Commodities in cultural perspective. University Press, Cambridge, pp 3–63

Krüger F (1998) Taking advantage of rural assets as a coping staretrgy for the urban poor: the case of rural-urban interrelationships in Botswana. Environ Urban 10:119–134

Lenz H (1973) El papel indígena mexicano. Serie Cultura Sep Setentas 65. Editorial Cultura Sep Setentas, México

López C (2003) The endurance of Mexican *Amate* paper: exploring additional dimensions to the sustainable development concept. PhD Thesis. CSTM-Twente University Press, The Netherlands

López-Austin A (1994) Tamoanchan y Tlalocan. Sección Obras de Antropología. Fondo de Cultura Económica, México

Mander M, Le Breton GL (2006) Overview of the medicinal plant industry in southern Africa. In: Diederichs N (ed) Commercialising medicinal plants: a southern African guide. Sun, Stellenbosch, South Africa, pp 1–9

Mathez-Stiefel S-L, Boillat S, Rist S (2007) Promoting the diversity of world-views: an ontological approach to bio-cultural diversity. In Haverkort B and Rist S (eds) Endogenous development and bio-cultural diversity. The interplay of worldviews, globalisation and locality. Compas Series on Worldviews and Science 6. Centre for Development and Environment (CDE), Bern, Switzerland, pp 67–82

McLaren J (1919) Arts and crafts of the Xhosas: a study based on philology. S Afr J Sci 15:441–449

Oviedo G, Maffi L (2000) Indigenous and traditional peoples of the world and ecoregion conservation. An integrated approach to conserving the world's biological and cultural diversity. WWF-World Wide Fund for Nature, Gland, Switzerland

Padoch C, Brondizio E, Costa S, Pinedo-Vasquez M, Sears RR, Siqueira A (2008) Urban forest and rural cities: multi-sited households, consumption patterns, and forest resources in Amazonia. Ecol Sci 13(2):2. http://www.ecologyandsociety.org/vol13/iss2/art2/

Peters CM (1996) The ecology and management of non-timber forest resources. World Bank Technical Paper no. 322

Peters CM, Gentry AH, Mendelsohn RO (1989) Valuation of an Amazonian rainforest. Nature 339:655–656

Posey DA (1988) The declaration of Belem. In: Posey DA, Overal W (eds) Proceedings of the first international congress of ethnobiology. D. Museu Paraense Goeldi, Belem

Posey DA (1999) Cultural and spiritual values of biodiversity. A complementary contribution to the global biodiversity assessment. In: Posey DA (ed) Cultural and spiritual values of biodiversity. UNEP and Intermediate Technology, London, pp 1–19

Pretty J, Adams B, Berkes F, de Athayde S, Dudley N, Hunn E, Maffi L, Milton K, Rapport D, Robbins P, Sterling F, Stolton S, Tsing A, Vintinner E, Pilgrim S (2009) The intersections of biological diversity and cultural diversity: towards integration. Conserv Soc 9:100–112

Quintanar-Isaías A, Zárate-Castrejón JL, López C, Salgado-Ugarte IH (2004) Anatomía e histoquímica de la corteza de cinco especies de Moraceae. Polibotánica 17:15–38

Sandell K (2006) Access under stress: the right of public access tradition in Sweden. In: McIntyre N, Williams DR, McHugh KE (eds) Multiple dwelling and tourism: negotiating place, home and identity. CABI, Wallingford, pp 278–294

Schreckenberg K (2004) The contribution of shea butter (*Vitellaria paradoxa* C.F. Gaertner) to local livelihoods in Benin. In: Sunderland T, Ndoye O (eds) Forest products, livelihoods and conservation: case studies of NTFP systems, vol 2. Center for International Forestry Research, Bogor, Indonesia, pp 91–113

Seeman EC (1990) Usos del Papel en el Calendario Ritual Mexica. Serie Etnohistoria 207, Instituto Nacional de Antropología e Historia INAH, México

Secretariat of the Convention on Biological Diversity (2004) Akwé: kon voluntary guidelines for the conduct of cultural, environmental and social impact assessment regarding developments proposed to take place on, or which are likely to impact on, sacred sites and on lands and waters traditionally occupied or used by indigenous and local communities. CBD Guidelines Series, Montreal, Canada

Shackleton SE (1990) Socio-economic importance of *Cymbopogon validus* in Mkambati game reserve, Transkei. S Afr J Bot 56:675–682

Shackleton SE, Campbell BM (2007) The traditional broom trade in Bushbuckridge, South Africa: helping poor women cope with adversity. Econ Bot 6:256–268

Shackleton CM, Shackleton SE (2004) Use of woodland resources for direct household provisioning. In: Lawes MJ, Eeley HAC, Shackleton CM, Geach BS (eds) Indigenous forests and woodlands in

South Africa: policy, people and practices. University of KwaZulu-Natal Press, Scottsville, South Africa, pp 195–197

Shackleton SE, Shackleton CM, Cunningham AB, Lombard C, Sullivan CA, Netshiluvhi TR (2002) Knowledge on *Sclerocarya birrea* subsp. *caffra* with emphasis on its importance as a non-timber forest product in South and southern Africa: a summary. Part 1: Taxonomy, ecology and role in rural livelihoods. S Afr For J 194:27–42

Shackleton CM, Shackleton SE, Buiten E, Bird N (2007) The importance of dry forests and woodlands in rural livelihoods and poverty alleviation in South Africa. For Pol Econ 9:558–577

Shackleton SE, Campbell B, Lotz-Sisitka H, Shackleton CM (2008) Links between the local trade in natural products, livelihoods and poverty alleviation in a semi-arid region of South Africa. World Dev 36:505–526

Shackleton CM, Parkin F, Chauke MI, Downsborough L, Olsen A, Weideman C, Brill G (2009a) Conservation, commercialisation and confusion: harvesting of *Ischreyolepis* in a coastal forest, South Africa. Environ Dev Sustain 11:229–240

Shackleton SE, Shackleton CM, Wynberg R, Sullivan C, Leakey R, Mander M, McHardy T, den Adel S, Botelle A, du Plessis P, Lombard C, Laird S, Cunningham T. O'Regan D (2009b) Livelihood trade-offs in the commercialisation of multiple-use NTFPs: lessons from marula (*Sclerocarya birrea* subsp. *caffra*) in southern Africa. In: Uma Shaanker R, Hiremath AJ, Joseph GC, Rai ND (eds) Non-timber forest products: conservation, management and policy in the tropics. ATREE, Bangalore, pp 139–173

Shanley P, Gaia G (2004) Poor man's fruit turns profitable: *Endopleura uchi* in managed groves near Belém, Brazil. Translated from: Shanley P, Gaia G (2004) A 'fruta do pobre' se torna lucrativa: a Endopleura uchi Cuatrec. em áreas manejadas próximo a Belém, Brasil. In: Alexiades MN, Shanley P (eds) Productos Forestales, Medios de Subsistencia y Conservación. Estudios de Caso sobre Sistemas de Manejo de Productos Forestales No Maderables, vol 3. Center for International Forestry Research, Bogor, Indonesia, pp 211–231

Shaw EM, Van Warmelo NJ (1981) The material culture of the Cape Nguni, Part 3: Subsistence. Ann S Afr Museum 58(3):376

Sheil D, Wunder S (2002) The value of tropical forest to communities: complications, caveats, and cautions. Conserv Ecol 6:9–25

Soga JH (1931) The Ama-Xosa: life and customs. Lovedale, Alice

Statistics South Africa (1996) Census 1996. http://www.statssa.gov.za/census

Sunderland T, Ndoye O (eds) (2004) Forest products, livelihoods and conservation. Case studies of non-timber forest systems, vol 2: Africa. Centre for International Forestry Research, Bogor, Indonesia.

Toledo VM (2002) Ethnoecology: a conceptual framework for the study of indigenous knowledge of nature. In: Stepp JR, Wyndham FS, Zarger RK (eds) Ethnobiology and biocultural diversity. International Society of Ethnobiology, University of George Press, Athens, Georgia, pp 511–522

Torres B (1982) El papel del amate, historia y significado. In: El Universo del Amate. Museo Nacional de Culturas Populares, Cultura SEP, México, pp 13–29

Van Oudtshoorn F (1992) Guide to grasses of South Africa. Briza, Pretoria

Vázquez-Yanes C (1998) *Trema micrantha* (L.) Blume (Ulmaceae): a promising neotropical tree for site amelioration of deforested land. Agrofor Syst 40:97–104

Viljoen R (1994) Fuelwood, social forestry and rural development. In Plant for Life/The Biomass Initiative Conference Report. Conference 28–29 Sept 1994

Von Hagen V (1945) La fabricación del papel entre los aztecas y los mayas. Editorial Nuevo Mundo, México

Wiersum KF, Shackleton C (2005) Rural dynamics and biodiversity conservation in southern Africa. In: Ros-Tonen AF, Dietz T (eds) Linking global conservation objectives and local livelihood needs: lessons from Africa. Edwin Mellen, UK, pp 67–92

Wollenberg E, Ingles A (eds) (1998) Incomes from the forest. Methods for the development and conservation of forest products for local communities. Center for International Forestry Research, Bogor, Indonesia

Wong JLG (2000) The biometrics of non-timber forest product resource assessment: a review of current methodology. FAO Non-wood Series, Rome, Italy

Woodley E, Crowley E, de Pryck J, Carmen A (2008) Cultural indicators of indigenous peoples' food and agro-ecological systems. FAO, International Indian Treaty Council (IITC)/Government of Norway, Christensen Fund

Chapter 6
From the Forest to the Stomach: Bushmeat Consumption from Rural to Urban Settings in Central Africa

Nathalie van Vliet, Robert Nasi, and Andrew Taber

Abstract Non-timber forest products are not used solely by people living in remote rural villages. They are also important components of urban livelihoods, in both the developing and developed world. This chapter illustrates this through examination of the urban demand for bushmeat in Central Africa. We consider the drivers of the trade and consumer preferences, as well as whether the offtake is sustainable. Overall, there is extensive trade with bushmeat being supplied to urban consumers in a variety of different markets. There is marked preference for certain species, independent of the availability or price of domestic alternatives. While trade in some species, especially large and long-lived ones, is unsustainable, trade in others has persisted for decades and there is little evidence of dwindling supply. Policy and management recommendations are complex because of the interplay of cultural preferences evolving within an urbanising population.

6.1 Introduction

Non-timber forest products (NTFPs) are used by millions of people, including rural and urban populations in countries of both the developed and developing world. Yet, for a number of reasons, most of the debates around poverty alleviation and conservation implications of the use of NTFPs are drawn from examples and information from rural settings. This undermines a coherent understanding of NTFPs and potentially results in biased policies which fail to address concerns and issues surrounding urban use of NTFPs.

N. van Vliet (✉)
Post Doctoral researcher, Department of Geography and Geology, University of Copenhagen, Oster Voldgade, 101350 Copenhagen K, Denmark
e-mail: vanvlietnathalie@yahoo.com

R. Nasi and A. Taber
Center for International Forestry Research (CIFOR), P.O. Box 0113 BOCBD, Bogor 16000, Indonesia
e-mail: r.nasi@cgiar.org; a.taber@cgiar.org

S. Shackleton et al. (eds.), *Non-Timber Forest Products in the Global Context*,
Tropical Forestry 7, DOI 10.1007/978-3-642-17983-9_6,

Urban use is extensive, including products such as medicinal plants, fuelwood, construction timber, cultural artifacts, and variety of plant and animal food products (e.g., Ruiz Peréz et al. 2000; Cocks 2006; Shackleton et al. 2006; van Vliet and Nasi 2008). Some of these are sourced within and near urban areas, and some are transported into urban areas from outlying rural regions. Importantly, the supply of urban consumer demand may be an important transitional livelihood strategy for households newly migrated to urban areas (Stoian 2005; Tacoli 2006). In being new migrants they lack a secure economic base within the city, but they still retain links to their former rural home and have the knowledge to harness rural resources (Krüger 1998). Consequently, becoming involved in trade in such resources is an attractive livelihood option, which also frequently has low barriers to entry. However, many dimensions of this urban demand remain unknown or insufficiently understood, in particular, the magnitude of demand relative to rural consumption patterns as well as the longevity of the demand, as consumer profiles change in an urban setting through exposure to increased media, education, modernisation, and market goods.

This chapter examines some of these issues, drawing on specific examples of bushmeat trade in Central and West Africa, from the hunter to the final consumer. We analyze the reasons for hunting at the village level and discuss the often made distinction between subsistence and commercial hunting. We describe the trade from rural to urban areas and analyze the contribution of urban consumers to overall bushmeat consumption (Fig. 6.1). Finally, at the consumer level, we examine the reasons for bushmeat consumption to understand the main drivers of the bushmeat market chain.

Fig. 6.1 Bushmeat meals advertised at a street restaurant in Libreville, Gabon (photo: Nathalie van Vliet)

For the purposes of this chapter, bushmeat is defined as non-domesticated terrestrial mammals, birds, reptiles, and amphibians harvested by humans for food (Nasi et al. 2008). While insects, crustaceans, grubs, molluscs, and fish can be locally important dietary items, it is the larger vertebrates that constitute the majority of the terrestrial wild animal biomass consumed by humans. Although the term "bushmeat" originated from Africa, it is now widely used across the tropics. The perception that many species used for bushmeat are being overexploited for human consumption, has resulted in the notion of a "bushmeat crisis". For example, the current annual harvest in the Congo Basin could exceed two million tonnes (Fa et al. 2003). Duikers (small forest antelopes), pigs, primates, and rodents are the most commonly hunted animals in that region, with duikers, both numerically (>75%) and in terms of biomass, being the most important bushmeat species. Primates rarely constitute more than 20% of the animals sold in bushmeat markets, and apes less than 0.5%. Rodents and other small, but fast-reproducing, species increase in proportion in the items on offer in long-established markets, presumably because slow-reproducing and large animals have been depleted in accessible forests (Cowlishaw et al. 2005).

6.2 Sustainability of Bushmeat Offtake

During the last decade, a growing number of researchers in Central African countries have tried to determine the effects of bushmeat hunting and the levels at which it becomes unsustainable, including in the Democratic Republic of Congo (Hart 2000; de Merode et al. 2004), Central African Republic (Noss 1998a, b, 2000), Gabon (Feer 1993, 1996; Lahm 1993; Starkey 2004; van Vliet and Nasi 2008), Cameroon (Dethier 1995; Delvingt et al. 1997; Muchaal and Ngandjui 1999; Ngandjui and Blanc 2000; Bousquet et al. 2001; Abugiche 2008), and Equatorial Guinea (Fa et al. 1995, 2005). This wealth of information provides evidence that the scale of hunting poses a threat to many tropical forest species (Milner-Gulland and Akçakaya 2001; Nasi et al. 2008). Hunting has been specifically identified as a threat for 84 mammalian species and subspecies from West and Central Africa (International Union for Conservation of Nature [IUCN] 2000, as cited in Bowen-Jones et al. 2002), and 60% of the mammal species might be hunted unsustainably (Fa et al. 2002). Most studies in tropical forest regions show that large game species are the first to disappear. Hunters tend to favour hunting of large mammal species because success supplies a large amount of meat, and larger animals tend to be the ones that have the most valuable horns, antlers, tusks, furs, skins, or other artifacts. Thus, the value returned per unit of time and effort expended is greatest for large animals while they are still in reasonable abundance in the forest. However, large species are rare compared to smaller ones and reproduce slowly; thus, they are especially vulnerable to overhunting and have reduced capacity to recover from population declines. However, small residual populations may survive in even highly disturbed or heavily hunted environments (Meijaard and Sheil 2008).

The long-term persistence of the bushmeat trade, documented in Africa over several centuries, suggests that the trade of some of the species can be sustainable (Cowlishaw et al. 2005), although increasing human populations, and hence demand for bushmeat, pose a limit at some scale. Vulnerable taxa (large bodied, slow-reproducing species) can only sustain low levels of hunting and are often depleted in hunted areas, but robust taxa (small size, fast-reproducing species), such as rodents and small antelope, appear resilient to hunting pressure and persist even in agricultural landscapes around the cities. For example, the Takoradi market in Ghana shows that large urban centers can be sustainably supplied with bushmeat over several decades by robust species from an agricultural landscape (Cowlishaw et al. 2005). Some species may even be locally advantaged by hunting and other land use practices which alter interspecific competition dynamics and the provision of favourable food or habitat (Bodmer et al. 1997; Cullen et al. 2000; Salas and Kim 2002; Hurtado-Gonzales and Bodmer 2004; Cowlishaw et al. 2005).

Despite the evidence that overhunting in some areas is leading to the depletion of certain species, the bushmeat trade cannot be simply banned given the essential roles that bushmeat plays as a main source of dietary protein and also for cultural identity (Pearce 2006), as well as providing important sources of income for some rural people (Lahm 1993; Wilkie and Carpenter 1999; Bakarr et al. 2001). Wildlife use has important livelihood aspects and serves multiple roles (Brown 2003). Wildlife products are often major items of consumption and have high medicinal and spiritual values in many human cultures (Scoones et al. 1992). Bushmeat, in particular, offers a number of benefits to forest-dwelling populations.

The bushmeat trade could be considered as a facet of the "tragedy of the commons" and be dealt with in the broader framework of adaptive management of renewable natural resources, integrated into complex market chains from rural to urban areas, such as those that exist for timber or fuelwood (Nasi et al. 2008). Nowadays, the recognition of the role of bushmeat and other wild animal products in the national and local economies provides a first essential step in promoting the sustainable use of this resource. The recommendations from the Bushmeat Liaison group of the Convention of Biological Diversity (CBD 2009) highlight that the notion of sustainability in the context of bushmeat trade must include sociological, cultural, and economic dimensions. Harvested populations should not be reduced to densities whereby they cease to fulfill their economic role in contributing to livelihoods and to provide the range of nutritional, medicinal, cultural, and social services to dependent populations.

6.3 Subsistence or Commercial Hunting? A Blurred Distinction

It is frequently espoused that subsistence hunting is legitimate and somewhat sustainable, but that commercial hunting is not (Bennett 2002; Robinson and Bennett 2000a, b). It is often suggested that unsustainable commercial hunting may negatively affect rural livelihoods if it results in lower returns for subsistence hunters (Bennett 2002; Davies 2002; Robinson and Bennet 2002). As a result, many conservation-oriented organisations advocate strict controls or bans on any commercial trade in bushmeat as a solution to the conservation aspects of the trade that would also benefit poor rural populations (Robinson and Bennett 2000b). In contrast, many development-oriented authors suggest that a regulated bushmeat trade, which maintains the supplies of appropriate species from wildlands and anthropogenic habitats, can contribute to economic growth in countries where there are few employment and economic options (Davies 2002; Brown 2003; Nasi et al. 2008).

However, for many tropical forest peoples, the distinction between subsistence and commercial use is blurred, with meat from the forest supplementing both diets and incomes. Bushmeat often represents both the primary source of animal protein and the main cash-earning commodity for the inhabitants of the humid forest regions of the tropics. Recent studies (Takforyan 2001; De Mérode et al. 2003; Kümpel et al. 2010) show that bushmeat sales within the village (as opposed to urban markets) can be significant; 30% in Cameroon and up to 90% in the Democratic Republic of Congo. This therefore contradicts the conventional wisdom that banning external market sales of bushmeat and restricting consumption to local subsistence use offer a "win–win" strategy to the benefit of both conservation and the poor (De Mérode et al. 2003).

In rural communities wildlife provides significant calories to residents, as well as essential protein and fats [for a comprehensive review of the importance and role of wildlife in nutrition see Hladick et al. (1989, 1996) and Froment et al. (1996)]. Even where there has been a change from a hunter-gatherer lifestyle to pastoralism or agriculture, hunting and gathering remain important to a high proportion of rural households in tropical forests and elsewhere. Hunting provides between 30% and 80% of the overall protein intake of rural households in Central Africa (Koppert et al. 1996) and nearly 100% of animal proteins. Some recent studies (De Mérode et al. 2003) show that bushmeat clearly plays an important food security role during the lean season. What is known of the nutritional composition of bushmeat species suggests that these provide an equivalent or even greater quality of food than domestic meats as they have less fat and more protein (Ntiamoa-Baidu 1997). These proteins cannot be substituted by proteins of vegetable origin, such as cassava or *gnetum* leaves, as they are poorer in amino-acids (Pagezy 1996). Where wildlife is still abundant, hunting is more profitable and therefore preferred to livestock (Fa and Brown 2009). Bushmeat is frequently an open access resource, so the cost of its production is always lower than the cost of raising livestock, even though at times considerable labour time is required to access and capture bushmeat (Kaschula and Shackleton 2009). Given the low productivity of domestic livestock in tropical forest conditions, and the high risks and investment costs associated with it, livestock husbandry is rarely a feasible option in remote rural areas. Most livestock are kept as a form of reserve banking, and to satisfy particular cultural needs (Fa and Brown 2009).

Many people also depend on wildlife resources as a buffer to see them through times of hardship (e.g., unemployment, illness of relatives, and crop failure), or to gain additional income for special needs (e.g., school fees, festivals, and funerals). This is the classical safety net role of NTFPs (Paumgarten 2005). Although

Fig. 6.2 Carcass of red duiker beeing transported by bike to Kisangani, DRC (photo: Casimir Nebesse)

bushmeat is most often sold only after the basic subsistence requirements are satisfied, many families use hunting to occasionally supplement short-term cash needs. For rural people, with limited or no access to capital, land, or livestock, the harvest of wildlife resources offers the best return for labour input (Shackleton et al. 2007). Indeed, the bushmeat market chain is beneficial to the primary producer who has low levels of investment, low risks, and a good profitability. Moreover, this activity is integrated in the household production system promoting an equilibrium between women's and men's labour: men are in charge of hunting, while women are involved in the trade. Bushmeat is an easily traded resource as it is transportable, has a high value/weight ratio, and is easily preserved at low cost (Fig. 6.2). Although cash income from the sale of wildlife products can be highly variable, the returns from hunting are generally higher than average local wages (Gally and Jeanmart 1996; Ntiamoa-Baidu 1997; Bennett and Robinson 2000).

6.4 The Bushmeat Market Chain: From the Forest to Urban Areas

Throughout tropical forest countries, many people benefit from wild meat. This includes those who eat it as part of a forest-dependent subsistence lifestyle, those who trade and transport it at all points along different supply chains, and those who consume it in restaurants and homes, often far from the forest.

Bushmeat is traded in six different settings or market types (Abernethy and Ntsame Effa 2001):

- Fixed markets, where sellers have a stand where they trade during recognised hours and days of the week. These markets are only found in stable population centers, and their operation depends on local authority tolerance.

- Regular trading locations, but without a physical stand or recognised hours. These markets are less regular than the fixed markets, but their location and approximate hours are fairly predictable to the local population. They are often found on port quays, near bus and train stations, or at road junctions in smaller towns and villages.
- Mobile delivery points. Here the location is locally known, but the hours and days of trade are very irregular, dependent on hunter success. These markets are often roadside locations, where Car-loads of meat will be brought when hunters return from a hunting trip. They are often supplied by commercial hunters who have regular clients (called subscribers), and delivery points are often in the quarters of larger towns and cities.
- Direct delivery to a client on order. Some hunters hunt on order, for a commercial client or restaurateur. Their meat is not offered for open sale, but delivered directly to the door, for a negotiated price.
- Village markets. These are held informally in villages where village resident hunters are hunting daily and other villagers know that any catch will be for sale around a certain hour of the day.
- Family roadside sales and village sales. These are points of sale trading small and irregular items for small amounts of supplementary family income. They are seen across Central Africa on all roadsides where traffic is frequent, but typically trade less than three items per day and often far less than this, maybe only one item per month.

Since the 1950s, a growing demand from urban areas, combined with larger populations more generally, has catalyzed the trade in wildlife resources. Much is increasingly being drawn from forest-agriculture mosaics into towns and cities as favoured or inexpensive sources of animal protein (Cowlishaw et al. 2005). From first harvest to final sale, the trade in bushmeat for local, national, or regional trade now forms an important part of the informal sector's "hidden economy". Access to markets is a key factor in realising economic values of wild products, including bushmeat. Moreover, the determination of people to access markets, if there is sufficient economic incentive to do so, should not be underestimated (Neumann and Hirsch 2000). If prices and profits are high enough, local traders will make use of any transport networks over considerable distances to get perishable goods to market. As a result, hunting and the bushmeat trade, although largely ignored in official trade and national statistics, play a crucial role in the economies of numerous Central African countries, but being part of the hidden economy, are not tapped as a source of government revenues (Fargeot 2009).

6.5 Contribution of Urban Areas to the Overall Bushmeat Consumption

Growing urban populations have been implicated as the primary driver of high levels of bushmeat offtake in Central Africa (Starkey 2004). The large quantity of bushmeat passing through urban markets is often cited as evidence of this

(BCTF 2002a; Bowen-Jones and Pendry 1999). Market studies, such as those of Wilkie and Carpenter (1999), Starkey (2004), Nasi et al. (2008), show that residents of the Congo Basin countries eat as much, if not more, meat as many residents of northern industrial countries (average of 47 kg/person/year vs. 30 kg/person/year), and that urban families eat less bushmeat than rural families. Wilkie and Carpenter (1999) estimate that rural consumption outweighs urban consumption in every country in Central Africa, often by a large margin. On the other hand, Chardonnet et al. (1995) suggest that while per capita urban consumption of bushmeat is low, aggregate urban consumption is higher than aggregate rural consumption due to the high and growing population density of urban areas. But Starkey (2004) shows that in Gabon, the capital city of Libreville accounts for 40% of the population, but likely accounts for less than 14% of aggregate bushmeat consumption. Given current trends in population distribution, urban demand for bushmeat may come to exceed rural demand. Wilkie and Carpenter (1999) also show that bushmeat constitutes the primary source of meat for most residents of the Congo Basin, and that the gross quantity of bushmeat consumed in forest and urban areas across the Congo Basin may exceed two million tonnes per year.

The assumption that urban consumption is the primary driver of high levels of bushmeat offtake in Central Africa is based on the large quantities of bushmeat passing through urban markets rather than on measured consumption data (e.g., ApeAlliance 1998; Bowen-Jones and Pendry 1999). These urban market surveys have often shown that the volume of bushmeat traded is large. For example, Steel (1994) estimated that the amount of bushmeat traded in six large bushmeat markets in Gabon was of the order of 1,000 tonnes/year. An inventory in 1995–1996 of the four main markets in the Cameroon capital, Yaoundé, indicated a similar amount, with estimated sales of 70–90 tonnes of bushmeat monthly (Bahuchet and Ioveva 1999). Estimates of the national value of the bushmeat trade range from USD 42 to 205 million across countries in West and Central Africa (Davies 2002). Similarly, Fa et al. (1995) suggested that the volume of bushmeat traded annually in Equatorial Guinea's two main markets might be of the order of 178 tons, which is considerable given the small population and surface area of the country (700,000 and 28,000 km^2, respectively). Fargeot and Diéval (2000) estimate annual consumption in Bangui, Central African Republic, to be of the order of 9,500 tonnes/year, of which at least half passes through formal markets. Since much bushmeat may pass through informal channels such as from rural hunters direct to urban consumers rather than through markets, formal bushmeat markets channel an unknown portion of total urban consumption (Bahuchet and Ioveva 1999; Trefon 1998; Trefon and de Maret 1999). Such findings have led to suggestions that the visible market trade in bushmeat may only be the tip of the iceberg and that the overall contribution of urban areas to bushmeat consumption may be far greater (Bakarr et al. 2001; Fa et al. 2002; Rose 2001).

6.6 Why Do Urban Consumers Eat Bushmeat?

In many rural areas, even if bushmeat is not the only source of protein available, it seems by far the cheapest source and as such plays an essential role in people's diets (Oates 1996; Auzel and Wilkie 2000; Dethier and Ghuirghi 2000; Bennett 2002; de Merode et al. 2004). What is less clear is whether in urban areas bushmeat is consumed primarily as a luxury good or as a necessity.

Rural consumption of bushmeat is linked to its availability. In villages, bushmeat is the cheapest and often the only type of animal protein available together with fish (Starkey 2004; Coad 2008; van Vliet 2008). In urban areas, consumers have access to other sources of protein such as meat from livestock and poultry species. Yet smoked bushmeat often provides protein to the poorest urban families who buy the less-expensive species in very small quantities daily. So, why do urban consumers buy bushmeat?

Preference for bushmeat compared to alternatives has an influence on bushmeat consumption. Schenck et al. (2006) analyzed taste choices in Gabon and suggested that consumers differentiate among bushmeat species and that wildlife cannot be treated as a generic food source. Besides, consumers with a stated preference for fish chose porcupine rather than chicken, but did not choose duiker more often than beef. Similarly, consumers who stated a preference for chicken avoided porcupine and showed no preference for blue duiker or beef. In Equatorial Guinea, the top three most-preferred foods are all fresh fish or bushmeat species including red snapper (*Lutjanus campechanus*), porcupine (*Atherurus africanus*), and blue duiker (*Cephalophus monticola*), whereas the top most consumed foods are frozen mackerel, frozen chicken, and frozen pork (Kumpel 2006). The availability of bushmeat in urban towns also influences consumption choices beyond preferences. In Nigeria, using a combination of taste tests and questionnaires, cane rat (*Cricetomys emini*) was rated the highest by consumers according to sensory quality, but lower than mutton and beef in terms of consumption, constrained by cost and availability (Ladele et al. 1996). This suggests that preferences are not fixed but adapt according to experience and circumstance. People already consume greater quantities of livestock and fish in those parts of West Africa where human population densities are greater and fewer wilderness areas remain to provide bushmeat (Brashares et al. 2004; Cowlishaw et al. 2005).

In urban areas, consumers typically choose the cheapest form of meat (Wilkie and Carpenter 1999; Wilkie et al. 1998) or the most available and not necessarily the most preferred. The price of bushmeat in comparison to other sources of protein also affects bushmeat consumption. Wilkie et al. (2005) showed that changes in the price of poultry or livestock do not influence the level of wildlife consumption by Gabonese households, while changes in fish prices change bushmeat consumption where fish and bushmeat are substitutes. In Gabon, the price and elasticity of demand have been shown to vary along a spatial gradient, with bushmeat becoming cheaper and demand more inelastic, compared with domestic meats and fish, the further one travels from the market that supplies these alternatives (Starkey 2004). In Ghana,

Brashares et al. (2004) provide good evidence that at times of low fish availability, the price of fish and the volume of bushmeat sold in rural markets both increase, suggesting that consumers treat bushmeat as a substitute for fish and vice versa.

These results suggest that availability and low prices of alternative proteins would help replace bushmeat consumption by other sources of protein. However, the incentives for bushmeat consumption are complex and do not depend only on availability and prices, especially for the wealthiest families who consume fresh bushmeat without concern for the price. In the northeastern Democratic Republic of Congo, wealthier households tend to consume more bushmeat (de Merode 1998), even when bushmeat is more expensive than other alternatives. In urban Gabonese towns, the wealthiest households consume less bushmeat per person per day than poorer households, but are less sensitive to prices and often choose fresh wild meat (not smoked) and of the more expensive species (porcupine, red river hog, or python) (Knights 2008).

Besides the economic factors that drive demand for bushmeat, cultural factors contribute some understanding in interpretation of bushmeat consumption patterns. East et al. (2005) used a study of consumption and preferences in Bata, Equatorial Guinea, to indicate that besides income, ethnicity and nationality are key determinants of consumption of bushmeat. In Bata, Equatoguineans are much more likely than other nationalities to buy bushmeat, while purchasers of fresh domestic meat are more likely to be Muslims from Cameroon and Nigeria (Kumpel 2006). The Islamic prescription that animals be killed in a specific fashion (halal) means that all bushmeat species are de facto forbidden for strict Muslims.

Some authors have also shown that a cultural preference for bushmeat encourages consumers to pay high prices for bushmeat (Trefon 1998; Bahuchet and Ioveva 1999; Trefon and de Maret 1999). For example, King (1994) suggests that in urban areas of western Cameroon the rate of consumption seems predominantly dictated by preference or taste rather than a lack of alternatives. Chicken, beef, pork, and fish are commonly available in urban restaurants and from street corner "chops stalls" at cheaper prices than bushmeat. In Gabon, familiarity with the taste of bushmeat due to childhood experience is clearly a major factor in determining preference (Starkey 2004). In northeast Gabon, wealthier families choose the type of meat according to preference (red-river hog meat, *Potamochoerus porcus*) rather than according to prices (Okouyi 2006). Angoué et al. (2000) suggest that in Gabon, bushmeat is associated with the village, with rituals and with ceremonies (Fig. 6.3). The desire to eat bushmeat can be explained by the taste, habit, tradition, prestige, ritual, and nostalgia. In northeast Gabon, bushmeat consumption is particularly high during men's circumcision ceremonies organised in the dry season (van Vliet and Nasi 2008). In some villages close to Makokou (Gabon) the number of animals killed for consumption during these ceremonies represents half of the total annual offtake, and big prey species (such as red river hog) are especially sought-after (van Vliet and Nasi 2008). The traditional role of bushmeat has also been shown in Equatorial Guinea, where some species are considered to have magical or medicinal properties that increase their value. One example is the blue duiker, which costs 1,000–2,000 CFA more per

Fig. 6.3 Skin of a civet used during the circumcision ceremonies in Ba-kota villages from North East Gabon (photo: Nathalie van Vliet)

carcass when alive, due to its perceived curative powers for sick children (Kumpel 2006). The consumption of some species is associated with particular ceremonies, like the bay duiker (*Cephalophus dorsalis*) consumed during burials in southeast Cameroon by the Anyang people (van Vliet pers. obs.).

Urban consumers in Central Africa perceive bushmeat as a local, natural, and healthy food compared to livestock, which is mostly industrial and imported. In Equatorial Guinea, Kumpel (2006) showed that fresh bushmeat is widely regarded as more healthy and nutritious than any frozen meat. When consumers were asked why they bought bushmeat, a large proportion stated it was because it was fresh or healthy. Similarly, in Gabon, bushmeat is also perceived as a "*healthy food, natural, fresh and without artificial additives*" (Starkey 2004).

Taboos on certain foods are widespread in parts of Central Africa (Angoué et al. 2000; Joiris 1997). In Gabon, a large variety of species are reported as being forbidden, but the most frequently mentioned are carnivores (particularly leopards), yellow-backed duikers (*Cephalophus silvicultor*), tortoises (*Kinxys* spp.), and primates (Starkey 2004). Ethnic groups from all areas of Gabon (Starkey 2004) and Democratic Republic of Congo (Gambalemoke pers. comm.) consider that women should not consume carnivore meat (especially when pregnant). Crowned guenon (*Cercopithecus pogonias*) and African palm civet (*Nandinia binotata*) are thought to make women infertile, and snakes and monitor lizards (*Varanus niloticus*) are traditionally only eaten by the elders (Kumpel 2006). Cultural taboos do not necessarily reduce the hunting level of taboo species (especially when non-specific hunting methods are used) but do reduce their trade value. Yellow-backed duikers are mentioned in hunter offtakes in villages near Makokou, but the meat is never observed in the bushmeat market of Makokou (Okouyi 2006; van Vliet 2008). The implications of breaking taboos can be very specific. The meat of white-bellied duiker (*Cephalophus leucogaster*), if eaten by pregnant women in northeast Gabon, is believed to cause diseases to the new born child (van Vliet 2008). Women who eat the meat of black-fronted duiker

(*C. nigrifrons*) will menstruate continuously (van Vliet 2008). It is believed that eating monitor lizards while pregnant can cause a woman to bear children with epilepsy (Starkey 2004).

6.7 Conclusions

The bushmeat meat trade has many dualities and contrasts; urban versus rural; conservation versus opportunities for sustainable use; and driven by need, profit, or culture. Consequently, the bushmeat trade continues to worry the conservation community and is of increased interest to researchers and practitioners coming from diverse disciplines. Bushmeat serves multiple roles at the hunter level and remains the main source of protein and income in most rural areas. The distinction between subsistence and commercial hunting is blurred because rural areas are well integrated in the cash economy despite their remoteness. Like many other NTFP products, bushmeat follows different supply chains before reaching the final consumers in villages or towns far from the forest. Urban consumption of bushmeat is seen as one of main drivers of the unsustainable use of wildlife for food. The general wisdom would suggest that consuming bushmeat in urban areas is not a necessity and should be the first to be targeted by conservation efforts. However, our analysis shows that the reasons behind bushmeat consumption are complex and integrate economic, cultural, and social reasons that should not be disregarded in efforts to promote the sustainable trade of bushmeat. More particularly, a clear understanding of consumer preferences (stated and actual) for both wildlife and alternatives is needed before any effort in supplying alternative protein sources is provided. In some cases, providing cheaper sources of alternatives (such as poultry or livestock) does not always lead to a reduction in bushmeat consumption, because these are not always seen as substitutes by consumers. This analysis further provides evidence that enforcement alone will not be successful in reducing the impacts of bushmeat trade on wildlife species, if the economic and cultural values of bushmeat are not taken into account. Bushmeat cannot be considered as a generic food. Indeed, depending on the species, state (smoked, fresh), and part of the animal, bushmeat follows different market chains and reaches different types of consumers. The poorest urban families often buy smoked bushmeat as the most available and cheapest source of protein, often from the less-expensive species, the less-preferred parts of the animal, and in very small quantities per day. Wealthier families consume fresh bushmeat of the most-preferred species disregarding prices and availability of other sources of protein.

Although the above discussion has been developed around the bushmeat trade, there are several generic conclusions for NTFPs. Firstly, there is vibrant trade from rural to urban areas which can have profound effects on rural livelihoods and natural resource supply. Therefore, when assessing the sustainability of bushmeat trade and it's contribution to household and regional economies, the examination of rural dimensions alone is insufficient. Secondly, as the population profile of developing nations swings towards a more urbanised one, there is no guarantee that demand for NTFPs will decline (see Chap. 5). This is because they serve multiple functions in livelihoods over and above the purely consumptive. There are cultural, spiritual, and taste preferences that override predictions and patterns of behaviour captured in economic models based on rational behaviour. Thirdly, despite long and continuous sustained heavy harvesting, some bushmeat species continue to thrive in natural and modified habitats. Thus, heavy harvesting pressure should not always be equated with a spectre of resource decline and possible local extinction, although that does occasionally occur. Each species and context needs to be examined, and predictions and interventions designed and implemented at the appropriate scale.

References

Abernethy KA, Ntsame Effa E (2001) *Projet Gibier*: Interim report to the Direction de la Faune et Chasse, Unpublished Technical Report

Abugiche S (2008) Impact of Hunting and Bushmeat Trade on Biodiversity Loss in Cameroon: A Case Study of the Banyang-Mbo Wildlife Sanctuary. PhD dissertation. Brandenburg University of Technology, Faculty of Environmental Sciences And Process Engineering Chair of General Ecology. Germany

Angoué C, Assoko Ndong A, Binot A, Carette S, de Maret P, Trefon T (2000) Gabon. In Les peuples des forêts tropicales aujhourd'hui: Volume III Région Afrique Central. In: Bahuchet S, de Maret P (eds) Programme Avenir des Peuples des Forêts Tropicales (APFT). Université Libre de Bruxelles, Brussels, Belgium

ApeAlliance (1998) The African bushmeat trade – a recipe for extinction. Fauna and Flora International, Cambridge

Auzel P, Wilkie DS (2000) Wildlife use in northern Congo: hunting in a commercial logging concession. In: Robinson JG, El B (eds) Hunting for sustainability in tropical forests. Columbia University Press, New York, pp 413–426

Bahuchet S, Ioveva K (1999) De la forêt au marché : le commerce de gibier au sud Cameroun. In L'homme et la forêt tropicale (Eds S. Bahuchet, D. Bley, H. Pagezy and N. Vernazza-Licht), pp. 533–558. Travaux Société Ecologie Humaine, Paris

Bakarr M, Fonseca G, Mittermeier R, Rylands A, Painemilla K (eds) (2001) Hunting and bushmeat utilisation in the African rainforest – perspectives toward a blueprint for conservation action (Advances in Applied Biodiveristy Science 2). Conservation International, Washington, DC

BCTF (2002a) Bushmeat – a wildlife crisis in Central Africa and around the world. Bushmeat Crisis Task Force

BCTF (2002b) Bushmeat Crisis Task Force Economics Factsheet. Bushmeat Crisis Task Force

Bennett BC (2002) Forest products and traditional peoples: economic, biological, and cultural considerations. Nat Resour Forum 26:293–301

Bennett EL, Robinson JG (2000) Hunting of wildlife in tropical forests. Implications for biodiversity and forest peoples. Biodiversity Series, Impact Studies, Paper no 76, The World Bank Environment Department, Washington, DC

Bodmer RE, Eisenberg JF, Redford KH (1997) Hunting and the likelyhood of extinction of Amazonian mammals. Conservation Biology, 11(2):460–466

Bousquet F, Lepage C, Bakam I, Takforyan A (2001) A spatially-explicit individual-based model of blue duikers population dynamics: multi-agent simulations of bushmeat hunting in an eastern Cameroonian village. Ecol Modell 138:331–346

Bowen-Jones E, Pendry S (1999) The threat to primates and other mammals from the bushmeat trade in Africa, and how this threat could be diminished. Oryx 33:233–246

Bowen-Jones E, Brown D, Robinson E (2002) Assessment of the Solution-orientated research needed to promote a more sustainable bushmeat trade in Central and West Africa. Defra, Wildlife and Countryside Directorate, Bristol, UK

Brashares JS, Arcese P, Sam MK, Coppolillo PB, Sinclair ARE, Balmford A (2004) Bushmeat hunting, wildlife declines, and fish supply in West Africa. Science 306:1180–1183

Brown D (2003) Is the best the enemy of the good? Livelihoods perspectives on bushmeat harvesting and trade – some issues and challenges. ODI, London

CBD (2009) Report of the liaison group meeting on bushmeat, October 15–17 2010, Buenos Aires

Chardonnet P, Frits H, Zorzi N, Feron E (1995) Current importance of traditional hunting and major contrasts in wild meat consumption in sub-Saharan Africa. The Wildlife Society, Bethesda, MD

Coad L (2008) Bushmeat hunting in Gabon: socio-economics and hunter behaviour, PhD Thesis, University of Cambridge

Cocks ML (2006) Wild resources and practices in rural and urban households in South Africa: Implications for bio-cultural diversity conservation. PhD Thesis, Wageningen University, The Netherlands

Cowlishaw G, Mendelson S, Rowcliffe JM (2005) Evidence of post-depletion sustainability in a mature bushmeat market. J Appl Ecol 42:460–468

Cullen LJ, Bodmer R, Padúa CV (2000) Effects of hunting in habitat fragments of the Atlantic forests, Brazil. Biological Conservation 95:49–56

Davies G (2002) Bushmeat and international development. Conserv Biol 16:587–589

de Merode E (1998) Protected areas and local livelihoods: contrasting systems of wildlife management in the Democratic Republic of Congo. PhD Thesis, University College London, UK

De Merode E, Homewood K, Cowlishaw G (2004) The value of bushmeat and other wild foods to rural households living in extreme poverty in Democratic Republic of Congo. Biol Conserv 118:573–581

De Merode, E., Homewood, K. and Cowlishaw, G. 2003. Wild resources and livelihoods of poor households in Democratic Republic of Congo. Wildlife Policy Briefing no.1. ODI.

Delvingt W, Dethier M, Auzel P, Jeanmart P (1997) La chasse villageaoise Badjoué, gestion coutumière durable ou pillage de la ressource gibier? Presses Agronomiques de Gembloux, Gembloux

Dethier M (1995) Etude Chasse. Projet Ecofac Composante Cameroun, Dja Reserve, Cameroon

Dethier M, Ghuirgui A (2000) Etude de la chasse villageoise dans le secteur Ouest (route Mambélé-Ndélé) de la zone d'intervention du Projet ECOFAC. Projet ECOFAC, Composante RCA

East T, Kümpel NF, Milner-Gulland EG, Rowcliffe JM (2005) Determinants of urban bushmeat consumption in Rio Muni, Equatorial Guinea. Biol Conserv 126:206–215

Fa J, Brown D (2009) Impacts of hunting on mammals in African tropical moist forests: a review and synthesis. Mamm Rev 39:231–264

Fa JE, Juste J, Perez del Val J, Castroviejo J (1995) Impact of market hunting on Mammalian species of Equatorial Guinea. Conserv Biol 9:1107–1115

Fa J, Peres C, Meeuwig J (2002) Bushmeat exploitation in tropical forests: an intercontinental comparison. Conserv Biol 16:232–237

Fa J, Currie D, Meeuwig J (2003) Bushmeat and food security in the Congo Basin: linkages between wildlife and people's future. Env Conserv 30:71–78

Fa JE, Ryan SF, Bell DJ (2005) Hunting vulnerability, ecological characteristics and harvest rates of bushmeat species in afrotropical forests. Biological Conservation 121:167–176

Fargeot C (2009) Le commerce de la viande de chasse en Afrique Centrale : étude d'un marché-porte. Le PK 12 à Bangui (RCA) In: Colloque SFER Chasse, Territoires et Déveleppement Durable. Outils d'Analyse, en Jeux et Perspectives, 25–26 mars 2009, Clermont-Ferrand

Fargeot C, Diéval S (2000) La consommation de gibier à Bangui, quelques données économiques et biologiques. Canopée 18:5–7

Feer F (1993) The potential for sustainable hunting and rearing of game in tropical forests. In: Hladik CM, Hladik A, Ragezy H, Linares OF, Koppert GJA, Froment A (eds) Tropical forests, people and food. UNESCO/MAB, Paris, pp 691–708

Feer F (1996) Les potentialités de l'exploitation durable et de l'élevage du gibier en zone forestière forestières de la région côtière du Cameroun: Yassa, Mvae et Bakola. In: Hladik CM, Hladik A, Ragezy H, Linares OF, Koppert GJA, Froment A (eds) L'alimentation en forêt tropicale, interactions bioculturelles et perspectives de développement. Volume I, Les ressources alimentaires: production et consommation. UNESCO/MAB, Paris, pp 477–496

Froment A, de Garine I, Binam Bikoi Ch Loung JF (eds) (1996) Bien Manger et Bien Vivre: Anthropologie alimentaire et développement en Afrique intertropicale: du biologique au social. L'Harmattan- ORSTOM, Paris

Gally M, Jeanmart P (1996) Etude de la chasse villageoise en forêt dense humide d'Afrique Centrale. Faculté Universitaire des Sciences Agronomiques de Gembloux, Gembloux, Mémoire de fin d'étude

Hart JA (2000) Impact and Sustainability of indigenous hunting in the Ituri forest, Congo-Zaïre: a comparison of unhunted and hunted duiker populations. In: Robinson JG, Bennet EL (eds) Hunting for sustainability in tropical forests. Columbia University Press, New York, pp 106–153

Hladick CM, Bahuchet S, de Garine I (eds) (1989) Se nourrir en forêt équatoriale. UNESCO-CNRS, Paris

Hladik CM, Hladik A, Pagezy H, Linares OF, Koppert GJA, Froment A (eds) (1996) L'alimentation en forêt tropicale, interactions bioculturelles et perspectives de développement. Vol 1: Les ressources alimentaires: production et consommation; Volume 2: Bases culturelles des choix alimentaires et stratégies de développement. UNESCO, Paris

Hurtado-Gonzales JL, Bodmer RE (2004) Assessing the sustainability of brocket deer hunting in the Tamshiyacu-Tahuayo Communal Reserve, northeastern Peru. Biological Conservation 116:1–7

Joiris D (1997) La nature des uns et la nature des autres: mythe et réalité du monde rural face aux aires protégées d'Afrique Centrale. Civilisations XLIV:94–103

Kaschula SA, Shackleton CM (2009) The magnitude of bush-meat offtake by rural communities in South Africa: a quantitative study from the Mt Frere region of the Eastern Cape. Env Conserv 36:192–200

King S (1994) Utilisation of wildlife in Bakossiland, West Cameroon. Traffic Bull 14:63–73

Knights K (2008) Who ate all the crocodiles? An investigation of trends and patterns in trade and consumption of bushmeat in Gabon. MSc Conservation Science, Imperial College London, London

Koppert G, Dounias E, Froment A, Pasquet P (1996) Consommation alimentaire dans trois populations forestières de la région côtière du Cameroun: Yassa, Mvae et Bakola. In: L'alimentation en forêt tropicale, interactions bioculturelles et perspectives de développement. Vol 1 Les ressources alimentaires: production et consommation. pp. 477–496

Krüger F (1998) Taking advantage of rural assets as a coping strategy for the urban poor: the case of rural-urban interrelationships in Botswana. Environ Urbanisation 10:119–134

Kumpel N (2006) Incentives for sustainable hunting of bushmeat in Rio Muni, Equatorial Guinea, PhD Thesis, Institute of Zoology/Imperial College London

Ladele AA, Joseph OA, Omotesho OA, IJaya TO (1996) Sensory quality ratings consumption pattern and preference for some selected meat types in Nigeria. Int J Food Sci Nutr 47:141–145

Lahm SA (1993) Ecology and economics of human/wildlife interaction in Northeastern Gabon. New York University, New York

Maisels F, Keming E, Kemei M, Toh C (2001) The extirpation of large mammals and implications for montane forest conservation: the case of the Kilum-Ijim Forest, North-west Province, Cameroon. Oryx 35:322–331

Meijaard E, Sheil D (2008) The persistence and conservation of Borneo's mammals in lowland rain forests managed for timber: observations, overviews and opportunities. Ecol Res 23:21–34

Milner-Gulland EJ, Akçakaya HR (2001) Sustainability indices for exploited populations. Trends Ecol Evol 16:686–692
Muchaal PK, Ngandjui G (1999) Impact of village hunting on wildlife populations in the western Dja Reserve, Cameroon. Conserv Biol 13:385–396
Nasi R, Brown D, Wilkie D, Bennett E, Tutin C, van Tol G Christophersen T (2008) Conservation and use of wildlife-based resources: the bushmeat crisis. Secretariat of the Convention on Biological Diversity, Montreal, and Center for International Forestry Research (CIFOR), Bogor. Technical Series no. 33
Neumann RP, Hirsch E (2000) Commercialisation of non-timber forest products: review and analysis of research. CIFOR, Bogor
Ngandjui G, Blanc CP (2000) Effects of hunting on mammalian populations in the western sector of the Dja reserve (Southern Cameroon). Game Wild Sci 17:93–113
Noss AJ (1998a) The impact of BaAka net hunting on rainforest wildlife. Biol Conserv 86:161–167
Noss AJ (1998b) The impacts of cable snare hunting on wildlife populations in the forest of the Central African Republic. Conserv Biol 12(2):390–398
Noss AJ (2000) Cable snares and nets in the Central African Republic. In: Robinson JG, Bennet EL (eds) Hunting for sustainability in tropical forests. Columbia University Press, New York, pp 106–153
Ntiamoa-Baidu Y (1997) Wildlife and food security in Africa. FAO Conservation Guide 33. Food and Agriculture Organisation of the United Nations, Rome
O'Brien TG, Kinnaird MF (2000) Differential Vulnerability of Large Birds and Mammals to Hunting in North Sulawesi, Indonesia and the Outlook for the Future. In J.G. Robinson and E.L. Bennett, eds. Hunting for Sustainability in Tropical Forests, pp. 199–213. New York: Columbia University Press
Oates J (ed) (1996) African primates action plan: African Primates Specialist Group, World Conservation Union (IUCN), Gland, Switzerland
Okouyi J (2006) Savoirs locaux et outils modernes cynégétiques: développement de la filière commerciale de viande de brousse à Makokou (Gabon). Université d'Orléans, Orléans
Pagezy H (1996) Importance des ressources naturelles dans l'alimentation du jeune enfant en forêt tropicale inondée. In: Hladik CM, Hladik A, Pagezy H, Linares OF, Koppert GJA, Froment A (eds) L'alimentation en forêt tropicale, interactions bioculturelles et perspectives de développement. Vol 1, Les ressources alimentaires: production et consommation. UNESCO, Paris, pp 569–588
Paumgarten F (2005) The role of non-timber forest products as safety-nets: a review of evidence with a focus on South Africa. Geojournal 64:189–197
Pearce F (2006) The protein gap. Cons Pract 6:117–123
Robinson JG, Bennet EL (2002) Will alleviating poverty solve the bushmeat crisis. Oryx 36:332–332
Robinson JG, Bennett EL (2000a) Hunting for sustainability in tropical forests. Columbia University Press, New York
Robinson JG, Bennett EL (2000b) Carrying capacity limits to sustainable hunting in tropical forests. In: Robinson JG, Bennett EL (eds) Hunting for sustainability in tropical forests. Columbia University Press, New York, pp 13–30
Rose A (2001) Social change and social values in mitigating bushmeat commerce. In: Bakarr M, de Fonseca G, Mittermeier R, Rylands A, Painemilla K (eds) Hunting and bushmeat utilisation in the African rainforest: perspectives towards a blueprint for conservation action. Conservation International, Washington, DC
Ruiz Peréz MR, Ndoye O, Eyeba A, Puntodewo A (2000) Spatial characterisation of non-timber forest products markets in the humid forest zone of Cameroon. Int Forest Rev 2:71–83
Salas LA, Kim JB (2002) Spatial factors and stochasticity in the evaluation of sustainable hunting of tapirs. Conservation Biology 16:86–96

Schenck M, Nsame Effa E, Starkey M, Wilkie D, Abernethy K, Telfer P, Godoy R, Treves A (2006) Why people eat bushmeat: results for two-choice taste tests in Gabon Central Africa. Hum Ecol 34:433–445

Scoones I, Melnyk M, Pretty J (1992) The hidden harvest: wild foods and agricultural systems: a literature review and annotated bibliography. IIED, SIDA and WWF, London, UK and Gland, Switzerland

Shackleton CM, McConnachie M, Chauke MI, Mentz J, Sutherland F, Gambiza J, Jones R (2006) Urban fuelwood demand and markets in a small town in South Africa: livelihood vulnerability and alien plant control. Int J Sust Dev World Ecol 13:1–11

Shackleton CM, Shackleton SE, Buiten E, Bird N (2007) The importance of dry forests and woodlands in rural livelihoods and poverty alleviation in South Africa. For Pol Econ 9:558–577

Starkey M (2004) Commerce and subsistence: the hunting, sale and consumption of bushmeat in Gabon, PhD Dissertation, Fitzwilliam College, University of Cambridge, UK

Steel L (1994) Étude sur le volume et la valeur du commerce de la viande de brousse au Gabon. WWF-Gabon. Libreville

Stoian D (2005) Making the best of two worlds: rural and peri-urban livelihood options sustained by non-timber forest products from the Bolivian Amazon. World Dev 33:1473–1490

Tacoli C (ed) (2006) Rural-urban linkages. Earthscan, London

Takforyan A (2001) Chasse villageoise et gestion locale de la faune sauvage en Afrique. Une étude de cas dans une forêt de l'Est-Cameroun. EHESS, Paris, France

Trefon T (1998) Urban threats to biodiversity in the Congo Basin. In: Besselink C, Sips P (eds) The Congo Basin: human and natural resources. Netherlands Committee for the IUCN, Amsterdam

Trefon T, de Maret P (1999) Snack nature dans les villes d'Afrique Centrale. In: Bahuchet S, Bley D, Pagezy H, Vernazza-Licht N (eds) L'homme et la forêt tropicale Editions de Bergier, Paris

van Vliet N (2008) Spatial and temporal variability within the "hunter-animal-village territory" system – towards a geographical approach to hunting sustainability in Central Africa- case study on duikers in north-east Gabon. PhD dissertation, Université Toulouse le Mirail: Faculté de Géographie

van Vliet N, Nasi R (2008) Hunting for livelihood in northeast Gabon: patterns, evolution, and sustainability. Ecol Soc 13:33, online

Wilkie DS, Carpenter JF (1999) Bushmeat hunting in the Congo Basin: an assessment of impacts and options for mitigation. Biodivers Conserv 8:927–955

Wilkie DS, Curran B, Tsombe R, GA MI (1998) Modeling the sustainability of subsistence farming and hunting in the Ituri forest of Zaıre. Conserv Biol 12:137–147

Wilkie D, Starkey M, Abernethy K, Nstame Effa EN, Telfer P, Godoy R (2005) Role of prices and wealth in consumer demand for bushmeat in Gabon, Central Africa. Conserv Biol 19:268–274

Part III
Systems for Sustainable Management of Non-timber Forest Products

Chapter 7
Harvesting Non-timber Forest Products Sustainably: Opportunities and Challenges

Tamara Ticktin and Charlie Shackleton

Abstract The growing knowledge and understanding of the contribution of non-timber forest products to local livelihoods is not matched with equal understanding of the sustainability of direct consumption or of market demand. There are tens of thousands of NTFP species, and there are only a few for which there have been detailed studies on their autecology and the ecological effects and sustainability of harvesting. Even amongst these, few are based on long-term data or have considered the impacts at all scales. Of the case studies to date, the results are mixed, with some clearly indicating overharvesting, and others still well within harvest limits (based on current data). In the absence of rigorous data, the likelihood of ecologically sustainable harvest systems will need to be inferred from broad patterns of the species attributes, harvest systems, and nature of demand.

7.1 Introduction

The sustainable harvest of NTFPs requires sustainability in overlapping arenas, namely the social, economic, political, and ecological. To understand the sustainability of NTFP harvesting from an ecological perspective, we need to know how, where, and when NTFPs are gathered, how this matches current and anticipated future demands, and what ecological impacts result from their harvest. However, the decisions harvesters make on how, where, and when to harvest NTFPs are shaped by cultural, political, and economic factors. Therefore, an ecological analysis of NTFP sustainability represents a key part of only a much broader analysis.

The aim of this chapter is to provide an overview of the ecological impacts of NTFP harvest and management. Despite the dependence of millions of people

---

T. Ticktin (✉)
Department of Botany, University of Hawai'i, Honolulu, HI, USA
e-mail: ticktin@hawaii.edu

C. Shackleton
Department of Environmental Science, Rhodes University, Grahamstown 6140, South Africa
e-mail: c.shackleton@ru.ac.za

S. Shackleton et al. (eds.), *Non-Timber Forest Products in the Global Context*,
Tropical Forestry 7, DOI 10.1007/978-3-642-17983-9_7,

worldwide on NTFPs for subsistence and trade (Kaimowitz 2003), this is an area that is surprisingly still understudied. In general, we know little about the harvest, management, and trade of most NTFPs, and even for the great majority of plant species that are known to be at risk due to trade, such as those listed in CITES appendices, little detailed ecological information exists. We start here by discussing the diversity of NTFP harvest systems, from wild harvest to cultivation. We then review what is known to date about NTFP harvest sustainability by discussing the impacts of harvesting at varying ecological scales and identifying some of the factors that influence them. We conclude by summarising some of the opportunities and challenges for sustainable harvest. Although NTFPs include a range of biological taxa (see Chap. 1), most of the examples we draw from are botanical.

7.2 NTFP Harvest Systems

Before we can understand the impacts of NTFP harvesting at various scales, it is necessary to first appreciate the diverse array of systems from which NTFPs are collected. It is clear that the potential and real impacts from collecting NTFPs from already intensively modified or managed systems (such as a tropical home-garden) will be different to those in largely natural or undisturbed systems (e.g., extensive forests with no or low human population densities).

Whilst there have been a number of attempts to classify the systems from which NTFPs are harvested, and the intensity, at the outset it is necessary to dispel any notion that NTFPs are collected mainly from wild or natural areas. NTFPs are found and harvested from lands spanning the full continuum of intensity of human use and disturbance, from extensive, low impact forests, wetlands, and mountains to cultivated fields and gardens around rural settlements, commercial plantations and enrichment plantings, urban allotments and vacant spaces in towns and cities (Fig. 7.1). Within these physical locations, disturbance and management regimes, NTFPs are selectively collected, nurtured as wildlings, or actively planted and tended. Moreover, human impacts on these systems, as well as on specific NTFP species, have spanned millennia (Denevan 1992), with both simplification through species removal as well as active or passive enrichment through species additions (e.g., Reid and Ellis 1995; Cunningham 1997; Tipping et al. 1999). Whilst the intensity of use and management may have changed through time, it is probably safe to say that there are few areas that have not been touched by human influence in one way or another, which may result in direct or indirect impacts (positive or negative) on particular NTFP species and populations. It is possible to speculate that perhaps the bulk of NTFP material used in significant quantities to support daily needs (as opposed to income generation) is found in areas of intermediate disturbance, such as fallows and grazing lands (McGregor 1995; Schreckenberg 1999; Pulido and Caballero 2006), where the suite of locally indigenous NTFPs are augmented by species brought in or facilitated by human actions.

Fig. 7.1 *Sclerocarya birrea* trees close to homesteads and maintained in fields in Bushbuckridge, South Africa (photo: Roger Leakey)

Table 7.1 The continuum of management systems for NTFP harvesting (adapted from Wiersum 1997)

	System	Management
Increasing human intervention	Wild populations	Uncontrolled collection from the wild
		Controlled collection from the wild
		Directed actions to stimulate growth or regeneration of wild occurring individuals or patches
		Wild land enrichment
	Maintained populations	Nurturing or planting of wildlings in human-dominated landscapes, homesteads, or gardens (e.g., agroforests)
		Maintaining adults in situ when clearing lands for cultivation or occupation
		Protection of sites or individuals (for utilitarian or cultural reasons (e.g., sacred forests)
	Farmed populations	Cultivation and regeneration of NTFPs as a secondary or supplementary crop in (or around) homesteads, gardens, or fields
		Planting of NTFPs for cultural reasons, such as to mark burial sites, temple trees, grave sites, etc.
	Domesticated populations	Establishment of fields or gardens in which the NTFP is actively planted, tended and is the dominant crop
		Selection through time for desirable traits (e.g., taste, size, growth rate)
	Urban populations	Maintenance or planting in urban areas

Wiersum (1997) presents a helpful overview of the continuum of indigenous forest management systems, which needs relatively little adaption to be functional for NTFP systems as well, which we present in Table 7.1. They are arranged along a continuum of increasing human intervention and hence impacts (both positive and negative). It is common, but with exceptions, that there is a positive relationship between increasing intervention and increasing security of tenure to the resource or the land on which it is located. Thus, wild populations have the least controls and whilst they may occur on State, private, or commons land, the harvesting regime is either typically open access or common property. This may graduate through usufruct or private tenure as access to the land or specific resources becomes more controlled through farming or domestication. Increasing security of tenure or access is also common with increasing marketability of specific NTFPs (van den Berg et al. 2007).

7.3 Ecological Impacts of NTFP Harvest

NTFP harvest systems can have impacts at multiple ecological scales, from individuals to ecosystems. Therefore, from an ecological perspective, NTFP harvest sustainability requires not only that NTFP populations are able to persist over the long-term, but also that harvest does not negatively affect community and ecosystem functions.

The most direct impact of NTFP harvest is on the vital rates, i.e., the survival, growth, and reproduction, of the harvested individual. For example, leaf harvest involves removal of photosynthetic material and nutrients and can significantly decrease reproductive output in many palms and ferns (e.g., Milton 1987; Endress et al. 2006). Leaf harvest can also increase growth due to the reallocation of stored resources, at least over a short term (Endress et al. 2006). Similarly, harvest of NTFPs such as exudates and bark can decrease vital rates by reallocating resources to wound healing and new resin or bark production. For example, in Ethiopia, *Boswellia papyrifera* trees that are heavily harvested for franckincense resin produce three times less healthy and filled seeds than unharvested trees (Rijkers et al. 2006).

Significant changes in some vital rates can in turn lead to impacts at the population level. The harvest of underground organs from perennial plants often involves mortality of the whole plant, and for many species, even low levels of harvest may result in a significant decline in long-term population growth rates (see review by Ticktin 2004). For example, this is the case for some populations of the Himalayan medicinal herb, *Nardostachys grandiflora*, where harvest is only sustainable if $<10\%$ of rhizomes are removed at intervals of at least 5 years (Ghimire et al. 2008). At the other end of the spectrum, many tree species can tolerate very high levels of fruit, seed, or flower harvest with little or no decrease in long-term population growth rates (Ticktin 2004). For example, the estimated sustainable harvest rate for marula fruits (*Sclerocarya birrea* subsp. *caffra*) in South Africa is 92% (Emanuel et al. 2005) (Fig. 7.1).

The potential to withstand harvesting depends to a large degree on how sensitive the long-term population growth rate (or the finite rate of population increase λ, estimated using matrix population models) is to changes in the vital rates altered by NTFP harvest. Meta-analyses have shown that the elasticity of vital rates varies broadly across life histories. For example, population growth rates of long-lived perennials tend to be very highly sensitive to changes in survival. In contrast, those of semelparous (reproduce only once) herbs and iteroparous (reproduce over several cycles) herbs of open habitats tend to be more sensitive to changes in growth and fertility in addition to survival (Franco and Silvertown 2004).

Life Table Response Experiments (LTRE, Caswell 2001) are a type of analysis that can be used to identify the vital rates that contribute most to observed differences in population growth rates between harvested and non-harvested NTFP populations. NTFP harvesting has the highest chance of sustainability when the vital rates with the highest elasticities have low life table response experiment values (Zuidema et al. 2007). For example, in several species of understory palms, leaf harvest causes significant declines in reproduction and growth. However, since these have low elasticity, harvesting has little effect on population growth rates (Zuidema et al. 2007). In contrast, when the vital rates with the highest elasticity values have high life table response experiment values, long-term population decline is much more likely because NTFP harvesting affects the vital rates most important for population growth. For example, for populations of *N. grandiflora* mentioned above to tolerate very low rates of harvest, survival of the largest individuals has the highest elasticity values and also has high life table response experiment values, since large adults suffer high mortality during rhizome harvest (Ghimire et al. 2008).

When harvest of NTFPs can result in adult mortality, resilience can be greatly increased if a part of the individual (stump or underground stem) is left, and the species is a good resprouter (Cunningham 2001; Botha et al. 2004). Alternatively, if the entire plant is removed, resprouters may be at a disadvantage (Raimondo and Donaldson 2003), since long-lived species that invest more in persistence or ability to resprout tend to have fewer seedlings and slower growth than non-sprouting species (Bond and Midgley 2001). They therefore tend to have very slow recovery rates.

NTFP harvest and selection can lead to evolutionary changes which may in turn have demographic impacts. For example, long-term harvesting of the largest individuals by harvesters may lead to an unconscious selection for smaller plants, and therefore to a significant decrease in plant size over time. One example is the Himalayan snow lotus (*Saussuria laniceps*), a heavily harvested medicinal plant which has decreased in size over time by an average of 9 cm (Law and Salick 2005). If smaller sized-plants have lower fitness, then both the effects of harvest and selection may decrease population viability.

On the other hand, NTFP harvesters often look for desirable characteristics, leading to increasingly higher frequencies of these characteristics in populations with increasing levels of human intervention (e.g., Shackleton et al. 2003; Leakey et al. 2004, Box 7.1). Increased human intervention can also lead to both increases and decreases in genetic diversity (Casas et al. 2007). However, heavy and uncontrolled harvest may lead to decreased genetic diversity, as has been illustrated with the medicinal plant, American ginseng (Cruse-Saunders and Hamrick 2004).

Box 7.1 Plant Management and Domestication in the Tehuacan Valley, Central Mexico

Alejandro Casas, Centro de Investigaciones en Ecosistemas, Universidad Nacional Autónoma de México, Morelia, Mexico, acasas@oikos.unam.mx

The Tehuacan Valley is one of the areas of Mesoamerica where plant management has been better documented. Archaeological studies have reconstructed a chronology of use and management since prehistory, including from some of the oldest remains of agriculture of the New World. Ethnobotanical studies have documented that people of this region currently use more than 1,600 plant species, most of them wild plants gathered from the forests, and nearly 500 species receive some type of management. Plant management systems include a high variety of agricultural systems of domesticated plants, mostly introduced, but nearly 50 species are native, with wild relatives occurring in local forests. Approximately 120 wild plant species are managed in situ, associated with agroforestry systems through various forms of tolerance (sparing useful plants in cleared areas), promotion (enhancing numbers of wild useful plants in disturbed forest), special care (protection against herbivores or competitors), seed sowing, and transplanting of vegetative parts or complete plants.

In situ management is directed at maintaining or increasing the availability of useful plants, but for some species it is also directed at increasing abundance of particularly good phenotypes through artificial selection. For instance, studies on management of the columnar cacti *Escontria chiotilla*, *Stenocereus* spp., and *Polaskia* spp. documented that phenotypes identified and favoured by local peoples are those with larger and sweeter fruits, special pulp colours, fewer spines, and thinner fruit peel. Recognition of variation in attributes and artificial selection have also been documented in trees such as *Leucaena esculenta*, *Sideroxylon palmeri*, *Ceiba aesculifolia* and in annual plants such as *Anoda cristata* and *Crotalaria pumila*.

Studies of columnar cacti indicate that morphological divergence between wild and managed populations is influenced by the intensity of artificial selection. Within a species, wild and cultivated populations are most divergent, whereas the in situ managed populations in agroforestry systems are intermediately divergent with respect to the others. Similarly, among species, morphological divergence between wild and managed populations is higher in species under higher artificial selection intensity and more easily isolated from their wild relatives. Population genetics studies have revealed the occurrence of high levels of gene flow among wild and managed populations when these populations coexist. Genetic differentiation between wild and managed populations has been identified to be generally slight, but higher in those species more intensely managed (*Stenocereus* spp.); this illustrates the role of humans and artificial selection in maintaining such differentiation. This work illustrates how traditional management systems for NTFP maintain high levels of genetic variation and should be considered in strategies for biodiversity conservation.

Further reading: Casas et al. (2006, 2007).

The participatory approach to domesticating trees that produce NTFPs in agroforestry systems (Leakey et al. 2003) is designed to minimise the problems of reduced genetic diversity (Leakey et al. 2004).

NTFP harvesting may result in ecological impacts at the community level by altering interactions between NTFP species and other organisms. For example, Moegenburg and Levey (2002) illustrated that high intensity harvest of acai palm fruit (*Euterpe oleracea*) in the Brazilian Amazon reduces avian frugivore diversity by 22%, although low intensity harvest has no effect. Enrichment plantings or agroforestry systems can have the opposite effect. For example, enhancement of acai populations can support more fruit-eating birds, but also changes the composition of avian community towards fruit eaters (Moegenburg and Levey 2002).

NTFP harvesting may also alter the structure and composition of plant communities. In India, dry deciduous forests subject to high intensity extraction of NTFPs have lower tree species richness and higher proportions of wind-dispersed versus animal-dispersed understory plants and seedlings than comparable areas of forest with lower intensities of NTFP harvest (Murali et al. 1996; Ganeshaiah et al. 1998). NTFP harvest may also increase the cover of invasive species, although this can also be mitigated by certain harvest practices (Cunningham 1993; Ticktin et al. 2006).

Finally, NTFP harvest can affect ecosystem-level processes, including nutrient dynamics and soil erosion processes (Witkowski and Lamont 1996; Siebert 2001). O'Hara (1999) illustrated that harvesting the leaves of the palm *Sabal mauritiiformis* in Belize, which are used for thatch, does not remove significant levels of limiting nutrients from harvest sites. However, she demonstrated that *S. mauritiiformis* appears to contribute significant sources of P, K, and Zn sources during certain seasons, and that the magnitude of the contributions of *S. mauritiiformis* to total ecosystem cycling is much greater for dense populations than for sparse populations. This suggests that although harvesting high density NTFP populations may be least damaging from a population perspective, it could have impacts from an ecosystem perspective. This highlights the need to concurrently carry out research at different ecological levels.

7.3.1 Relationships Among NTFP Management, Habitat, and Sustainability

Regardless of the ecological scale at which it is assessed, the sustainability of NTFP harvest is highly dependent on several key factors. These include the plant part(s) harvested, the life history characteristics of the species, the nature and intensity of harvest and management practices, and the larger socioeconomic, political, and ecological context in which the products are harvested. In Table 7.2 we present some general trends in the relationships between these factors and potential for sustainable harvest. The arrows in the table indicate that characteristics listed in the columns for "high" and "low" potential for sustainable harvest represent two ends of a continuum.

Table 7.2 Factors affecting the potential for sustainable NTFP harvest[a]

Category	Attributes	Potential for sustainable harvest		
		High	Medium	Low
Ecological	Plant part harvested	Fruit, seeds, short-lived leaves, dead wood	Exudates, phloem sap, long-lived leaves	Whole plants, roots, bulbs, bark, apical meristems
	Distribution and habitat specificity	Widespread, broad; high life history plasticity	→	Restricted, highly specific; low life history plasticity
	Population size and growth rates	Large populations, fast growth	→	Small populations, slow growth
	Reproduction	High rates of sexual and/or vegetation reproduction; continuous recruitment	→	Monocarpic or irregular and periodic sexual reproduction only; low recruitment
	Pollination, dispersal	Abiotic and/or generalist relationships	→	Specialist relationships
	Resilience to natural disturbance	High (e.g., high resprouting, fire tolerance, seedbank and/or good recruitment after disturbance)	→	Low
	Ecological integrity of landscape	Presence of necessary pollinators, dispersers, or other organisms that foster persistence of NTFP	→	Low abundance of lack of other organisms on which NTFP depend
Socio-political	NTFP uses	Single or non-competing uses; harvest of selected size-classes only	→	Multiple conflictive uses and harvest of different or all size classes
	NTFP management	Highly tended wild or maintained populations; farmed or domesticated; secure tenure	→	Uncontrolled collection from wild; open-access resource
	Governance systems	Recognised, respected, implemented	→	Open access, competing claims on

(*continued*)

Table 7.2 (continued)

Category	Attributes	Potential for sustainable harvest		
		High	Medium	Low
				resources by locals and outsiders
	Land use context	No major competing land-uses with NTFP harvest	→	Many competing land uses (e.g., logging, livestock grazing, fire, agriculture...)
Economic	Seasonality of harvest	Short season with high abundance	→	Available all year round
	Commercialisation	Local markets, low to medium value	→	High value, external markets
	Substitutability	Many species can provide the same or similar product	→	Only one or few species offer the same product

[a]Adapted and modified from Cunningham (2001) and Peters (1994)

In addition, many of the social and ecological factors that affect the potential for sustainability are interrelated (Fig. 7.2). For example, the structure and composition of the habitats in which NTFPs occur are shaped by the physical environment and by the natural and anthropogenic disturbance regimes to which they are subjected. These habitat types in turn both shape and are shaped by NTFP harvest practices and their ecological impacts. NTFP harvest practices are of course also governed by cultural, socio-economic, and political factors. Assessing the sustainability of NTFP harvest and the resilience of populations to changes requires an understanding of the interactions among these factors (Fig. 7.2).

Harvest strategies can vary widely both across and within human communities, with important consequences for NTFP productivity and sustainability (Box 7.2). Many studies have illustrated that the ecological impacts of harvesting can vary significantly according to seasonal timing of harvest, timing of harvest in the plant life-cycle, and the frequency, intensity, and methods of harvesting (Ticktin 2004; Ghimire et al. 2004, 2008; Endress et al. 2006; Guedge et al. 2007). For example, the restriction of harvest to specific size-classes, established either by traditional rules, the market, or cultural preference, can play a key role in ensuring sustainability (Nantel et al. 1996; Endress et al. 2006; Shackleton et al. 2009) (Fig. 7.2).

Many NTFP species grow or are planted in a diversity of habitats and types of production systems. Differences in abiotic factors across these systems, such as light, nutrients, and water availability, can also alter the ecological impacts of harvesting since they can allow for faster recuperation after harvesting. For example, in Mexico, populations of the terrestrial bromeliad *Aechmea magdelenae*,

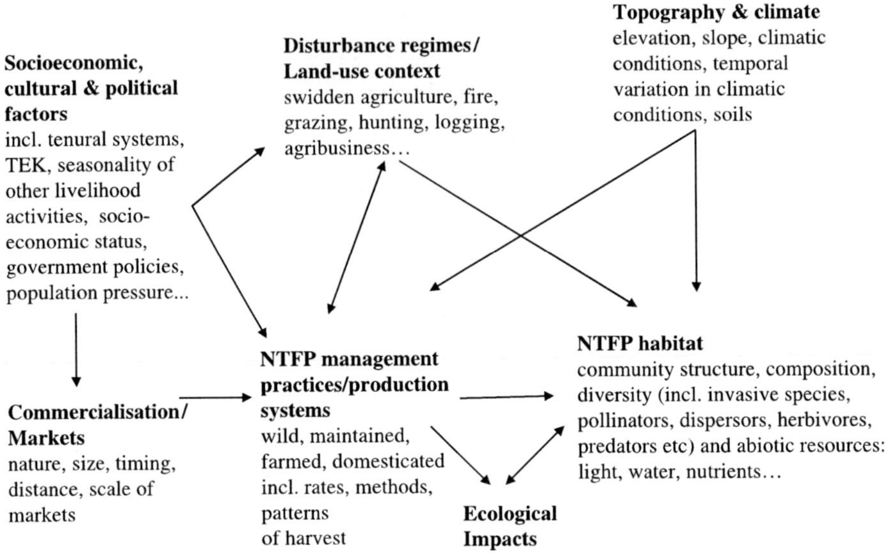

Fig. 7.2 Relationships among some of the social and ecological factors that affect NTFP sustainability

Box 7.2 Heterogeneity in Ethnoecological Knowledge and Management of Medicinal Plants in the Himalayas of Nepal

Suresh Kumar Ghimire, *Yildiz Aumeeruddy-Thomas*, and *Doyle McKey*, Central Department of Botany, Tribhuvan University, Kirtipur, Katmandu, Nepal, Sureshkghimire@yahoo.com; Centre d'Ecologie Foncionnelle et Evolutive, Université de Montpellier II, UMR 5175 Montpellier, France, yildiz.thomas@cefe.cnrs.fr; Centre d'Ecologie Foncionnelle et Evolutive, Université de Montpellier II, 5175, Montpellier, France, doyle.mckey@cefe.cnrs.fr

A large number of medicinal plants (MPs) are threatened in the Himalayas due to over-exploitation for trade, and knowledge of sustainability of their use is urgently needed. In a long-term study in Shey-Phoksundo National Park and Buffer Zone, northwestern Nepal, we assessed the factors affecting sustainable use of Himalayan MPs by integrating local knowledge and practices with research in conservation ecology. We first evaluated how local knowledge and harvesting patterns of MPs varied among different users and social groups in Tibetan and mixed Tibeto–Nepalese societies, and then incorporated our results into the design of studies to assess the ecological impacts of MP harvest.

(continued)

We found three groups of MP users in the area (1) traditional healers (*amchi* – who practice Tibetan Medicine in the national park and *dhami*, who practice a shamanistic healing system in the buffer zone); (2) commercial collectors, who live in the park buffer zone; and (3) other agro-pastoralists, who live both inside the park and the buffer zone. These groups hold a rich body of local knowledge relating to the use and ethnoecology of MPs. Management of MPs is based on this broad ethnoecological knowledge, as well as on socio-cultural, economic, and political considerations. Local knowledge and management practices varied substantially within and between social groups, and this variation was related to variation in peoples' activities, their cultural and historical linkages to the environment, their levels of specialisation in relation to MPs, and the extra-local socio-economic factors that govern people's activities. Among the three classes of users, traditional healers (particularly *amchi*) possessed substantially higher knowledge on the diversity, natural history, and environmental requirements of MPs. Knowledge of commercial collectors was most detailed for plants that are traded. In terms of harvesting patterns, we found two general approaches (1) selective or choosy harvesting, conducted by *amchi* in the national park, and (2) non-selective and massive harvesting, typical of exploitation of MPs in the buffer zone. In the case of commercial collectors, specialised ethnoecological knowledge of traded species did not necessarily lead to good harvesting practices. Knowledge of commercial collectors is less integrated culturally but this knowledge base also indicates the potential for improving management, provided the market is well organised.

We assessed the impact of local practices of rhizome harvesting described above on the population dynamics of two high-altitude perennial medicinal herbs, *Nardostachys grandiflora* (Valerianaceae) and *Neopicrorhiza scrophulariiflora* (Scrophulariaceae). We found that *N. scrophulariiflora* is more vulnerable to harvesting than *N. grandiflora* due to differences in strategies for vegetative reproduction and to the harvesting practices associated with these strategies. The scope for harvesting sustainability even varied within a species in relation to environmental variation that affected the rate of recovery of populations from the loss of individuals. Our results illustrate that for the same plant part, differences in harvesting approaches, in addition to differences in plant growth strategies and habitat conditions, can lead to responses that vary both within and between plant species. This work illustrates how current human management shapes the structure and dynamics of MPs at the population level and clarifies how studies of local knowledge and practices may contribute to formulating new hypotheses in ecology as well as designing more sustainable practices.

Further reading: Ghimire et al. (2004, 2005, 2008).

whose leaves are harvested for their silky fibre and used in handicrafts, receive more light in secondary forests and recuperate four times faster after ramet harvest than those growing in the old growth forests (Ticktin and Nantel 2004). In Benin, harvest of foliage for fodder and bark for medicine from African mahogany (*Khaya senegalensis*) reduces fruit production by 75% in the drier northern part of the country, but in the more moist southern region the effect of harvesting is not significant (Gaoue and Ticktin 2008). Clearly, any guidelines for sustainable management must take these kinds of differences into account.

Life history strategies for a given species can vary across habitat types as well. For example in Mexico, life history strategies for xa'an palm (*Sabal yapa*), whose leaves are harvested for thatch, varies among fallows, home gardens, and crop fields so that leaf harvest is sustainable in all habitats, but it is achieved in different ways (Martínez-Ballesté et al. 2005). Those species whose life history strategies show high plasticity may be most resilient to harvest (Fig. 7.2).

Differences in biotic interactions across habitat types or within habitats over time can also affect harvest sustainability. For instance, the ability for American ginseng to withstand harvesting is reduced with increasing populations of deer, which browse on this species (McGraw and Furedi 2005). At the same time, NTFP species subject to harvesting patterns that mimic biotic interactions for which they have developed evolutionary responses such as resprouting, may be most tolerant to harvesting (Siebert 2000).

Changes in anthropogenic disturbance regimes to the landscapes where NTFPs grow can affect their distribution and densities and alter their resilience to harvesting (Fig. 7.2). For example, although current leaf harvest rates for xa'an palms are sustainable, increasing intensity of shifting agriculture and decreasing fallow times can diminish leaf supply (Pulido and Caballero 2006). In addition, many NTFPs are subject to multiple pressures and may be more threatened by other land-uses than by harvesting (Fig. 7.3). For example, in Eastern Amazonia, declining abundance of

Fig. 7.3 Fuelwood harvesters, Peddie district, Eastern Cape, South Africa (photo: Charlie Shackleton)

NTFPs due to logging and fire has resulted in a lack of forest products to meet even subsistence needs of local communities (Shanley et al. 2002).

Generations of observation, experimentation, and adaptation by local harvesters often leads to the development of detailed traditional ecological knowledge (TEK) and highly sophisticated traditional management practices for maintaining NTFPs and other culturally and economically important resources and landscapes. TEK systems usually include social institutions for governing resource use including systems of tenure and taboos or other restrictions on the nature, timing, location, or amount of harvest, including stories and legends (Turner et al. 2000; Colding and Folke 2001; Shankaar et al. 2004). NTFP populations managed by knowledgeable harvesters may show high growth rates under high harvesting pressure, while populations of the same species managed by others may decline under much lower levels of harvesting (Ticktin 2004). However, understanding traditional or local management systems can be complex, and harvesters may not consider their actions, however sophisticated, as any kind of "management" per se. This highlights the importance of long-term research with local harvesters and of combining ethnographic research with ecological studies, especially including participant observation. While TEK alone may not hold all the answers, it can and should play a key role in developing sustainable management plans (Box 7.2, 7.3).

Box 7.3 Combining Traditional Ecological Knowledge (TEK) with Western Science in Forest Inventories in British Columbia

Wendy Cocksedge, Centre for Non-Timber Resources, Royal Roads University, 2005 Sooke Road, Victoria, BC V9B 5Y2, Canada, Wendy.cocksedge@royalroads.ca

Although critically important to livelihoods of local communities, NTFPs are often undervalued and overlooked within conventional land and resource management plans. Better integration of NTFP resource information into conventional resource inventories is needed both for sustainability of the ecosystems, particularly in light of cumulative effects such as logging, development, insect epidemics, and climate change, and facilitating compatible management for multiple resource values.

An NTFP inventory can be considered simply a focused vegetation inventory. The focus is the inclusion of parameters of plant quality in the inventory; that is, the usability of the plant by local harvesters. The inclusion of quality attributes is necessary to be able to assess available abundance. For example, soapberry (*Shepherdia canadensis*) is found throughout much of the interior of British Columbia, but only a small subset of this area produces berries of sufficient quantity and quality to be of value to harvesters.

Combining traditional knowledge with western science methods to augment conventional (Western) resource inventories reflects the requirements

(continued)

of the community and other users, such as wildlife, by incorporating the specific morphological features required, while maintaining a landscape-level approach. This method is being tested in British Columbia. Local experts, including Aboriginal and non-Aboriginal harvesters, have identified priority species for the first round of study, and have defined quality for each species based on their experience. Using this information, standardised ratings were developed that use specific rating scales (e.g., 0–5) of quality, from low to high. These ratings can then be used by ecologists within other inventory projects, as usually it is only a few more items to complete on a plot card. As the ecologists are usually unfamiliar with the use of the species, each rating requires specific measureable criteria, such as the estimated volume of berries per square metre or the diameter and length of a floral greenery branch. Each rating is specific to the use of the plant; for example, Oregon grape (*Mahonia aquifolium*) would have three separate ratings associated with the cover value, one for each of the branches (floral greenery), the rhizomes (medicinal), and the berries (edible).

Attribute tables can then be developed by correlating high quality areas to their associated site and stand values, such as specific soil types, moisture levels, elevation, and stand age. These attributes can be used within predictive maps or models, and as this inventory is done within conventional systems, it is possible to extrapolate the predictive models to larger or alternative areas, assuming similar ecosystems or plant usage.

By developing a better understanding of the conditions required for high quality NTFP presence, we can enhance the quantity and quality of these species by compatibly managing for both tree and understory values.

Further reading: Cocksedge (2006) and Cocksedge et al. (2010)

7.3.2 Temporal Dynamics of NTFP Harvest Systems

The sustainability of NTFP harvesting is also closely linked to dimensions of the seasonal availability of the resource and how that dovetails with other livelihood activities. For example, a highly seasonal resource, which offers only a short duration harvesting season, potentially promotes (but does not guarantee) better prospects for sustainable use than does a resource that is available year round. This is because the short duration NTFP requires greater organisation and mobilisation of household labour to harvest significant proportions in the time available: for example, collection of mopane worms (*Imbrasia belina*) in Botswana. In contrast, an NTFP with year round availability can be harvested at any time which can be coordinated to fit in with periods of low labour demand for other activities. Thus, the temporal dynamics relate to both the availability of the resource as well as the availability of household labour. When these are in synchrony, the greater is the possibility for significant proportions of the resource to be harvested.

The temporal patterns of availability relate to seasonal and intra-seasonal scales. Typically, reproductive organs used as NTFPs (e.g., flowers, fruits, or seeds) are available only for specific periods within the annual or seasonal cycle. However, for some species the period of availability may be reasonably predictable from one year to the next such that local knowledge that they are available in specific months accumulates. In many instances local traditions have evolved around the harvest, either to control the actual period of harvest or to celebrate its arrival. For example, traditional controls on thatch grass harvesting in parts of southern Africa require the local chief to declare the harvesting season open (Mwalukomo 2007). Traditionally, no harvesting is permitted before this is done, but in areas where traditional authority is waning, this practice is often disregarded. A similar system of the local traditional authority declaring an open season applies to harvesting of in-shore fisheries and gastropods around several Indonesian islands (Evans et al. 1997). This is mirrored in western Europe and North America where wildlife or conservation authorities issue permits for hunting, fishing, or collecting a range of resources only during set periods of the year. With respect to celebrations, a good example is the requirement that households must offer a portion of the first marula (*Sclerocarya birrea* subsp. *caffra*) beer brew of the season to the local tribal leader at a community gathering (Shackleton et al. 2002).

However, the cues for the harvesting of other species may be a specific suite of intra-seasonal conditions, rather than just time of year. For example, masting species, or species that require particular temperature or moisture conditions, or species that emerge, flower or seed only after fire or flooding, or insect and small mammal population eruptions. These more specialised requirements or conditions may be familiar to local harvesters, but precisely when they will be manifest are not. Consequently, there is a strong element of unpredictability about when the NTFP will be available and in what quantity. As a result, some harvesting households may be unable to respond in time because they are otherwise occupied when the NTFP peak emerges, or not be able to optimise the volume harvested or income earned (e.g., Nevenimo et al. 2008). There is typically an inverse relationship between availability and price in the local market place (e.g., Muñiz-Miret et al. 1996), and consequently being able to predict NTFP supply before it peaks results in better incomes per unit effort.

At the opposite end of the spectrum are those NTFPs that are not dependent upon seasonal climatic conditions or intra-seasonal cues. For example, fuelwood, the bark of trees used for medicine, and some types of bushmeat. However, although the resource supply may be relatively unseasonal, frequently the species composition of the harvest may change with season. For example, *Lysiloma spp.* constitutes over 43% of the fuelwood volume in markets in Masaya (Nicaragua) in the dry season, but only 16% in the wet season (McCrary et al. 2005).

As mentioned above, the seasonality of NTFP harvest is not solely a result of the resource availability, but may also be a reflection of harvesting calendars. These are a consequence of (1) changing demands on labour during the year and (2) the relative importance of other livelihood activities in contributing food or income to the household. It is well recorded that in many regions the contribution of NTFPs to household consumption increases during the non-agricultural season, due to both reduced food availability from crops and increased availability of labour, which is no longer occupied in the fields (e.g., Mahapatra et al. 2005). The predictable production cycles for land preparation, planting, and harvesting require some or all of the household labour to be involved in these arable production activities. Another example is the seasonality of casual wage labour. If the returns for such casual wage labour are deemed to be greater than from NTFP harvesting or agriculture, such harvesting or production will decline or cease (e.g., Conelly 1994). However, the opposite also applies. Casual traders in NTFPs will increase their activities at times when market conditions improve. These may be predicable periods so that they can prepare in advance, such as craft producers or traders in marine resources trying to catch seasonal tourist peaks (e.g., Shackleton et al. 2007), or traders in ceremonial products increasing production and trade at times just before the ceremonial period (e.g., grass brooms – Shackleton and Campbell (2007)) (Fig. 7.4). Less predictable peaks in harvesting may also occur in times of

Fig. 7.4 Transporting raw materials for twig (*Athrixia phylicoides*) and grass (*Festuca costata*) brooms from the Drakensburg escarpment to Bushbuckridge in the lowveld of South Africa (photo: Sheona Shackleton)

stress or shock to a particular household (such as death or retrenchment of a breadwinner) or to a community as a whole (through drought, floods, and diseases to crops or livestock). This relates to the well-recognised safety net value of NTFPs (Shackleton and Shackleton 2004; McSweeney 2005; Paumgarten 2005).

7.4 Opportunities and Challenges for Sustainable Harvest

This review has illustrated some of the complexities involved in understanding the ecological impacts of NTFP harvest. The diversity of NTFP harvest systems and the many factors that influence their ecological impacts, both positive and negative, provide opportunities as well as challenges for sustainable harvesting.

Although it is clear that many non-cultivated NTFP populations are overharvested and declining (Ticktin 2004), the literature also suggests that there is much scope for sustainability. When collection from wild populations is controlled and populations are tended through management practices, a diversity of species can withstand fairly heavy rates of harvest. These include some species harvested for parts predicted to have medium or high ecological impacts, such as leaves, bark, and whole plants (Ticktin et al. 2002; Martínez-Ballesté et al. 2005; Endress et al. 2006; Guedge et al. 2007; Zuidema et al. 2007). However, an important challenge is that many wild-harvested NTFPs are now open-access resources. Controlled harvesting usually requires some kind of informal or formal tenural arrangements, and these can be especially difficult to maintain or develop in socio-ecological contexts characterised by decreasing land-bases (e.g., due to changing land-use practices), increasing harvest pressure, changing cultural practices, among other factors.

Intensification of NTFP management through farming or cultivation provides important options to increase production and access. While clearing natural vegetation to grow NTFPs in monocultures or low diversity production systems can clearly have negative ecological impacts, NTFPs grown in diverse home gardens, fallows, agroforestry systems, or enrichment plantings offer much potential for sustainability. For example, agroforestry systems that involve low intensity management and retain the canopy cover of native trees can conserve a high diversity of plants, mammals, birds, and insects (Bhagwat et al. 2008). At the same time, constraints to intensification of NTFP management signify that cultivation of many threatened NTFPs is not economically feasible (Schwippman et al. 2002) and that it requires secure tenure, which limits landless people from participating. In addition, although cultivation can help save threatened species, it can have negative consequences for conservation on a larger scale. This is because maintaining controlled harvesting from wild populations can provide harvesters with important economic incentives to conserve the forests or other habitat in which NTFPs grow (Shackleton 2001), which is lost when the bulk of the supply is farmed under restricted access or tenure.

Although there is now a large and growing literature on the sustainability of NTFP harvest and management, there still remain important knowledge gaps. Most studies to date have been short-term, yet the few longer term studies that exist illustrate that harvest impacts may only become apparent after multiple years (Endress et al. 2006). The lack of long-term data is particularly important for NTFP populations that may respond to harvesting in a non-linear fashion, such that decline is only evident once some threshold is reached. There still is relatively little information on the ecological impacts of harvesting some of the most frequently used plant parts such as bark, exudates, and underground organs. Many studies on the ecological impacts of NTFP harvesting have focused on populations in natural habitats, but most NTFPs are collected or grown in a variety of habitats subject to multiple anthropogenic disturbances. Approaches that consider the diversity of habitats in which NTFPs are harvested, tended, and cultivated, as well as the interconnections among them, can provide better insight on current sustainability and how future changes in landscape use or climate may affect it (Ticktin 2005; Pulido and Caballero 2006). Similarly, more studies at the community and ecosystem levels can provide insight on NTFP cultivation systems that can best conserve biodiversity while maintaining productivity.

Finally, regardless of where a given NTFP species lies in the continuum from wild harvest to domestication, it is clear that local management practices play a fundamental role in determining harvest sustainability. In addition, the ways in which the responses to harvest of numerous species vary significantly over space and time suggest that in many cases harvest limits may have little meaning outside the specific conditions in which they were determined. Therefore, participatory research with local harvesters to understand local knowledge and practices, including their drivers and ecological impacts, and to promote and support local experimentation and adaptive management strategies, provide key opportunities for identifying and promoting sustainable NTFP harvests.

References

Bhagwat SA, Willis KJ, Birks HJB, Whittaker RJ (2008) Agroforestry: a refuge for biodiversity? Trends Ecol Evol 23:261–264

Bond WJ, Midgley JJ (2001) Ecology of sprouting in woody plants, the persistence niche. Trends Ecol Evol 16:45–51

Botha J, Witkowski ETF, Shackleton CM (2004) The impact of commercial harvesting on *Waburgia salutaris* ('pepper-bark tree') in Mpumalanga South Africa. Biodivers Conserv 13:1675–1698

Casas A, Cruse J, Morales E, Otero-Arnaiz A, Valiente-Banuet A (2006) Maintenance of phenotypic and genotypic diversity of *Stenocereus stellatus* (Cactaceae) by indigenous peoples in Central Mexico. Biodivers Conserv 15:879–898

Casas A, Otero-Arnaiz A, Pérez-Negrón E, Valiente-Banuet A (2007) In situ management and domestication of plants in Mesoamerica. Ann Bot 100:1101–1115

Caswell H (2001) Matrix population models, 2nd edn. Sinauer Associates, Sunderland, Massachusetts

Cocksedge W (ed) (2006) Incorporating non-timber forest products into sustainable forest management: an overview for forest managers. Royal Roads University, McGregor Model Forest, Canadian Model Forest Network. Victoria, BC. http://cntr.royalroads.ca/publications

Cocksedge W, Titus B, Mitchell D (2010) Benefits and barriers to compatible management in BC; results of a survey. J Ecosyst Manage 11:9–26

Colding J, Folke C (2001) Social Taboos: invisible systems of local resource management and biological conservation. Ecol Appl 11:584–600

Conelly WT (1994) Population pressure, labour availability and agricultural disintensification: the decline of farming ion Rusinga Island, Kenya. Hum Ecol 22:145–163

Cruse-Saunders J, Hamrick JL (2004) Genetic diversity in harvested and protected populations of American ginseng *Panax quinquefolius* L. (Araliaceae). Am J Bot 91:540–548

Cunningham AB (1993) African medicinal plants. Setting priorities at the interface between conservation and primary healthcare. People and Plants Working Paper 1. UNESCO, Paris

Cunningham AB (1997) Landscape domestication and cultural change: human ecology of the Cuvelai-Etosha region. Madoqua 20:37–48

Cunningham AB (2001) Applied ethnobotany: people, wild plant use and conservation. Earthscan, London

Denevan G (1992) The pristine myth: the landscape of the Americas in 1492. Ann Assoc Am Geogr 82:369–385

Emanuel PL, Shackleton CM, Baxter JS (2005) Modelling the sustainable harvest of *Sclerocarya birrea* subsp. *caffra* fruits in the South African lowveld. For Ecol Manage 214:91–103

Endress BA, Gorchov DL, Berry EJ (2006) Sustainability of a non-timber forest product: effects of alternative leaf harvest practices over 6 years on yield and demography of the palm *Chamaedorea radicalis*. For Ecol Manage 234:181–191

Evans SE, Gill ME, Retraubun AS, Abrahamz J, Dangeubun J (1997) Traditional management practices and the conservation of the gastropod (*Trochus niloticus*) and fish stocks in the Maluku Province (Eastern Indonesia). Fish Res 31:83–91

Franco M, Silvertown J (2004) A comparative demography of plants based on elasticities of vital rates. Ecology 85:532–538

Ganeshaiah KN, Shaanker RU, Murali KS, Shankar U, Bawa KS (1998) Extraction of non-timber forest products in the forest of Bilingiri Rangan Hills, India. 5. Influence of dispersal mode on species response to anthropogenic pressures. Econ Bot 52:316–319

Gaoue OG, Ticktin T (2008) Effects of bark and foliage harvest on *Khaya senegalensis* (Meliaceae) reproductive performance and morphology in Benin. J Appl Ecol 45:31–40

Ghimire SK, McKey D, Aumeeruddy-Thomas Y (2004) Heterogeneity in ethnoecological knowledge and management of medicinal plants in the Himalayas of Nepal: implications for conservation. Ecol Soc 9:6 http://www.ecologyandsociety.org/vol9/iss3/art6/

Ghimire SK, McKey D, Aumeeruddy-Thomas Y (2005) Conservation of Himalayan medicinal plants: harvesting patterns and ecology of two threatened species, *Nardostachys grandiflora* DC. and *Neopicrorhiza scrophulariiflora* (Pennell) D.Y. Dong. Biol Conserv 124:463–475

Ghimire SK, Gimenez O, Pradel R, McKey D, Aumeeruddy-Thomas Y (2008) Demographic variation and population viability in a threatened Himalayan medicinal and aromatic herb *Nardostachys grandiflora*: matrix modelling of harvesting effects in two contrasting habitats. J Appl Ecol 45:41–51

Guedge NM, Zuidema PA, During H, Foahom B, Lejoly J (2007) Tree bark as a non-timber forest product: the effect of bark collection on population structure and dynamics of *Garcinia lucida* Vesque. For Ecol Manage 240:1–12

Kaimowitz D (2003) Not by bread alone . . . forests and rural livelihoods in sub-Saharan Africa. In: Oksanen T, Pajari B, Tuomasjukka T (eds) Forestry in poverty reduction strategies: capturing the potential. EFI Proceedings No. 47. European Forest Institute, Joensuu, pp 45–64

Law L, Salick J (2005) Human-induced dwarging of Himalayan snow lotus Saussuria laniceps (Asteraceae). Proc Natl Acad Sci USA 102:10218–10220

Leakey RRB, Schreckenberg K, Tchoundjeu Z (2003) The participatory domestication of West African indigenous fruits. Int For Rev 5:338–347

Leakey RRB, Tchoundjeu Z, Smith RI, Munro RC, Fondoun JM, Kengue J, Anegbeh PO, Atangana AR, Waruhiu AN, Asaah E, Usoro C, Ukafor V (2004) Evidence that subsistence farmers have domesticated indigenous fruits (*Dacryodes edulis* and *Irvingia gabonensis*) in Cameroon and Nigeria. Agrofor Syst 60:101–111

Mahapatra AK, Albers HJ, Robinson EJZ (2005) The impact of NTFP sales on rural households' cash income in India's dry deciduous forest. Environ Manage 35:1–8

Martínez-Ballesté A, Martorell C, Martínez-Ramos M, Caballero J (2005) Applying retrospective demographic models to assess sustainable use: the Maya management of xa'an palms. Ecol Soc 10:17

McCrary JK, Walsh B, Hammett AL (2005) Species, sources, seasonality, and substitutability of fuelwood commercialisation in Masaya, Nicaragua. For Ecol Manage 205:299–309

McGraw JB, Furedi MA (2005) Deer browsing and population viability of an understory plant. Science 307:920–922

McGregor J (1995) Gathered produce in Zimbabwe's communal areas: changing resource availability and use. Ecol Food Nutr 33:163–193

McSweeney K (2005) Natural insurance, forest access, and compounded misfortune: Forest resources in smallholder coping strategies before and after Hurricane Mitch, northeastern Honduras. World Dev 33:1453–1471

Milton S (1987) Effects of harvesting on four species of forest ferns in South Africa. Biol Conserv 41:133–146

Moegenburg SM, Levey DJ (2002) Prospects for conserving biodiversity in Amazonian extractive reserves. Ecol Lett 5:320–324

Muñiz-Miret N, Vamos R, Haraoka M, Montagnini F, Mendelsohn RO (1996) The economic value of managing açai palm (*Euterpe oleracea* Mart.) in the floodplains of the Amazon estuary, Pará, Brazil. For Ecol Manage 87:163–173

Murali KS, Shankar U, Shaanker R, Ganeshaiah KN, Bawa KS (1996) Extraction of non-timber forest products in the forests of Biligiri Rangan Hills, India. 2. Impact of NTFP extraction on regeneration, population structure, and species composition. Econ Bot 50:252–269

Mwalukomo H (2007) The role of traditional leaders in environmental governance in the context of decentralisation: a case study of grass utilisation in QwaQwa, Eastern Free State. M.Sc thesis, University of the Witwatersrand, Johannesburg, 129 pp

Nantel P, Gagnon D, Nault A (1996) Population viability analysis of American ginseng and wild leek harvested in stochastic environments. Conserv Biol 10:608–621

Nevenimo T, Johnston M, Binifa J, Gwabu C, Angen J, Moxon J, Leakey R (2008) Domestication potential and marketing of *Canarium indicum* nuts in the Pacific: producer and consumer surveys in Papua New Guinea (East New Britain). For Trees Livelihood 18:253–269

O'Hara JL (1999) An ecosystem approach to monitoring non-timber forest products harvest: the case study of Bayleaf Palm (*Sabal mauritiiformis*) in the Rio Bravo conservation and management area, Belize. PhD Thesis, Yale University, Conn, USA

Paumgarten F (2005) The role of non-timber forest products as safety-nets: a review of evidence with a focus on South Africa. GeoJournal 64:189–197

Peters CM (1994) Sustainable harvest of non-timber plant resources in tropical moist forest: an ecological primer. Bronx, NY, USA

Pulido MT, Caballero J (2006) The impact of shifting agriculture on the availability of non-timber forest products: the example of *Sabal yapa* in the Maya lowlands of Mexico. For Ecol Manage 222:399–409

Raimondo D, Donaldson JS (2003) Responses of cycads with different life histories to the impact of plant collecting. Biol Conserv 111:345–358

Reid RS, Ellis JE (1995) Impacts of pastoralists on woodlands in south Turkana, Kenya: livestock mediated tree recruitment. Ecol Appl 5:978–992

Rijkers T, Ogbazghi W, Wessel M, Bongers F (2006) The effect of tapping for frankincense on sexual reproduction in *Boswellia papyrifera*. J Appl Ecol 43:1188–1195

Schreckenberg K (1999) Products of a managed landscape: non-timber forest products in the parklands of the Bassila region, Benin. Glob Ecol Biogeogr 8:279–289

Schwippman U, Leaman D, Cunningham AB (2002) Impact of cultivation and gathering of medicinal plants on biodiversity: global trends and issues. In: Biodiversity and the ecosystem approach in agriculture, forestry and fisheries. FAO, Rome

Shackleton CM (2001) Re-examining local and market orientated use of wild species for the conservation of biodiversity. Environ Conserv 28:70–278

Shackleton SE, Campbell B (2007) The traditional broom trade in Bushbuckridge, South Africa: helping poor women cope with adversity. Econ Bot 61:256–268

Shackleton CM, Shackleton SE (2004) The importance of non-timber forest products in rural livelihood security and as safety-nets: evidence from South Africa. S Afr J Sci 100:658–664

Shackleton SE, Shackleton CM, Cunningham T, Lombard C, Sullivan CA, Netshiluvhi TR (2002) Knowledge on *Sclerocarya birrea* subsp. *caffra* with emphasis on its importance as a non-timber forest product in South and southern Africa: a summary. Part 1: Taxonomy, ecology and role in rural livelihoods. S Afr For J 194:27–41

Shackleton CM, Botha J, Emanuel PL (2003) Productivity and abundance of *Sclerocarya birrea* subsp. *caffra* in and around rural settlements and protected areas of the Bushbuckridge Lowveld, South Africa. For Trees Livelihood 13:217–232

Shackleton CM, Timmermans HG, Nongwe N, Hamer N, Palmer R (2007) Direct-use values of non-timber forest products from two areas on the Transkei Wild Coast. Agrekon 46:135–156

Shackleton CM, Chauke MI, Downsborough L, Olsen A, Parkin F, Weideman C, Brill G (2009) Conservation, commercialisation and confusion: harvesting of *Ischreyolepis* in a coastal forest, South Africa. Environ Dev Sustain 11:229–240

Shankaar RU, Ganeshaiah KN, Krishnan S, Ramya R, Meera C, Aravind NA, Kumar A, Rao D, Vanaraj G, Ramachandra J, Gauthier R, Ghazoul J, Poole N, Chinnappa Reddy BV (2004) Livelihood gains and ecological costs of non-timber forest product dependence: assessing the roles of dependence, ecological knowledge and market structure in three contrasting human and ecological settings in south India. Environ Conserv 31(3):242–253

Shanley P, Luz L, Swingland IR (2002) The faint promise of a distant market: a survey of Belem's trade in non-timber forest products. Biodivers Conserv 11:615–636

Siebert SF (2000) Abundance and growth of *Desmoncus orthacanthos* Mart. (Palmae) in response to light and ramet harvesting in five forest sites in Belize. For Ecol Manage 137:83–90

Siebert SF (2001) Nutrient levels in rattan foliage and cane, and implications for harvesting. Biotropica 33:361–363

Ticktin T (2004) The ecological consequences of harvesting non-timber forest products. J Appl Ecol 41:11–21

Ticktin T (2005) Applying a metapopulation framework to the management and conservation of a non-timber forest species. For Ecol Manage 206:249–261

Ticktin T, Nantel P (2004) Dynamics of harvested populations of the tropical understory herb *Aechmea magdalenae* in old-growth *versus* secondary forests. Biol Conserv 120:461–470

Ticktin T, Nantel P, Ramírez F, Johns T (2002) Effects of variation on harvest limits for nontimber forest species in Mexico. Conserv Biol 16:691–705

Ticktin T, Whitehead AN, Fraiola H (2006) Traditional gathering of native *hula* plants in alien-invaded Hawaiian forests: adaptive practices, impacts on alien invasive species, and conservation implications. Environ Conserv 33:185–194

Tipping R, Buchanan J, Davies A, Tisdall E (1999) Woodlands biodiversity, paleo-human ecology and some implications for conservation management. J Biogeogr 26:33–43

Turner NM, Ignace B, Ignace R (2000) Traditional ecological knowledge and wisdom of aboriginal people in British Columbia. Ecol Appl 10:1275–1287

Van den Berg D, Wiersum KF, van Dijk H (2007) The role and dynamics of community institutions in the management of NTFP resources. For Trees Livelihood 17:183–197

Wiersum KF (1997) Indigenous exploitation and management of tropical forest resources: an evolutionary continuum in forest-people interactions. Agric Ecosyst Environ 63:1–16

Witkowski ETF, Lamont BB (1996) Nutrient losses from commercial picking and cockatoo removal of *Banksia hookeriana* blooms at the organ, plant and site levels. J Appl Ecol 33:131–140

Zuidema P, de Kroon H, Werger MJA (2007) Testing sustainability by prospective and retrospective demographic analyses. Ecol Appl 17:118–128

Chapter 8
Timber and Non-timber Forest Product Extraction and Management in the Tropics: Towards Compatibility?

Manuel R. Guariguata, Carmen García-Fernández, Robert Nasi, Douglas Sheil, Cristina Herrero-Jáuregui, Peter Cronkleton, Ousseynou Ndoye, and Verina Ingram

Abstract Tropical forests have the potential to satisfy multiple demands for goods and services. Yet integrated management approaches across multiple goods remain elusive. Here we consider selective harvesting of timber and non-timber forest product (NTFP) extraction. We analyze the current status of this combination and speculate on prospects and challenges regarding (1) resource inventory, (2) ecology and silviculture, (3) conflict in the use of multipurpose tree species, (4) wildlife conservation and use, (5) tenure, and (6) product certification. Convincing conclusions are hampered by the relative paucity of comprehensive studies and lessons learned on what has worked and what has not in the context of integrated management for timber and NTFPs. Interventions for enhancing the compatibility of timber and NTFP extraction must be scaled in relation to the size of the area being managed, applied timber harvesting intensities, and the dynamics of multiactor, forest partnerships (e.g., between the private sector and local communities). In addition, training and education issues may have to be recrafted with multiple-use management approaches inserted into tropical forestry curricula.

M.R. Guariguata (✉), R. Nasi, P. Cronkleton, and V. Ingram
Center for International Forestry Research (CIFOR), P.O. Box 0113 BOCBD, Bogor 16000, Indonesia
e-mail: m.guariguata@cgiar.org

C. García-Fernández
Bioversity International, Via dei Tre Denari 472/a, 00057 Maccarese, Rome, Italy

D. Sheil
Institute of Tropical Forest Conservation (ITFC), Bwindi Impenetrable National Park (at Ruhija), P.O. Box 44, Kabale, Uganda

C. Herrero-Jáuregui
Departamento de Ecología, Facultad de Biología, Universidad Complutense de Madrid, Madrid 28040, España

O. Ndoye
Food and Agriculture Organisation of the United Nations (FAO), P.O. Box 281, Yaoundé, Cameroon

S. Shackleton et al. (eds.), *Non-Timber Forest Products in the Global Context*, Tropical Forestry 7, DOI 10.1007/978-3-642-17983-9_8,

8.1 Introduction

Tropical forests have the potential to satisfy multiple demands for timber and non-timber forest products (NTFPs) and marketed and nonmarketed ecosystem services, while including industrial and nonindustrial actors – both for present and future generations (Kant 2004). To accommodate these requirements, sustainable forest management (SFM) emerged in the early 1990s (Poore 2003), and multiple-use forestry became entrenched within SFM as a way to achieve socially and environmentally driven development models in the tropics (Panayotou and Ashton 1992). Yet, clearly defined multiuse approaches to natural forest management remain elusive (García-Fernández et al. 2008). The application of reduced impact logging (RIL) guidelines (reviewed in Putz et al. 2008) largely contributed to an increase in the area of natural forest under SFM from less than one million ha in 1988 (Poore et al. 1989) to about 36 million ha by 2005 (ITTO 2006). However, this quest for sustainability was largely disconnected from other forest goods and services, including NTFPs which are still treated in relative isolation (Lawrence 2003). Clearly, SFM is about more than RIL, and there is now renewed interest in developing multiple-use tropical forestry (e.g., Sist et al. 2008; Shanley et al. 2008).

Diversified forest demands can be met either by spatially segregating uses for particular goods and services (Vincent and Binkley 1993; Binkley 1997; Zhang 2005) or by managing forest stands to meet multiple objectives from the same area. The latter model is widespread across the tropics (Sayer and Byron 1996; Poore 2003; Nittler and Tschinkel 2005), but whether or not multiple-use of forest goods and services facilitates sustainability still generates much debate. Skeptics question the extent to which economic returns from NTFPs and other values are sufficient to outweigh the financial costs of modifying and/or applying RIL norms (Barreto et al. 1998; Pearce et al. 2003) and silvicultural practices needed for sustaining timber production over the long term (e.g., enrichment planting, Schulze 2008; liberation thinning, Wadsworth and Zweede 2006). Advocates emphasise that by incorporating many forest goods and services, including the voices of different stakeholders, a social and financial edge can be gained over timber-dominated models (Ashton et al. 2001; Hiremath 2004; Wang and Wilson 2007).

This chapter examines one of the possible combinations for multiple use: extraction and management of valuable timber and NTFPs. We focus on mechanised, selective logging as this remains the dominant and most profitable option in natural tropical forests and excludes agroforests, regenerating fallows, and/or planted forests (Toledo et al. 2003; Belcher et al. 2005; Michon et al. 2007). We emphasise six topics: (1) resource inventory, (2) ecology and silviculture, (3) conflict in the use of multipurpose tree species, (4) wildlife conservation and use, (5) tenure and access rights, and (6) product certification. Other factors (e.g., seasonality, legal frameworks, gender) may cut across the above topics and thus we provide an indicative list of these (Table 8.1) highlighting the way they may affect compatible management outcomes of timber and NTFPs.

Table 8.1 An indicative list of factors (left column) and the way these may affect compatible extraction and management of timber and non-timber forest products (NTFPs) in tropical forests

Factors	Compatibility influenced by
Biophysical	
Seasonality	– Production peaks for a given NTFP
Habitat overlap	– Extent of spatial segregation of timber and NTFPs due to edaphic/disturbance factors
Growth habit and product type	– Lianas, shrubs, epiphytes, palms; or fruits, foliage, resin, bark, vis-à-vis timber – Relative timber/NTFP values
Silvicultural practices	– Application thinning, liana removal, reduced impact logging norms, enrichment planting, site preparation – Whether the NTFP benefits from felling gaps
Length of timber rotation cycles	– Time to recover to preharvest levels
Preharvest timber inventories and marking of future crop trees	– NTFP growth habit (if it is an arborescent palm or a tree, rather than understory plants)
Social/institutional/legal	
Access to NTFP resources	– Extent of protection of NTFPs from logging and/or logging damage
Local knowledge	– Interaction between loggers and NTFP harvesters
Gender	– Who is involved in collecting NTFPs and local decision making during sales
Seasonality	– How it influences labour availability for harvesting timber and/or NTFPs
Property rights	– Modes of access (legal vs. customary, cooperative members vs. open access, determined by gender) – Extent to which some users are excluded – How management plans for timber respect property boundaries
Local governance	– Degree of organisation among producers – Extent of differences between established mechanisms to distribute revenues from timber and NTFPs
Training and education	– Degree to which NTFPs are incorporated into forestry curricula, and loggers and forest managers are aware of NTFP values
Legal frameworks	– Extent to which government-designed management plans for timber harmonise NTFP issues or vice versa – Enforcement of hunting bans or NTFP theft
Income diversification	– Extent to which timber and NTFP diversify income sources
Market chains	– Extent to which market chains for timber and NTFPs are complementary or divergent

8.2 Examining the Compatibility of Timber and NTFP Management

8.2.1 Resource Inventory

Based on a global assessment, Vantomme (2003) concluded that national statistics on NTFPs, including data on the resource base, are absent for all but a few internationally traded products (where data are usually limited to export quantities). It is therefore not surprising that little effort may have been directed at integrating inventories of NTFPs into timber censuses. When implemented, these inventories concentrate more on tallying the *presence* of locally important NTFPs than on estimating yields for guiding management. In the Congo Basin, NTFPs, including bushmeat and evidence of bushmeat hunting, are routinely recorded in timber inventories; however, in most cases, this information (e.g., Van Vliet and Nasi 2008) is rarely used in informing the design of multiuse management plans. Mapping the presence of locally important NTFP species before logging may, nevertheless, be necessary to ensure that they are maintained in forests managed primarily for timber. In Indonesian Borneo, for example, the palm *Eugeissona utilis*, one very important emergency forest food for the Punan hunter gatherers, grows along ridge tops and is often damaged when opening skid trails (Sheil et al. 2008). In this context, local knowledge is potentially critical in informing NTFP inventories alongside timber (Cunningham 2001; Lawrence et al. 2005; Shanley and Stockdale 2008).

Even in cases where timber and NTFPs have high commercial value, the cost-effectiveness of implementing integrated inventories of timber and NTFPs may depend on the extent of biological similarity between both types of product. Despite early efforts (e.g., Pineda 1996) in the community forestry concessions of Petén, Guatemala, in designing integrated inventory protocols for timber and NTFPs, including the fronds of high-valued *xate* (*Chamaedorea* spp.) understory palms, their implementation has been limited to date (Louman et al. 2008). Timber in the Petén is harvested from annual compartments of fixed area under decades-long rotations, while *xate* palms take only 4–6 months to regain preharvest yields. Because of its wide distribution across the entire forest, *xate* can therefore be harvested more frequently and over larger areas than within annual logging blocks; hence a different inventory protocol was designed (outlined in Godoy et al. 2009). Moreover, the size of plots used for timber inventory was insufficient for concurrent, reliable estimates of sustainable harvest rates of *xate* leaves that were needed to

Fig. 8.1 Logging damage to NTFP-bearing trees can be minimised by flagging them during preharvest timber inventorying. Photograph by Juan Carlos Licona from the Bolivian Amazon

fulfill FSC certification standards currently enjoyed by this NTFP. In contrast, arborescent palms or other NTFP-bearing trees are more amenable for integrated timber–NTFP inventories since little deviation is needed from common practice. For example, the management potential of both timber and NTFPs derived from palms and trees (fruit, seed oils, latex) in Amazonian floodplain forests was determined through standard, tree inventory assessment (Fortini et al. 2006). Another advantage of shared biological similarity is that, in the case of arborescent life forms, logging damage to NTFP-bearing trees can be easily minimised by marking them during routine, preharvest timber inventories (Guariguata et al. 2009, Fig. 8.1).

8.2.2 Ecology and Silviculture for Timber and NTFP Management

From a biophysical standpoint, the compatibility of management for timber and NTFP harvesting may be positively or negatively affected by the wide range of logging intensities applied across the forested tropics (Putz et al. 2001; Sist and Ferreira 2007), direct postlogging impacts such as increased tree mortality rates (Gourlet-Fleury et al. 2004; Schulze and Zweede 2006), overall changes in forest structure (Jonkers 1987; Johns et al. 1996), increased levels of solar radiation (Webb 1999; Pereira et al. 2002), the presence of disturbed or otherwise compacted soil (Hendrison 1990; Pinard et al. 2000), and the ecological attributes of the NTFP in question. For example, climbing palms (many of which are high-value NTFPs such as *Desmoncus* spp. and rattans such as *Calamus* spp.) usually benefit from logging-related canopy opening (Siebert 1993, 2000; Asseng Ze 2008). Similarly, understory NTFPs may survive better and elevate their reproductive activity after logging gaps are created (e.g., Costa and Magnusson 2003), although not all of these may benefit from high-light environments (e.g., Ocampo Sánchez 2004). Postlogging, tree fruit production due to improved crown illumination is also possible (Johns 1988; Guariguata and Sáenz 2002; but see Fonseca et al. 2009). However, any beneficial postlogging effects on NTFP growth and yield are expected to be both localised and possibly short lived, in the context of the long rotation cycles in selectively logged tropical forests (not less than 30–40 years) if no silvicultural treatments are further applied.

The few published works on the effects of selective logging on NTFP yields point to compatibility at the stand level, at least under experimental conditions. In lowland Nicaragua, Salick et al. (1995) reported that the density of locally useful woody plants was comparable in both logged and unlogged plots. Similarly, 9 years after RIL was applied in a Costa Rican montane forest, the harvestable biomass of nonvascular epiphytes (a locally important NTFP) equaled that of adjacent unlogged plots (Romero 1999). In the eastern Amazon, Menton et al. (2009) concluded that smallholder (~90 ha) forests that were selectively logged under RIL norms showed, after 18 months, no discernible difference in harvest yields of both game and tree fruits when compared to smallholder forests where no logging took place. The authors hypothesised that low NTFP harvest rates and minimal logging impacts both accounted for the observed compatibility of timber and NTFP management at the landscape scale (although they reported high interhousehold variation). Similar studies along these lines are necessary to better inform the design of compatible management interventions for timber and NTFPs. However, RIL may not always favour compatibility. For example, both number and size of logging gaps in forests logged under RIL norms in Amazonian forests may be insufficient for light-demanding timber trees, including those with concurrent NTFP value, to regenerate (Schulze 2008; Schulze et al. 2008a).

Some operational norms applied in forests managed for timber may facilitate NTFP management objectives. For example, lianas in tree crowns can reduce tree fruiting (Wright et al. 2005) including timber species (Fonseca et al. 2009). Liana cutting, applied primarily as a way to reduce logging damage to residual trees and to improve worker's safety (Putz et al. 2008), could be extended in managed forests to enhance fruit production in NTFP-bearing trees as suggested for Brazil nut (*Bertholletia excelsa*, Lecythidaceae; Kainer et al. 2006, 2007). Silvicultural treatments after selective tree harvest such as removing tree neighbors from future crop trees (De Graaf et al. 1999; Wadsworth and Zweede 2006) and stand refinement and soil scarification in logging gaps (e.g., Peña-Claros et al. 2008) may be adequate for either natural regeneration or enrichment planting of light-demanding NTFPs. Harvest systems typically applied in Asian dipterocarp forests such as shelterwood cutting (which remove or reduce the overstory) also appear ecologically and economically amenable for managing timber and light-demanding NTFPs (Ashton et al. 2001). However, existing silvicultural treatments may require adjustment. For example, the current Indonesian regulation on timber cutting (TPTI) requires companies to slash all undergrowth and climbers every year for 5 years in each compartment following timber extraction in order to control weeds and promote the regeneration of timber species. High-value or else locally important NTFPs are usually slashed as well (e.g., rattans, food and medicinal plants; Sheil et al. 2006). Meijaard et al. (2005) suggested revoking this policy as it is largely perceived as both technically and socially questionable. In contrast to the above examples, very little seems to be reported on how silviculture of NTFPs affects timber values. Trauernicht and Ticktin (2005) showed in southern Mexico how the planting the understory *xate* palm *Chamaedorea hooperiana* under natural forest cover led to a reduction of the density of saplings of timber species (probably due to slashing during site preparation).

8.2.3 Conflict of Use

Conflict of use arises when the same tree species provides both timber and NTFP values. And it exacerbates when different stakeholders are involved in the extraction of each (Laird 1999; Menton 2003; Shanley and Luz 2003, Fig. 8.2). Herrero-Jáuregui et al. (2009) observed that 47% of all timber species currently

Fig. 8.2 A medicinal plant collector extracting the bark of *Tabebuia* sp. at a sawmill in Eastern Brazil. Across this region, this species shows acute "conflict of use" between timber and non-timber values to different stakeholders (photo: Murilo Serra)

traded in the Amazonian state of Pará in Brazil also have documented non-timber uses. As expected, the greater the resources' values, the greater the conflict. Four species scored specially high in this respect: *Dipteryx odorata*, *Hymenaea courbaril* (both Fabaceae), *Tabebuia serratifolia*, and *T. impetiginosa* (both Bignoniaceae). Medicinal plant collectors greatly value the bark of *T. impetiginosa* and *H. courbaril* trees, whereas the oil from the seeds of *D. odorata* is widely extracted for cosmetic and medicinal purposes. In the particular case of *T. impetiginosa* and *H. courbaril*, conflict of use is acute because both species regenerate poorly due to their light-demanding attributes, low population densities, and low growth rates (Schulze 2008). In particular, the long-term population persistence and the capacity of local people to collect bark of *T. impetiginosa* for local medicinal and public health purposes (Gómez-Castellanos et al. 2009) may disappear over time if postlogging enrichment planting is not implemented (Schulze et al. 2008a, b).

Conflict of use is also widespread in Central Africa. In Cameroon, out of the 23 top timber species being exported, over half of these also have NTFP value (Ndoye and Tieguhong 2004). In both Cameroon and Central African Republic, the three most exploited timber species, *Triplochiton scleroxylon* (Sterculiaceae), *Entandrophragma cylindricum* (Meliaceae), and *Milicia excelsa* (Moraceae), are also sources of medicine and food (Tieguhong and Ndoye 2007). In Cameroon and the Democratic Republic of Congo, the Forestry Laws have clarified logging companies' obligations toward local people with provisions to avoid timber exploitation obstructing villagers in exercising their user rights. To meet this objective, local communities and timber companies work together to reach agreements on maintaining tree species with conflict of use. Others have suggested that government agencies assign harvest quotas for those timber species with high NTFP value and compensate timber companies for any forgone revenue (Tieguhong and Ndoye 2007).

An alternative intervention for minimising conflict of use includes legal protection from logging when both the NTFP economic and social value equals or exceeds the timber value. Such protection is currently extended for the Brazil nut tree in Brazil, Peru, and Bolivia, due to its pivotal role in sustaining rural livelihoods (Ortiz 2002; Peres et al. 2003). However, the extent of conflict of use is often culturally and geographically specific, thus complicating any necessary steps toward legal protection at broad spatial scales. For example, in the Pokola-Kabo-Loundoungou forest concession in Congo, five species extracted for timber were noted as having no NTFP value; yet they were commonly used as NTFPs in southwestern and eastern Cameroon. Conversely, one of the most commercially valuable timber species (*E. cylindricum*) is used as a medicine in central and eastern Cameroon, but not in the south west (N'Zala 2002).

Another intervention is the spatial separation of management units for either timber or NTFPs (e.g., da Silva et al. 2002). The feasibility of this option will depend, among other factors, on the nature of the NTFP in question and its habitat requirements. For example, the locally valuable, multipurpose tree *Carapa guianensis* (Meliaceae) shows higher adult densities in seasonally flooded than *terra firme* forests in the southwestern Brazilian Amazon (Klimas et al. 2007). Management objectives for either timber or the high-quality oil extracted from its seeds can be spatially segregated, for example, if seed harvest intensities are anticipated to be high. In this case, allocating flooded forest areas only for seed collection may be a sensible alternative. Yet, areas destined for tree seed collection need to be extensive enough to compensate for interannual and/or intertree variability in seed production, a typical trait of many tropical forest trees (Wright et al. 2005). A related issue to consider in multipurpose tree species is the nature of the relationship between individual size and NTFP yields. For example, if fruit production peaks at intermediate (instead of large) diameter classes (e.g., Soehartono and Newton 2001; Kainer et al. 2007), the largest (i.e. less fecund) individuals are amenable to harvesting or otherwise setting aside during multiuse planning.

8.2.4 Wildlife Conservation and Use

Most vertebrate species can persist in selectively logged forests as long as most indirect effects such as hunting, forest fragmentation, and forest fires are controlled (Johns 1997; Meijaard et al. 2005; Azevedo-Ramos et al. 2006). Yet, the indirect effects are widespread and pervasive (Laurance and Peres 2006), particularly hunting for bushmeat. For example, per capita wildlife harvest rates in settlements adjacent to logging concessions are much higher than those away from concessions

(Robinson et al. 1999; Auzel and Wilkie 2000; Thibault and Blaney 2003). Furthermore, bushmeat is usually hunted by outsiders at the expense of those who own prior legitimate claims to forest wildlife (Poulsen et al. 2009). Regulating or banning hunting in timber concessions is now a widely agreed measure by national governments, researchers, concessionaires, nongovernmental organisations, and the international community (Bennett and Robinson 2001; Meijaard et al. 2005; Nasi et al. 2008).

The compatibility of timber harvesting with the survival of wildlife may be contingent on other interrelated measures. One is to put pressure on timber concessionaires to control the activities of their own employees, such as banning them from purchasing bushmeat from forest villagers and/or providing workers and families with alternative protein sources. Some companies are implementing such regulations in Congo, Gabon, and Cameroon, in partnership with nongovernmental organisations and usually with external funding support (Aviram et al. 2003). Another measure is allowing rural communities to sell bushmeat for local consumption in, or nearby, urban centers. An example is the *Congolaise Industrielle des Bois* timber concession in North Congo, where communal hunting areas were created for abundant and ecologically resilient species (i.e. with high intrinsic rates of population increase such as ungulates and rodents; Bennett and Robinson 2000) while prohibiting the hunting of legally protected species (Elkan et al. 2006). These kinds of initiatives are likely to promote compatible timber and wildlife uses although further interventions may be needed. Some have suggested, for example, to locate sawmills in existing cities to avoid urbanisation in, or adjacent to, logging concessions (Poulsen et al. 2009). Besides terrestrial vertebrates, the effects of selective logging on aquatic wildlife for human consumption appear less studied. In the hilly landscapes of Borneo, many locally important fish species are known to be sensitive to disturbance due to enhanced stream sediment levels after logging roads are built (Meijaard et al. 2005). Locating logging roads away from streams and minimising their width may help to reduce sediment loads into streams and rivers.

8.2.5 Tenure and Access Rights

In addition to the topics discussed above, moving toward compatibility of timber and NTFP management requires understanding of who has rights and responsibilities for management decisions for both types of product. Rather than individual and comprehensive ownership rights, forest property in many contexts is an overlapping "bundle of rights" including those to access and harvest the resource, to manage it and exclude others, and to sell or transfer resource rights to others (Schlager and Ostrom 1992). Typically, local stakeholders hold only a partial set of rights while others have rights over the same resource or property (Meinzen-Dick and Mwangi 2008). The type of right held and the presence of multiple rights holders will influence the compatibility of integrated management approaches for timber and NTFPs, and the prospects for enforcing norms and agreements. For example, the community forest concessions in Petén, Guatemala, were superimposed by the government on preexisting rights regarding to NTFPs such as the fronds of *xate* understory palms. Rights over *xate* are largely held by stakeholders outside community concession organisations, and conflicts between some community concessionaires and outside harvesters are not uncommon (Nittler and Tschinkel 2005). In northern Bolivia, conflict and confusion over tenure rights have resulted from industrial timber concessions being superimposed on customary properties such as forest estates called *barracas* and agroextractive communities who are dependent on Brazil nut extraction (de Jong et al. 2006). In both cases, even if selective timber harvesting and NTFP extraction are biophysically compatible, the potential for excluding legitimate rights holders from forest benefits may undermine the prospects of an integrated management regime from a social standpoint.

8.2.6 Product Certification

The proliferation of different certification standards, the presence of different groups of harvesters for either timber or NTFPs and the inherent diversity of NTFPs (Shanley et al. 2002), currently hampers the development of cost-efficient, harmonised labeling procedures for timber and NTFPs in a given forest. Lack of consumer awareness about the environmental and social aspects of NTFP extraction, compared to timber, may also become a barrier (Shanley et al. 2008). Furthermore, NTFP certification is usually product specific (food, personal care, or medicine), and its standards focus on issues of "product quality", "organic production", "good agricultural practices", and/or "source of origin" (Pierce and Laird 2003). In contrast, tropical timber certification through Forest Stewardship Council (FSC 1996) standards is holistic and granted at the level of the forest stand without guarantees that each timber species is extracted sustainably (Schulze et al. 2008b). The many differences in information needs, economic value, and management procedures for timber and NTFP certification suggest that much work lies ahead in moving toward compatibility from a certification standpoint.

Another constraint to compatible certification approaches is that knowledge on population density, regeneration rates, and optimum management practices for most NTFPs is scant (see Chap. 7). This knowledge is needed for delineating management standards, including sustainable harvest regimes. Although guidelines for timber management in tropical forests date back to more than a century of research and development (Dawkins and Philip 1998), formal management principles for NTFPs have a more recent history (Peters 1996; Wong et al. 2001; Stockdale 2005; Medicinal Plant Specialist Group 2007), while informal principles need further integration into certification procedures and forest management (Shanley and Stockdale 2008). Overall, few such principles have yet been validated or widely adopted.

A key variable influencing the harmonisation of certification procedures for NTFPs and timber is whether or not a given NTFP involves human consumption. For example, organic certification of Brazil nuts gathered from the Bolivian Amazon needs to follow strict international standards for collection, handling, and storage (e.g., European Union Regulation 2092/91; SIPPO-Swiss Import Promotion Programe 2005). Some Brazil nut cooperatives in Bolivia now prohibit members from harvesting timber in organically certified Brazil nut stands. The fact that organic labeling drives Brazil nut certification in Bolivia may explain the fact that, for example, FSC-certified timber concessions have not yet attempted to certify Brazil nuts under the current FSC standards developed for this NTFP (FSC norms for Brazil nut are also perceived as too complicated to implement; Pacheco and Cronkleton 2008). In contrast, FSC standards for management of *xate* fronds in Guatemala were recently appended to those of (FSC) timber in three different concessions covering about 190,000 ha of forest (Smartwood 2007). This is expected to facilitate and harmonise the auditing process while reducing the costs of applying multiple certification schemes. The fronds of *xate* are used in floral arrangements and are not consumed by humans thus facilitating compatibility of certification of timber and *xate* under FSC principles.

8.3 Conclusions

Given the growing demands on tropical forests regarding the many goods and services they provide, effective guidelines for multiple-use management systems are essential. Here we used timber as the primary output upon which trade-offs and management challenges could be identified when adding NTFPs as a secondary output (Panayotou and Ashton 1992); hence, our discussion above and below needs to be interpreted in this context. One obvious outcome from our analysis is that compatible management of timber and NTFPs is inherently multifactorial and context dependent (Table 8.1). In some situations, compatibility is possible and in others, it may prove difficult to achieve. This conclusion remains, however, speculative due to the paucity of published studies on integrated management approaches for timber and NTFPs across the tropics from which generalisations can be made. We provide some suggestions for moving ahead.

García-Fernández et al. (2008) hypothesised that governance conditions relating to land-devolution policies, effective collective institutions, and the design of multistakeholder management models are important enabling factors for multiple forest use to succeed in the tropics. An example offering partial support comes from Guatemala. When the community forestry concessions were created in the lowlands of the Petén during the early 1990s, multiple objectives (production of timber and NTFPs) were explicitly defined from the outset. Since then, concerted efforts by the national authority responsible for overseeing the implementation of sustainable management, forest managers, researchers, and organisations providing technical assistance have contributed toward compatibility of timber and NTFP management, in particular, by enforcing good management practices for the harvest of timber and commercially valuable *xate* fronds. The lessons learned from the Petén (see Carrera et al. 2004; Nittler and Tschinkel 2005) may be useful in other tropical locations where concurrent management of timber and NTFPs is sought. Others have argued (e.g., Sands 2005, p. 159) that a major predictor of success in implementing multiple forest use is ownership, or else direct oversight, by governments (whereas private companies would favour specialisation over a single commodity). Testing pilot management systems for timber and NTFPs in forests where governments exert a direct role may prove fruitful. The Central Africa regional norms (FAO-GTZ-COMIFAC 2008) developed for managing NTFPs indicate how countries can incorporate NTFPs in policy, legal, fiscal, and institutional frameworks and provide a working model vis-à-vis timber production.

We suggest three ways to promote compatibility of timber and NTFP extraction and management. One is to improve "passive" or "opportunistic" compatibility situations. For example, by enforcing the mitigation of logging impacts on the NTFP resource base (Tieguhong and Ndoye 2007; Guariguata et al. 2009). Another alternative is to explicitly enhance both timber and NTFP values. Recent calls for researching and implementing silvicultural intensification in forests managed under RIL guidelines to ensure long-term timber production in tropical forests (Fredericksen and Putz 2003; Peña-Claros et al. 2008; Schulze et al. 2008d) may open avenues for concurrent management of locally important NTFPs. Third is assessing biophysical, social, regulatory, and institutional aspects, so that trade-offs are minimised among stakeholders regarding timber and NTFPs (Purnomo et al. 2005; Lawrence 2007; Lynam et al. 2007). Ros-Tonen et al. (2008) provide examples from the Brazilian Amazon where partnerships between local forest users, the private sector, nongovernmental organisations, and the civil society could facilitate the insertion of NTFPs into timber-oriented models (see Chap. 9). The practicalities and effectiveness of the above proposals will also depend on the scale of management, timber harvesting intensities, and mode of extraction (e.g., RIL vs. conventional logging), and the NTFP harvesting intensities among others, for example, from extensive industrial timber concessions where NTFPs are allowed to be harvested by local people (e.g., Guariguata et al. 2009) to small, multiuse forests that are managed communally or by individual families (e.g., Rockwell et al. 2007a, b; Menton et al. 2009).

Finally, tropical forestry training and education institutions may have to be recrafted. Otherwise, we run the risk of perpetuating a timber bias when NTFP management plans are drafted. Although there are documented initiatives in Brazil at training tropical foresters in bridging the gap between timber and NTFP use, ecology, and management (Pinto et al. 2008; Shanley and Medina 2005), there is apparently little happening in other tropical locations. In closing, given the millions of hectares of natural forest allocated for timber production in the Amazon (Schulze et al. 2008c) and Congo basins (Nasi et al. 2006), and the equally vast area of natural tropical forest under control of rural communities (Sunderlin et al. 2008), there are plenty of opportunities, for designing and validating integrated management approaches that include timber and NTFPs.

References

Ashton MS, Mendelsohn R, Singhakurama BMP, Gunatilleke CVS, Gunatilleke IAUN, Evans A (2001) A financial analysis of rain forest silviculture in southwestern Sri Lanka. For Ecol Manage 154:431–441

Asseng Ze A (2008) Gestion durable des produits forestiers non ligneux dans la concession forestière de Pallisco. Etude Pilote sur les Techniques d'Exploitation Forestiere no. 24. Organisation des Nations Unies pour l'alimentation et l'agriculture, FAO, Rome

Auzel P, Wilkie DS (2000) Wildlife use in Northern Congo: hunting in a commercial logging concession. In: Robinson JG, Bennett EL (eds) Hunting for sustainability in tropical forests. Columbia University Press, New York, pp 413–426

Aviram R, Bass M, Parker K (2003) Extracting hope for bushmeat: case studies of oil, gas, mining and logging industry efforts for improved wildlife management. Bushmeat Crisis Task Force. http://www.bushmeat.org/docs.html. Accessed 23 Feb 2009

Azevedo-Ramos C, de Carvalho O, do Amaral BD (2006) Short-term effects of reduced-impact logging on eastern Amazon fauna. For Ecol Manage 232:26–35

Barreto P, Amaral P, Vidal E, Uhl C (1998) Costs and benefits of forest management for timber production in eastern Amazonia. For Ecol Manage 108:9–26

Belcher B, Michon G, Angelsen A, Ruiz-Pérez M, Asbjornsen H (2005) The socioeconomic conditions determining the development, persistence, and decline of forest garden systems. Econ Bot 59:245–253

Bennett EL, Robinson JG (2000) Hunting for sustainability: the start of a synthesis. In: Robinson JG, Bennett EL (eds) Hunting for sustainability in tropical forests. Columbia University Press, New York, pp 499–519

Bennett EL, Robinson JG (2001) Hunting of wildlife in tropical forests. Environment Department Paper no. 76. The World Bank, Washington, DC

Binkley CS (1997) Preserving nature through intensive plantation forestry: the case of forestland allocation with illustrations from British Columbia. For Chron 73:553–559

Carrera F, Stoian D, Campos JJ, Morales J, Pinelo G (2004) Forest certification in Guatemala, in: Proceedings of the symposium on forest certification in developing and transitional societies. Yale School of Forestry and Environmental Studies, New Haven, USA, pp 363–405

Costa FRC, Magnusson WE (2003) Effects of selective logging on the diversity and abundance of flowering and fruiting understory plants in a Central Amazonian forest. Biotropica 35:103–114

Cunningham AB (2001) Applied ethnobotany. Earthscan, London

da Silva DA, Campos JJ, Villalobos R, Louman B, Gonçalvez L (2002) Manejo forestal diversificado en una comunidad ribereña de la Amazonía brasileña: consideraciones sociales y silviculturales. Revista Forestal Centroamericana 38:78–84

Dawkins HC, Philip MS (1998) Tropical moist forest silviculture and management: a history of success and failure. CAB International, Wallingford, UK

De Graaf NR, Poels RLH, Van Rompaey RSAR (1999) Effect of silvicultural treatment on growth and mortality of rainforest in Surinam over long periods. For Ecol Manage 124:123–135

de Jong W, Ruiz S, Becker M (2006) Conflicts and communal forest management in Northern Bolivia. For Pol Econ 8:447–457

Elkan PW, Elkan SW, Moukassa A, Malonga R, Ngangoue M, Smith JLD (2006) Managing threats from bushmeat hunting in a timber concession in the Republic of Congo. In: Laurance WF, Peres CA (eds) Emerging threats to tropical forests. University of Chicago Press, Chicago, pp 393–415

FAO-GTZ-COMIFAC (2008) Directives sous-regionales relatives a la gestion durable des produits forestieres non-ligneux d'origine vegetale en Afrique Centrale. Document GCP/RAF/398/GER. Yaounde, Cameroun

Fonseca MG, Vidal E, dos Santos FAM (2009) Intraspecific variation in the fruiting of an Amazonian timber tree: implications for management. Biotropica 41:179–185

Fortini LB, Rabelo FG, Zarin DJ (2006) Mixed potential for sustainable forest use in the tidal floodplain of the Amazon River. For Ecol Manage 231:78–85
Fredericksen TS, Putz FE (2003) Silvicultural intensification for tropical forest conservation. Biodivers Conserv 12:1445–1453
FSC (1996) FSC Principles and criteria for forest stewardship. Document FSC-STD-01-001 (version 4-0). Bonn, Germany
García-Fernández C, Ruiz Pérez M, Wunder S (2008) Is multiple-use forest management widely implementable in the tropics? For Ecol Manage 256:1468–1476
Godoy CE, Madrid J, Manzanero M (2009) Guía y formato para la elaboración de planes de manejo integrados de recursos forestales no maderables (RFNM). Rainforest Alliance, Petén, Guatemala
Gómez-Castellanos JR, Prieto JM, Heinrich M (2009) Red lapacho (*Tabebuia impetiginosa*) – a global ethnopharmacological commodity? J Ethnopharmacol 121:1–13
Gourlet-Fleury S, Favrichon V, Schmitt L, Petronelli P (2004) Consequences of silvicultural treatments on stand dynamics at Paracou. In: Gourlet-Fleury S, Guehl J-M, Laroussinie O (eds) Ecology and management of a neotropical rainforest – lessons drawn from Paracou, a long-term experimental research site in French Guiana. Elsevier, Paris, pp 254–280
Guariguata MR, Sáenz G (2002) Post-logging acorn production and oak regeneration in a tropical montane forest, Costa Rica. For Ecol Manage 167:285–293
Guariguata MR, Licona JC, Mostacedo B, Cronkleton P (2009) Damage to Brazil nut trees (*Bertholletia excelsa*) during selective timber harvesting in Northern Bolivia. For Ecol Manage 258:788–793
Hendrison J (1990) Damage controlled logging in managed rain forest in Suriname. Agricultural University, Wageningen, The Netherlands
Herrero-Jáuregui C, García-Fernández C, Sist P, Casado M (2009) Conflict of use for multipurpose tree species in the state of Pará, eastern Amazonia, Brazil. Biodivers Conserv 18:1019–1044
Hiremath AJ (2004) The ecological consequences of managing forests for non-timber products. Conserv Soc 2:211–216
ITTO (2006) Status of tropical forest management 2005. ITTO Technical Series no. 24. Yokohama, Japan
Johns AD (1988) Effects of selective timber extraction on rain forest structure and composition and some consequences for frugivores and folivores. Biotropica 20:31–37
Johns AD (1997) Timber production and biodiversity conservation in tropical rain forests. Cambridge University Press, UK
Johns JS, Barreto P, Uhl C (1996) Logging damage during planned and unplanned logging operations in the eastern Amazon. For Ecol Manage 89:59–77
Jonkers WBJ (1987) Vegetation structure, logging damage and silviculture in a tropical rain forest of Suriname. Agricultural University, Wageningen, The Netherlands
Kainer KA, Wadt LHO, Gomes-Silva DAP, Capanu M (2006) Liana loads and their association with Bertholletia excelsa fruit and nut production, diameter growth and crown attributes. J Trop Ecol 22:147–154
Kainer KA, Wadt LHO, Staudhammer CL (2007) Explaining variation in Brazil nut fruit production. For Ecol Manage 250:244–255
Kant S (2004) Economics of sustainable forest management. For Pol Econ 6:197–203
Klimas CA, Kainer KA, Wadt LHO (2007) Population structure of *Carapa guianensis* in two forest types in the southwestern Brazilian Amazon. For Ecol Manage 250:256–265
Laird S (1999) The management of forests for timber and non-wood forest products in central Africa. In: Sunderland T, Clark L, Vantomme P (eds) Non-wood forest products of Central Africa: current research issues and prospects for development and conservation. FAO, Rome
Laurance WF, Peres CA (eds) (2006) Emerging threats to tropical forests. University of Chicago Press, Chicago
Lawrence A (2003) No forest without timber? Int For Rev 5:87–96

Lawrence A (2007) Beyond the second generation: towards adaptiveness in participatory forest management. CAB Reviews: Perspectives in Agriculture, Veterinary Sciences, Nutrition and Natural Resources. doi: 10.1079/PAVSNNR20072028

Lawrence A, Phillips OL, Reategui Ismodes A, Lopez M, Rose S, Wood D, Farfan AJ (2005) Local values for harvested forest plants in Madre de Dios, Peru: towards a more contextualised interpretation of quantitative ethnobotanical data. Biodivers Conserv 14:45–79

Louman B, García-Fernández C, Sabogal C, Ehringhaus C, Salazar M, Villacrés D (2008) Capacidades técnicas y desafíos del manejo forestal comunitario. In: Sabogal C, de Jong W, Pokorny B, Louman B (eds) Manejo forestal comunitario en America Latina: experiencias, lecciones aprendidas y retos para el futuro. Centro para la Investigación Forestal Internacional (CIFOR), Bogor, Indonesia, pp 75–112

Lynam T, de Jong W, Sheil D, Kusumanto T, Evans K (2007) A review of tools for incorporating community knowledge, preferences and values into decision making in natural resources management. Ecol Soc 12(1):5. http://www.ecologyandsociety.org/vol12/iss1/art5/

Medicinal Plant Specialist Group (2007) International Standard for Sustainable Wild Collection of Medicinal and Aromatic Plants (ISSC-MAP). Version 1.0. Bundesamt für Naturschutz (BfN), MPSG/SSC/IUCN, WWF Germany, and TRAFFIC (BfN-Skripten 195)

Meijaard E, Sheil D, Nasi R, Augeri D, Rosenbaum B, Iskandar D, Setyawati T, Lammertink M, Rachmatika I, Wong A, Soehartono T, Stanley S, O'Brien T (2005) Life after logging: reconciling wildlife conservation and production forestry in Indonesian Borneo. Center for International Forestry Research, Bogor, Indonesia

Meinzen-Dick R, Mwangi E (2008) Cutting the web of interests: pitfalls of formalising property rights. Land Use Pol 26:36–43

Menton M (2003) Effects of logging on non-timber forest product extraction in the Brazilian Amazon: community perceptions of change. Int For Rev 5:97–105

Menton M, Merry FD, Lawrence A, Brown N (2009) Company-community logging contracts in Amazonian settlements: impacts of livelihoods and NTFP harvests. Ecol Soc 14(1):39. http://www.ecologyandsociety.org/vol14/ iss1/art39/

Michon G, de Foresta H, Levang P, Verdeaux F (2007) Domestic forests: a new paradigm for integrating local communities' forestry into tropical forest science. Ecol Soc 12(2):1. http://www.ecologyandsociety.org/vol12/iss2/art1

N'Zala D (2002) Conservation et gestion durable des écosystèmes des forêts tropicales humides de l'Afrique centrale. Etude de cas d'aménagement forestier exemplaire en Afrique centrale: La concession de Pokola-Kabo-Loundoungou, République du Congo. Document de travail FM/19F. Service de la mise en valeur des ressources forestières, Département des Forêts. Rome, FAO

Nasi R, Cassagne B, Billand A (2006) Forest management in Central Africa: where are we? Int For Rev 8:14–20

Nasi R, Brown D, Wilkie D, Bennett E, Tutin C, van Tol G, Christophersen T (2008) Conservation and use of wildlife based resources: the bushmeat crisis. CBD Technical Series no. 33. Secretariat of the Convention on Biological Diversity, Montreal, Canada

Ndoye O, Tieguhong JC (2004) Forest resources and rural livelihoods: the conflict between timber and non-timber forest products in the Congo Basin. Scand J For Res 19:36–44

Nittler J, Tschinkel H (2005) Community forest management in the Maya Biosphere Reserve of Guatemala. Protection through profits. Submitted to the United States Agency for International Development (USAID) and the Sustainable Agriculture and Natural Resource Management (SANREM) Collaborative Research Support Program (CRSP), University of Georgia

Ocampo Sánchez RA (2004) Ipecacuana, *Psychotria ipecacuanha* (Brotero) Stokes: un producto no maderable cultivado bajo el bosque en Huetar Norte, Costa Rica. In: Alexiades M, Shanley P (eds) Productos Forestales, Medios de Subsistencia y Conservación, vol 3. CIFOR, Bogor, Indonesia, pp 257–273

Ortiz E (2002) Brazil Nut (*Bertholletia excelsa*). In: Shanley P, Pierce AR, Laird SA, Guillén A (eds) Tapping the green market: certification and management of non-timber forest products. Earthscan, London, pp 61–74

Pacheco P, Cronkleton P (2008) Developing standards for Brazil nuts in Bolivia. In: Shanley P, Pierce A, Laird S, Robinson D (eds) Beyond timber: certification and management of non-timber forest products. Center for International Forestry Research, Bogor, Indonesia, pp 41–42

Panayotou T, Ashton PS (1992) Not by timber alone: economics and ecology for sustaining tropical forests. Island, Washington DC

Pearce D, Putz FE, Vanclay J (2003) Sustainable forestry: panacea or folly? For Ecol Manage 172:229–247

Peña-Claros M, Fredericksen TS, Alarcón A, Blate GM, Choque U, Leaño C, Licona JC, Mostacedo B, Pariona W, Villegas Z, Putz FE (2008) Beyond reduced-impact logging: silvicultural treatments to increase growth rates of tropical trees. For Ecol Manage 256:1458–1467

Pereira R Jr, Zweede J, Asner GP, Keller M (2002) Forest canopy damage and recovery in reduced-impact and conventional selective logging in eastern Pará, Brazil. For Ecol Manage 168:77–89

Peres CA, Baider C, Zuidema PA, Wadt LHO, Kainer KA, Gomes-Silva DAP, Salomão RP, Simões LL, Franciosi ERN, Cornejo F, Gribel R, Shepard GH, Kanashiro M, Coventry P, Yu DW, Watkinson AR, Freckleton RP (2003) Demographic threats to the sustainability of Brazil nut exploitation. Science 302:2112–2114

Peters CM (1996) The ecology and management of non-timber forest resources. World Bank Technical Paper no. 322. The World Bank, Washington, DC

Pierce A, Laird S (2003) In search of comprehensive standards for non-timber forest products in the botanicals trade. Int For Rev 5:138–147

Pinard MA, Barker MG, Tay J (2000) Soil disturbance and post-logging forest recovery on bulldozer paths in Sabah, Malaysia. For Ecol Manage 130:213–225

Pineda PA (1996) Diseño y aplicación de un inventario forestal diversificado (maderables y no maderables) en Petén. M.Sc. Thesis. Centro Agronómico Tropical de Investigación y Enseñanza (CATIE), Costa Rica

Pinto LFG, Shanley P, Gomes APC, Robinson D (2008) Experience with NTFP certification in Brazil. For Trees Livelihood 18:37–54

Poore D (2003) Changing landscapes: the development of the international tropical timber organisation and its influence on tropical forest management. Earthscan, London

Poore D, Burgess J, Palmer J, Rietbergen S, Synott T (1989) No timber without trees: sustainability in the tropical forests. Earthscan, London

Poulsen JR, Clark CJ, Mavah G, Elkan PW (2009) Bushmeat supply and consumption in a tropical logging concession in Northern Congo. Conserv Biol. doi:10.1111/j.1523-1739.2009.01251.x

Purnomo H, Mendoza GA, Prabhu R (2005) Analysis of local perspectives on sustainable forest management: an Indonesian case study. J Environ Manage 74:111–126

Putz FE, Blate G, Redford KH, Fimbel R, Robinson J (2001) Tropical forest management and conservation of biodiversity: an overview. Conserv Biol 15:7–20

Putz FE, Sist P, Fredericksen T, Dykstra D (2008) Reduced-impact logging: challenges and opportunities. For Ecol Manage 256:1427–1433

Robinson JG, Redford KH, Bennett EL (1999) Wildlife harvest in logged tropical forest. Science 284:595–596

Rockwell CA, Kainer KA, Staudhammer CL, Baraloto C (2007a) Future crop tree damage in a certified community forest in southwestern Amazonia. For Ecol Manage 242:108–118

Rockwell CA, Kainer KA, Marcondes N, Baraloto C (2007b) Ecological limitations of reduced-impact logging at the smallholder scale. For Ecol Manage 238:365–374

Romero C (1999) Reduced impact logging effects on commercial non-vascular pendant epiphyte biomass in Costa Rica. For Ecol Manage 118:117–125

Ros-Tonen MAF, van Andel T, Morsello C, Otsuki K, Rosendo S, Scholz I (2008) Forest-related partnerships in Brazilian Amazonia: there is more to sustainable forest management than reduced impact logging. For Ecol Manage 256:1482–1497

Salick J, Mejia A, Anderson T (1995) Non-timber forest products integrated with natural forest management, Rio San Juan, Nicaragua. Ecol Appl 5:878–895

Sands R (2005) Forestry in a global context. CABI, Wallingford, UK

Sayer JA, Byron N (1996) Technological advance and conservation of resources. Int J Sustain Dev World Ecol 3:43–53

Schlager E, Ostrom E (1992) Property rights regimes and natural resources: a conceptual analysis. Land Econ 68:249–262

Schulze M (2008) Technical and financial analysis of enrichment planting in logging gaps as a potential component of forest management in the eastern Amazon. For Ecol Manage 255:866–879

Schulze M, Zweede J (2006) Canopy dynamics in unlogged and logged forest stands in the eastern Amazon. For Ecol Manage 236:56–64

Schulze M, Grogan J, Uhl C, Lentini M, Vidal E (2008a) Evaluating ipê (*Tabebuia*, Bignoniaceae) logging in Amazonia: sustainable management or catalyst for forest degradation? Biol Conserv 141:2071–2085

Schulze M, Grogan J, Vidal E (2008b) Forest certification in Amazonia: standards matter. Oryx 42:229–239

Schulze M, Grogan J, Vidal E (2008c) Technical challenges to sustainable forest management in concessions on public lands in the Brazilian Amazon. J Sustain For 26:61–76

Schulze M, Grogran J, Landis MR, Vidal E (2008d) How rare is too rare to harvest? Management challenges posed by timber species occurring at low densities in the Brazilian Amazon. For Ecol Manage 256:1443–1457

Shanley P, Luz L (2003) The impacts of forest degradation on medicinal plant use and implications for health care in eastern Amazonia. Bioscience 53:573–584

Shanley P, Medina G (eds) (2005) Frutíferas e plantas úteis na vida amazônica. Center for International Forestry Research, Bogor, Indonesia

Shanley P, Stockdale M (2008) Traditional knowledge, forest management, and certification: a reality check. For Trees Livelihood 18:55–67

Shanley P, Pierce AR, Laird SA, Guillen A (2002) Tapping the green market: certification and management of non-timber forest products. Earthscan, London

Shanley P, Pierce A, Laird S, Robinson D (2008) Beyond timber: certification and management of non-timber forest products. Center for International Forestry Research, Bogor, Indonesia

Sheil D, Puri R, Wan M, Basuki I, van Heist M, Liswanti N, Rumiyati RI, Samsoedin I (2006) Recognising local people's priorities for tropical forest biodiversity. Ambio 35:17–24

Sheil D, van Heist M, Liswanti N, Basuki I, Wan M (2008) Biodiversity, landscapes and livelihoods: a local perspective. In: Moeliono M, Wollenberg E, Limberg G (eds) The decentralisation of forest governance: politics, economics and the fight for control of forest in Indonesian Borneo. Earthscan, London, pp 61–90

Siebert SF (1993) The abundance and site preferences of rattan (*Calamus exilis* and *Calamus zollingeri*) in two Indonesian national parks. For Ecol Manage 59:105–113

Siebert SF (2000) Abundance and growth of *Desmoncus orthacantos* Mart. (Palmae) in response to light and ramet harvesting in five forest sites in Belize. For Ecol Manage 137:83–90

SIPPO-Swiss Import Promotion Programe (2005) Guidance manual for organic collection of wild plants. SIPPO, Zurich, Switzerland

Sist P, Ferreira FN (2007) Sustainability of reduced-impact logging in the eastern Amazon. For Ecol Manage 243:199–209

Sist P, García-Fernández C, Fredericksen T (2008) Moving beyond reduced-impact logging towards a more holistic management of tropical forests. For Ecol Manage 256:7–9

Smartwood (2007) Adenda a los estándares interinos de Rainforest Alliance/Smartwood para la certificación del manejo de xate (*Chamaedorea* spp.) como NTFP en Guatemala. Documento FM-32-GUATEMALA. http://www.rainforest-alliance.org/forestry/documents/smartwood-guatemalaFMinterimstandardnov08spa.pdf. Accessed 12 Feb 2009

Soehartono T, Newton AC (2001) Reproductive ecology of *Aquilaria* spp. in Indonesia. For Ecol Manage 152:59–71

Stockdale M (2005) Steps to sustainable and community-based NTFP management. Non-Timber Forest Products-Exchange Programme for South and Southeast Asia, The Philippines

Sunderlin WD, Hatcher J, Liddle M (2008) From exclusion to ownership? Challenges and opportunities in advancing forest tenure reform. Rights and Resources Initiative, Washington, DC

Thibault M, Blaney S (2003) The oil industry as an underlying factor in the bushmeat crisis in Central Africa. Conserv Biol 17:1807–1813

Tieguhong JC, Ndoye O (2007) The impact of timber harvesting on the availability of non-wood forest products in the Congo Basin. Forest Harvesting Case Study 23. Food and Agricultural Organisation of the United Nations, Rome

Toledo VM, Ortiz-Espejel B, Corté L, Moguel P, Ordoñez MJ (2003) The multiple use of tropical forests by indigenous peoples in Mexico: a case of adaptive management. Ecol Soc 7(3):9. http://www.consecol.org/vol7/iss3/art9/

Trauernicht C, Ticktin T (2005) The effects of non-timber forest product cultivation on the plant community structure and composition of a humid tropical forest in southern Mexico. For Ecol Manage 219:269–278

Van Vliet N, Nasi R (2008) Mammal distribution in a Central African logging concession area. Biodivers Conserv 17:1241–1249

Vantomme P (2003) Compiling statistics on non-wood forest products as policy and decision-making tools at the national level. Int For Rev 5:156–160

Vincent R, Binkley CS (1993) Efficient multiple-use forestry may require land-use specialisation. Land Econ 69:370–376

Wadsworth FH, Zweede JC (2006) Liberation: acceptable production of tropical forest timber. For Ecol Manage 233:45–51

Wang S, Wilson B (2007) Pluralism in the economics of sustainable forest management. For Pol Econ 9:743–750

Webb EL (1999) Growth ecology of *Carapa nicaraguensis* Aublet. (Meliaceae): implications for natural forest management. Biotropica 31:102–110

Wong J, Thornber K, Baker N (2001) Resource assessment of non-wood forest products: experience and biometric principles. NWFP Series 13. Food and Agriculture Organisation of the United Nations, Rome

Wright SJ, Jaramillo MA, Pavon J, Condit R, Hubbell SP, Foster RB (2005) Reproductive size thresholds in tropical trees: variation among individuals, species, and forests. J Trop Ecol 21:307–315

Zhang Y (2005) Multiple-use forestry vs. forestland-use specialisation revisited. For Pol Econ 7:143–156

Chapter 9
Pro-poor Governance of Non-timber Forest Products: The Need for Secure Tenure, the Rule of Law, Market Access and Partnerships

Mirjam A.F. Ros-Tonen and Koen Kusters

Abstract It is now well appreciated that the sustainability and livelihood contributions of NTFPs depend upon the manner in which they are managed and governed. Yet, governance systems are complex, context specific and constantly changing, especially as markets develop. Therefore, the attributes of good governance need to be identified, along with the enabling conditions that allow them to evolve and persist. In this chapter, we review some key conditions for good and pro-poor NTFP governance, with a focus on secure rights, equitable rule of law, market access and the building of partnerships. Through examination of case studies we conclude that many of the NTFP governance challenges can be met through building partnerships.

9.1 Introduction

This chapter deals with aspects of non-timber forest products (NTFPs) governance. We define NTFPs as "all plant and animal products from forested landscapes, including human-modified ones" (Ros-Tonen and Wiersum 2005, p 147). This definition excludes ecotourism and environmental services, although, as we argue below where we consider the potential of Payments for Environmental Services (PES) for popverty alleviation, these could be considered non-tangible NTFPs (Chap. 1). Forest governance is understood in this chapter as being the process of implementing and monitoring the allocation of forest land and resources and making the relevant policy (Macqueen and Bila 2004). It encompasses decisions on how and to what ends forests are managed, who are involved in these decisions, and what is done to enforce forest laws and policies on the ground. It also refers to the regulatory and institutional frameworks for the conservation, use and trade

M.A.F. Ros-Tonen (✉) and K. Kusters
Amsterdam Institute of Social Science Research, University of Amsterdam, Nieuwe Prinsengracht 130, 1018 VZ, Amsterdam, The Netherlands
e-mail: m.a.f.ros-tonen@uva.nl; k.kusters@uva.nl

S. Shackleton et al. (eds.), *Non-Timber Forest Products in the Global Context*, Tropical Forestry 7, DOI 10.1007/978-3-642-17983-9_9,

Fig. 9.1 Working with multiple parties is crucial in pro-poor NTFP governance (photo: Mirjam Ros-Tonen)

of forest resources and the principles that guide the interactions between those taking part in the design and implementation of different kinds of arrangements. In contrast to centralised government by the national state, governance also includes governments at other levels as well as private and civic actors such as companies, communities and non-governmental organisations (NGOs). Forest governance has become increasingly hybrid, multilevel and cross-sectoral (Lemos and Agrawal 2006; Ros-Tonen et al. 2008) and hence increasingly "interactive" (Kooiman and Bavinck 2005, p 17).

In this chapter, we seek to identify the conditions for good NTFP governance (Fig. 9.1). In addition to general conditions that enable the conservation and sustainable use of the resource, we thereby pay specific attention to conditions that determine whether NTFP exploitation can potentially contribute to poverty alleviation. Following Sunderlin et al. (2005), we interpret poverty alleviation as a continuum between poverty mitigation (forest resources as a safety net or gap filler in times of economic hardship) and poverty elimination (forest resources as a means to help people escape from poverty by generating savings, investment, asset accumulation and permanent increases in income) (see Chap. 3).

Below we first review some general issues related to good forest governance. We then specify and illustrate these issues using several cases. Based on the review, we extract the main lessons and challenges to be faced in NTFP governance. Our focus on NTFP resource use is not to deny that forests can and should be governed for other purposes, such as the maintenance of their supporting, regulating and cultural services.

9.2 NTFP Governance: Some General Issues

Although the etymology of the word governance dates back to the ancient Greeks, where the original meaning of the word *kubernan* was "steering a vessel" (Jessop 1998), scientific attention to governance is relatively new. In several languages, the word was, for a long time, synonymous with government or the modes and manner of governing. At the end of the 1980s, other meanings emerged in response to the need for a new concept to cover actions by actors other than the State. Since then the term governance has been increasingly used in relation to global, corporate and local governance (Jessop 1998), and it became a buzz word in international donor circles when the World Bank launched the "good governance" norm in 1989 (World Bank 1989; Kooiman and Bavinck 2005).

If UNESCAP's (n.d.) eight characteristics of good governance are applied to the forest sector (cf. Brown et al. 2002; Mayers et al. 2002), it follows that forest governance should be:

- *Accountable* – meaning that all actors involved in forest governance (governmental institutions, private sector or civil society organisations) are held responsible by those affected by their decisions.
- *Transparent* – meaning that forest policies and regulations are clear to all who will be affected by them, and that information about them is freely available to all stakeholders in an accessible and understandable form.
- *Responsive* – meaning that it meets the livelihood needs of forest-dependent people.
- *Equitable and inclusive* – meaning that it stops marginalising the forest-dependent poor and enables them to maintain and improve their well-being based on equitable shares of forest benefits.
- *Effective and efficient* – meaning that it promotes efficient use of forest resources for both conservation and sustainable use, and puts effective arrangements in place to include the various stakeholders.
- *Following the rule of law* – meaning that it applies forest laws and regulations impartially, without excluding forest-dependent people from access to, or trade in, forest resources.
- *Participatory* – meaning that all relevant stakeholders are directly or indirectly involved in forest decision-making processes that affect them.
- *Consensus oriented* – meaning that it is based on a shared and negotiated vision of the societal role of forests and the role of each stakeholder in terms of rights, responsibilities and use.

It may be impossible to comply with all these aspects of good forest governance, so the question becomes what are the key conditions for pro-poor NTFP governance? In addressing this question, we follow the four main challenges in forest governance put forward by Mayers and Vermeulen (2002a):

- Strengthen rights, capabilities and local decision-making so that poor people can take action to improve local livelihoods.

Fig. 9.2 Market Santarém, Brazilian Amazonia. Removing market barriers is key to poverty alleviation (photo: Mirjam Ros-Tonen)

- Reduce poor people's vulnerability by cutting the regulatory burden on them and promoting the rule and legitimacy of law.
- Enable market opportunities by removing the barriers to market entry (Fig. 9.2), appropriate valuation of forest resources, ensure that markets for environmental services benefit poor people, support associations and finance local forest businesses.
- Work in partnership by supporting participatory processes, promoting inter-agency learning and action, and making private sector and NGOs partners in poverty reduction.

In the rest of this chapter, we elaborate on these issues and specify and clarify them for NTFPs. We then summarise some lessons learned for the governance of NTFP resources.

9.3 Strengthening Rights

Strengthening property rights is arguably the main element of the first challenge put forward by Mayers and Vermeulen (2002a). Following Schlager and Ostrom (1992), we use property rights in this chapter in the sense of a bundle of rights that refer to access, withdrawal, management, exclusion and alienation. It should be noted that property rights refer not only to land but also to trees and the products derived from them, and that tenure rights to land and to trees can be separate (Feder and Feeny 1991). Property rights for local people have been promoted on four grounds. First, local ownership or long-term access and control rights to forest assets may effectively contribute to poverty reduction because they enable local people to capitalise on forest resources through deals with businesses (Mayers and

Vermeulen 2002b) or participation in certification, PES and REDD (Reduced Emissions from Deforestation and Degradation) schemes (Arnold 2001; Scherr et al. 2003; Skutsch et al. 2008). Second, and related to this, well-defined and secure property rights have been promoted as a key condition to achieve sustainable management of natural resources as they would ensure that managers can reap the benefits of management, protect the resources from overexploitation, and promote long-term investment. Third, community-based organisations (CBOs), NGOs and researchers have drawn attention to the fact that State authorities often deny local people's access to forest resources. Property rights are therefore regarded as a justice issue (e.g., Zerner 2000). Fourth, secure property rights are expected to help protect areas that are used by local people from appropriation by outsiders. For example, if NTFP management systems are located on State lands, conflicts may arise with government-sanctioned land uses such as forest concessions or plantations and this can lead to tenure insecurity and jeopardise the continuation of these production systems (e.g., Pagdee et al. 2006). Similarly, insecure property rights lead to the exclusion of people from the exploitation of forest resources, and to conflicts over forest land resulting from competing claims. Both exclusion and conflicts threaten the livelihoods of the rural poor (e.g., Hobly 2007).

In the last few decades, several countries have developed legal instruments to grant rights to forest resources to local communities. Based on the extrapolation of official tenure data for 24 of the top 30 forested countries that together hold 93% of the world's natural forests, White and Martin (2002) suggested that 22% of the forests in developing countries are currently reserved for, or owned by, community and indigenous groups. Many of these arrangements imply rights to extract or cultivate NTFPs. Well-known examples are the Certificates of Ancestral Domain Claim (CADC) in the Philippines (Lynch and Talbott 1995), the extractive reserves in Brazil (Schwartzman 1989; Allegretti 1990), and the Joint Forest Management schemes in India (Kumar 2002). Such tenure arrangements are generally expected to promote sustainable forest management while enabling local communities to capitalise on available resources.

Although these efforts to reform property rights for local people are promising (Scherr et al. 2003), the frequently quoted figures presented by White and Martin (2002) may present an overly optimistic picture for the following reasons: (1) many of these property rights involve a range of restrictions, for example, on commercial timber (e.g., Boaz 2004); (2) the highest quality forests usually remain in the hands of governments or large-scale enterprises (Scherr et al. 2003); (3) property rights are often designed on the basis of unrealistic assumptions, lack proper implementation and involve onerous procedures (e.g., Palis 2004); (4) the formalisation of property rights may threaten security if it fails to address the complexity of the existing system and competing claims (Neumann 1997); (5) granting property rights may continue or even increase existing inequalities which prejudice the least powerful, such as women and minorities (Ben White quoted in World Bank 2007, p 115) and (6) clear ownership rights are insufficient if poor people are unaware of their rights or if these rights are not backed up by local institutions (Mayers and Vermeulen 2002a; Shackleton et al. 2002). In addition, the relationship between tenure and sustainable forest use is ambiguous. First, having secure property rights to a forest area does not mean the holder of the property rights will leave the area forested. Second, deforestation can be a way to claim land and secure tenure (Schneider 1995; Chomitz 2007). Third, as demonstrated by the case of Krui in Sumatra (Box 9.1), *perceived* tenure security, which is affected by reduced outside threats and increased support of NGOs and research institutions, can be more important in shaping land-use decisions than the formal legal tenure status.

Box 9.1 The Recognition of User Rights in the Krui Area of Sumatra, Indonesia

Source: Kusters et al. (2007).

The planted agroforests in the Krui region, on the southwest coast of Sumatra, have been subject to tenure disputes. The communities in the area have been cultivating their lands with agroforests for centuries and therefore have a strong and longstanding ownership claim. Their land use became threatened when the central State classified the area as State land and started supporting industrial plantation development. As a result, farmers stopped planting trees and the future of the agroforests was at stake.

Under pressure from research institutions, NGOs and CBOs, the Indonesian Minister of Forestry (MoF) signed a decree in 1998 to provide the Krui farmers with tenure security. The decree, commonly referred to as the KdTI decree, enabled community leaders to apply for legal user rights (termed concession rights in the decree) for the agroforests located within the State zone. In the KdTI area, farmers were to be allowed to plant and harvest NTFPs (in particular the valuable *damar* resin), but they would not be allowed to convert the tree stands. In the words of Schlager and Ostrom (1992), the KdTI decree gave communities the opportunity to apply for the rights of access, withdrawal, management and exclusion. Rights of ownership, however, were not transferred. Application would provide communities with a de jure collective right, within which the de facto customary system based on individual and alienable plots would be maintained.

The decree was welcomed by NGOs and researchers as a breakthrough as it was the first time the Indonesian government had acknowledged local user rights on State forest land (Fay et al. 1998). However, a study conducted in 2005 revealed that none of the Krui communities had ever formally applied for their rights. Community leaders had been unwilling to apply as it would have implied recognition of State ownership of the land, while the communities would only accept a return of full ownership rights. This should be seen in the context of the decentralisation process after the fall of the authoritarian regime of Suharto in 1998, which influenced the expectation level of local communities.

Even though none of the communities applied for concession rights, the recognition of the local rights in the KdTI decree with the designation of the distinctive use zone has de facto helped to safeguard the agroforest from

(continued)

claims by outsiders. The decree represented the State's acceptance of the existing land-use practises and thereby altered the government's approach to the use and development of the State forest zone. The government stopped supporting expansion of industrial activities in the area. As such, the threat of appropriation decreased which, in combination with support from NGOs and research institutions, helped to restore farmers' perception of security. Between 1998 and 2005 the perception of security has been sufficient for farmers to make long-term investments in complex agroforestry systems.

9.4 Reducing Vulnerability by Promoting the Rule of Law

An important feature of good governance is the rule of law. In the context of forest use this means that forest laws and regulations are applied impartially, without excluding forest-dependent people from access to, or trade in, forest resources. This is often not the case (Mayers and Vermeulen 2002a; Shackleton et al. 2002). In a lot of countries, forest-dependent people are facing overregulation with regard to the use and trade of forest resources, particularly when they live close to conservation areas. More powerful forest users, in contrast, encounter fewer restrictions particularly with regard to logging or forest clearing for farming. Furthermore, many of them often succeed in undermining the rule of law through tax evasion or illegal forest exploitation due to the understaffing of forest agencies (Mayers and Vermeulen 2002a). Even where land ownership and authority have been devolved to local communities, as in Botswana and Namibia, the State continues to control natural resource use by setting wildlife quotas, renewing tourism and hunting concessions with the private sector and prohibiting subsistence hunting without a permit (Shackleton et al. 2002). Such a situation may adversely affect the poverty-reducing potential of NTFP production by denying people's access to forest resources or hampering their rights to exploit these. The need for a clear and coherent governance framework is illustrated by the study of Adano and Witsenburg (2004, 2005) in the Marsabit District in northern Kenya where both the forest's safety net function for the poor and the conservation of the Marsabit Forest Reserve are threatened as a result of deficient rule of law (Box 9.2).

Although the discussion in this section has focused mainly on formal State law, it should be noted that the rule of law should also apply to arrangements under customary law in situations of legal pluralism.

Box 9.2 Deficient Rule of Forest Law in Kenya

Source: Adano and Witsenburg (2004, 2005).

In Kenya, the problem of conflicting land-use practises between forests, agriculture and urban development is a major challenge facing the forest

(continued)

sector today (Mbugua 2001), as are the underlying power imbalances and conflicts of interests over forest-based resources. Forested areas face a serious threat of conversion to agricultural land and human settlement (Bubb et al. 2004). The mandate to prevent such conversion rests with the government, but the lack of a coherent governance framework hinders this task. Forest conservation and management are scattered across eleven statutes and twenty government agencies, while over 50 NGOs were engaged in environmental activities in Kenya in 1999 (Mugabe and Krhoda 1999; Seymour and Mugabe 2000). Management responsibility is vested in the Forest Department for all gazetted forests, while the Kenya Wildlife Service is responsible for forests in national parks and game reserves, County Councils for forests on trust lands and private individuals and companies manage forests on private lands. Decline of Kenya's economy, low budgetary allocations to government ministries and misappropriation of public resources (including corruption) decreased the management capacity of the Forest Department, in particular. The conservation of forests and forest resources is further hindered by regular excisions of protected forest areas endorsed by the government such as the excision of over 680 km^2 (about 10%) of the forest cover in 2001 (East African 2001, Kenya Forests Working Group KFWG 2002).

Whereas the government supports non-forest uses, the rural poor face more and more restrictions in the use of forest resources. On Marsabit Mountain in northern Kenya, the poorest households are the ones that sell forest products to earn income. Households on the mountain use the forest for watering and grazing livestock and for domestic purposes, and they convert forest into arable land. Along with this they also harvest forest and vegetation resources, such as fuel wood for cooking and making charcoal, poles for construction, grasses for thatching houses, fodder for livestock, medicinal plants and honey. The use of these products increases during dry or stress periods, which illustrates the forest's function as a safety net during times of economic hardship.

Access to such resources outside the settlements and from gully ridges, farm edges and woodland is subject to customary rights which prevailing conservation approaches fail to recognise (Mbugua 2001). Access to products from the National Park or the Forest Reserve, which provide wood fuel for 80% of the households, is formally only allowed with a monthly permit licence, issued upon payment of a prescribed fee. In addition, periodic bans are imposed in times of stress. Only a few products, such as medicinal herbs, may be acquired without a permit. Households furthermore appeared to pay varying permit prices, suggesting hitches in the prices charged. The command and control approach to forest access and the dual endorsement of forest permits raises concerns about the dominance of power and mistrust confronting resource users. The weak regulation of the protected areas created open access problems associated with the exploitation of the natural forest and is one of the main threats to the conservation of forest resources. The resource

(continued)

control measures also have unequal effects on households, since it is the relatively resource-poor households that derive a significant share of their income from forest resources and recur to these resources in times of scarcity.

9.5 Enabling Market Opportunities

Scherr et al. (2003) analysed the role of forest markets in rural livelihoods and argued in favour of a level playing field for low-income forest producers in forest markets. They highlighted the fact that conventional forestry, as well as approaches which are more sensitive to the livelihood needs of forest-dependent people such as social forestry and integrated conservation and development projects, marginalised these peoples' position on forest markets and at best supported subsistence activities. They point to new market opportunities resulting from the transition towards increased community ownership and control of forests, a growing demand for forest products versus increasing scarcity of accessible timber, and emerging niche markets for certified forest products and environmental services (which if supplied by forested landscapes can be considered non-tangible NTFPs). For small forest producers to benefit from these new potentials, enabling policies would be required to remove market barriers, reduce the regulatory burden on them and find new financial mechanisms and incentives. Stimulating producers' associations and strategic business partnerships would be another way of enhancing the participation of small producers in profitable forest markets.

How the existence of market barriers and a lack of enabling conditions can hinder promising NTFP trade is illustrated by the bamboo producers' case in the Philippines (Box 9.3). In spite of being a substantial source of income based on sustainable exploitation, the bamboo gatherers think it offers little future due to the illegal nature of their activities. As individuals they cannot meet the bureaucratic requirements of getting a permit. Removing this barrier could be a solution, but the bamboo gatherers themselves think the solution mainly has to do with becoming an organised producer association or cooperative.

Box 9.3 Tackling the Illegality of a Promising Source of NTFP Income in the Philippines

Source: Kusters et al. (2001).

In a community of forest migrants in the Sierra Madre mountains, called Puerta, located in the northeast of the island of Luzon, 50 of the 53 households engage in the cutting of the bamboo species buho (*Schizostachyum lumampao*). This species grows abundantly in the residual forest of the Sierra Madre and has a market in the lowland villages where the culms are, for example,

(continued)

used for the drying of tobacco leaves. Only 9% of the culms are cut on privately owned land, while the rest comes from public (residual) forest. All buho is extracted in a sustainable manner, because only mature culms are cut.

To be legally allowed to cut and sell buho, an official licence in the form of an Ordinary Minor Forest Products Permit is needed. This licence is granted by the Department of Environment and Natural Resources (DENR). The permit requires the payment of an application fee and forest charges. The DENR grants such a permit for buho cutting only to individuals or associations who participate in public bidding for a concession area. Since there is no licence holder in the study area, all cutting activities are prohibited de jure. De facto, however, there is a situation of open access for cutting buho in most of the residual forest. The DENR officials tolerate small-scale transport and trading of buho that was cut in the public forest, but the transport of illegally cut buho by truck is not tolerated and faces the risk of confiscation. Therefore, transport by truck usually takes place at night in an atmosphere of uncertainty and secrecy.

At the time of the research, buho selling was the most important source of income for the households after the selling of corn. On average it provided 13% of total income. Despite this economic importance, the buho gatherers thought there was only limited future potential. The uncertainty associated with the formal illegality of cutting and selling of buho played an important role in the perceived lack of prospects. Despite the low level of organisation among the gatherers in the village at the time of research, the buho cutters felt that organising individual gatherers into a cooperative could be an important step towards finding a solution. They reasoned that an organised group of gatherers could apply for a permit and that this would legalise the activities and facilitate greater trade and transport. Furthermore, an organised group of gatherers, together supplying a large amount of legal buho, might attract big buyers (who currently run the risk of confiscation when they buy large amounts of illegal buho), and improve their market position vis-à-vis other buho suppliers. In addition, the organisation of bamboo gatherers could pave the way to demarcated and protected extraction areas.

As regards enabling market opportunities, an important question is whether conventional NTFPs have sufficient potential to reduce poverty (Chap. 3), and whether governance efforts should not be directed towards new promising markets. A lot of NTFPs fulfil an important safety net and gap filling function to the poor, but this does not automatically imply that they have potential to lift people out of poverty (Arnold and Ruiz Pérez 2001; Belcher et al. 2005). Quantifying the value of several forest goods and services in the Amazon region, Verweij et al. (2009) also support the conclusion that returns from NTFPs and ecotourism should not be overestimated. This is largely due to contextual factors, such as poor tenure security, infrastructure and market access, but it also has to do with seasonality and the low densities at which some NTFP resources occur and their irregular distribution (Boot 1997; Ros-Tonen and Wiersum 2005).

In terms of poverty alleviating potential, more is currently expected from markets for certified forest products (including timber and non-timber forest products) and payment for environmental services (PES) such as carbon storage, hydrological services, biodiversity conservation, preserving landscape beauty or pollination services (Angelsen and Wunder 2003; Sunderlin et al. 2005; Verweij et al. 2009). As regards certification however, there are still a lot of obstacles that need to be removed before NTFP extractors can capitalise on these markets (Pierce et al. 2003). Most experience in the field of PES is still experimental (Sunderlin et al. 2005) and markets and effective financial mechanisms for environmental services still have to be developed (Verweij et al. 2009). Skutsch et al. (2008) also point out that, in relation to carbon payments under REDD schemes, several governance challenges still need to be met at local, national and international levels. These include questions related to (customary) ownership rights, the fair distribution of carbon benefits among various stakeholders, the institutional infrastructure to govern REDD, accounting methods to register carbon gains and losses, and transparent criteria for payments. Last but not least, there is the risk that benefits from REDD and PES, as in the case of other high-value NTFPs such as wildlife, remain in the hands of more powerful groups in society, with there being little interest in handing over control to local communities (Nelson and Agrawal 2008).

A concrete case in point is provided by Hall (2008), who reviewed the *Proambiente* PES programme for small farmers in Brazilian Amazonia (Box 9.4). He is moderately optimistic about the potential of such a scheme to cut down greenhouse gas emissions through reduced deforestation. However, he also makes it clear that, in addition to secure funding and extension support to farmers, a lot of governance challenges need to be dealt with before such experiments can be scaled up so as to contribute substantially to people's livelihoods.

Box 9.4 Governance Challenges as Regards Payments for Environmental Services (PES)

Source: Hall (2008).

Proambiente stands for "Programme for the socio-environmental development of rural family production" and was designed to pay small farmers for environmental services in twelve core areas or "poles" in Brazilian Amazonia. The environmental services which are subject to payment include (1) reduction or avoidance of deforestation, (2) carbon sequestration, (3) recuperation of ecosystem hydrological functions, (4) soil conservation, (5) preservation of biodiversity and (6) reduction of forest fire risks. Farmers can provide these services through the adoption of more sustainable farming systems such as agroforestry, extractivism and forest and pasture management. In order to apply for the payments, the beneficiaries should present a sustainable

(continued)

development plan for the community, a resource utilisation plan and a community agreement and also take care of auditing and certification. Provided independent monitoring demonstrates positive outcomes, about half a monthly minimum salary (USD 95) would be paid to a participating household.

After applying the scheme on a pilot scale, the idea was to scale it up into a national programme. However, four years of implementation demonstrated moderate achievements and several bottlenecks. The scheme became operational in ten of the twelve designated "poles" and benefited 42% of the participating households which were paid USD 325 on average. In addition to limited funds and deficient extension support to the farmers, a lot of the impediments to successful upscaling are related to governance. These include (1) lack of implementing capacity, (2) poor embedding of the economic value of environmental services in federal legislation, (3) incompatibility of various government policies and lack of cooperation between the many institutions involved (including private sector and civil society) and (4) a bias towards communities organised in labour unions and farmer associations, excluding non-unionised communities.

The Brazilian government is currently addressing these challenges since, in spite of limited achievements thus far, the scheme is regarded as being a promising complement to conventional conservation measures, with the potential to promote reduced emissions and deforestation as well as to combat poverty.

9.6 Alliances and Partnerships

The last challenge for good and pro-poor governance put forward by Mayers and Vermeulen (2002a) concerns the need to build partnerships between local forest dwellers and actors from other sectors (government, private sector and/or civil society). Partnerships are perceived here as being "more or less formal arrangements between two or more parties from various sectors around (at least partly) shared goals, in the expectation that each party will gain from the arrangement" (Ros-Tonen et al. 2007, p 5). The main idea behind the need for partnerships as an avenue via which to reach good and pro-poor governance of NTFP resources is, as Sunderlin et al. (2005, p 1388) put forward, that "forest-dependent people who live in or near forests tend to be politically weak or powerless". By bringing together power, assets, knowledge and skills with actors at other levels of scale, local forest dwellers can compensate for the lack of political and economic leverage. As mentioned above, this lack of power comes to the fore mainly in insecure property rights to forest resources, the skewed rule of law, and limited capacity to seize market opportunities.

In addition to being a way of obtaining greater leverage, partnerships are also needed to deal with the larger number and wider variety of actors in forest management. The past decades have shown an evolution from centrally guided forest management to decentralised governance, as a result of which several actors other than the State have a say in the allocation and use of forest resources. Several factors contributed to the inclusion of other actors in forest governance, with the most important at global level being (1) neo-liberal policy reforms and the declining role of the State, (2) the good governance debate in the 1990s, (3) the tendency towards global environmental action and governance, (4) globalisation and the corresponding improvement of transnational communication and information flows and (5) increasing reliance on the market and the role of corporate actors. At national and subnational levels, factors favourable to the shift from centralised government to multi-actor governance include (1) decentralisation policies, (2) the claims for and devolution of property rights to indigenous and other local communities and (3) the democratisation wave in the 1980s which resulted in strong civil society development and engagement in many places (Lemos and Agrawal 2006; Ros-Tonen et al. 2007).

As a result of these developments, forest policymaking and implementation shifted from the traditional centralised command and control approach to a network approach, in which government, corporate and civil society actors at different levels of scale collaborate on the basis of (at least partly) shared beliefs and dependency. The term 'forest governance' was coined to include the notion of democracy and the involvement of non-state actors in decision-making regarding the allocation and use of scarce forest resources.

The increasing number of actors also implies the need to deal with competing claims and diverging interests. In a context of declining forest resources, no single actor can be held responsible for managing forests and NTFP resources that are under pressure. Partnerships are a way of coping with this increased complexity of actors and claims. This holds true particularly for NTFP governance due to the limited means, power and market access of the producers involved.

Box 9.5 provides an example of partnerships around extractive reserves in Brazilian Amazonia which aim to improve both tenure security and the livelihoods of NTFP gatherers. This case shows that the partnership is a promising approach towards trying to secure the political empowerment of extractivists and their rights to the forest, but that there are still immense challenges to be met with regard to their economic empowerment (Brown and Rosendo 2000). In spite of considerable technical and financial support, it seems to be hard to improve the living standard and well-being of NTFP extractors living in isolated conditions, even if proper governance arrangements are in place.

Box 9.5 Multi-sector Partnerships Around Extractive Reserves Brazilian Amazonia

Sources: Brown and Rosendo (2000), Rosendo (2007) and Ros-Tonen et al. (2008).

Partnerships around extractive reserves in Brazilian Amazonia demonstrate a lot of the governance aspects dealt with in this chapter. Extractive

(continued)

reserves are protected areas, established by the government, that guarantee the rights of local populations to use natural resources for livelihoods (Allegretti 1990; Schwartzman 1989). This approach was proposed by the rubber tappers or *seringueiros* of the state of Acre in the mid-1980s as a means of securing rights to land. After joining forces with researchers, NGOs, the World Bank and other external agencies, the rubber tapper movement succeeded in getting this approach adopted by the Brazilian government as part of its environmental policy in 1990 (Hall 1997). Since then, the federal government has established about fifty extractive reserves in Brazil, covering an area of approximately 10 million ha (http://www.ibama.gov.br). In addition to these, there are 21 extractive reserves in Rondônia, covering a combined area of about one million hectares, which were created by the State government. This was done under a large-scale environmental management project funded by the World Bank, referred to as the Rondônia Natural Resources Management Project (PLANAFLORO), which was implemented between 1993 and 2001.

Improving livelihoods is conditional for the long-term viability of extractive reserves. Failure to do so might cause families to abandon the reserves in search of better opportunities elsewhere. In Rondônia, efforts to improve the economic viability of the reserves have benefited from alliances with NGOs that provided much needed capital and technical assistance. In particular partnerships with WWF and donors like the G7 Pilot Programme for the Protection of the Brazilian Rain forest (PPG7) have enabled the initiation of several income-generating projects since 1993. Examples include the diversification of NTFP extraction to other products such as açaí fruit and palm hearts of *Euterpe oleracea*, ecotourism, the production of a rubber-coated textile known as ecological leather and community-based logging. These activities provide necessary supplementary income to the low and seasonal income from the extraction of rubber, Brazil nuts (*Bertholletia excelsa*) and copaiba oil (*Copaifera* spp.). The partnership also aimed to reduce the extractors' dependency on intermediaries who capture a significant share of the profits. One attempt included the establishment of alternative marketing networks that provide better prices for the NTFPs. Another way was to provide the communities with storage facilities, working capital and transportation infrastructure to set up community-run trading posts that could supply the residents with essential goods for which they previously depended on intermediaries. In practise, however, the partnerships have been more successful in securing property rights than in improving the livelihood conditions of the extractors, who continue to live in deprived conditions.

9.7 Conclusions

In this chapter, we have reviewed some key conditions for good and pro-poor NTFP governance, following the challenges put forward by Mayers and Vermeulen (2002a) with regard to secure rights, equitable rule of law, market access and the building of partnerships.

With regard to rights we noted that secure tenure is considered key to reducing poverty, enhancing justice and promoting sustainable forest management and protection. It is clear that the lack of property rights increases poor peoples' vulnerability, thus endangering the poverty-alleviating function of NTFP extraction. However, the relationship between secure tenure and sustainable forest management is not always clear-cut: secure property rights do not by definition ensure the maintenance of tree cover. Whether this should lead to the decision to restrict land-use options to people should, under good governance conditions, be negotiated with all stakeholders, including policymakers, conservationists, private sector actors and local communities. The latter are often the weakest party, both in terms of political and economic power, and would need to organise and enter into partnerships with civil society organisations to have their claims heard. As the cases of extractive reserves in Brazilian Amazonia and the damar agroforests in the Krui area in Sumatra have shown, partnerships may help to ensure that ownership claims are heard and granted. In addition, shared research such as Participatory Action Research (PAR) as well as Adapted Collaborative Management (ACM) may enhance the skills and empowerment of local people and complement the work of NGOs.

Although securing property rights may safeguard the role of forests as a safety net in times of economic stress (poverty mitigation), it may not be enough for poverty elimination, as the PES scheme and extractive reserves cases in Brazilian Amazonia have shown. An important governance challenge to be met in this respect is the equitable rule of law. As made clear in the Marsabit case from Kenya, forest-dependent people often face restrictions to forest use that more powerful actors may be able to circumvent.

Overcoming restrictions to forest use as well as reducing the bureaucratic burden on the forest-dwelling poor are important conditions for meeting the third governance challenge: that of enabling market opportunities. The case of bamboo gatherers in the Philippines demonstrated how a potentially sustainable and lucrative activity lacks future prospects because it is difficult to acquire a permit on an individual basis with which the gatherers could operate legally.

In relation to enabling market opportunities, we raised the question of whether efforts should focus on conventional NTFPs or on promising new markets for non-tangible NTFPs (ecological services). With regard to these new and largely still experimental markets under PES and REDD schemes, a lot of governance challenges still have to be met. These challenges, as illustrated by the *Proambiente* case in Brazil, relate to the institutional frameworks governing such schemes, the implementing capacity, coherence with other policies and fair distribution of benefits among stakeholders, to name but a few.

Many of the NTFP governance challenges reviewed can be met through building partnerships. Such partnerships have become necessary, firstly because shifts in governance transferred authority from the central State to lower levels of government, the private sector, and local forest users. This increased the array of actors involved in forest governance and the need to deal with competing interests and claims. Secondly, the need to engage in partnerships is even more compelling for the forest-dwelling poor, with a view to increasing their limited political and economic leverage.

As illustrated by the cases presented in this chapter, partnerships can do a lot to create the conditions for the good governance of NTFP resources. They may help improve tenure security, the equitable rule of law, political empowerment of the rural poor in forest areas and access to markets through company–community deals and multi-sector partnerships involving private sector actors.

Governance for poverty reduction based on the marketing of NTFPs however faces huge challenges. This applies particularly to markets for "non-tangible" NTFPs due to the complexity of the governance structures required to make such markets work for local people. We argue that the building of partnerships between local forest users and actors from other sectors and geographical scales can serve as the starting point to tackle these governance challenges.

Acknowledgements The authors are grateful to Carol Colfer, Darcy Mitchell and Freerk Wiersum for useful comments on an earlier draft of this paper.

References

Adano WR, Witsenburg K (2004) Surviving pastoral decline: pastoral sedentarisation, natural resource management and livelihood diversification in Marsabit District, N Kenya. PhD thesis, University of Amsterdam, Amsterdam

Adano WR, Witsenburg K (2005) Global forest management decisions and local use of forest resources in Kenya: exploring the link. In: Ros-Tonen MAF, Dietz T (eds) African forests between nature and livelihood resources: interdisciplinary studies in conservation and forest management. The Edwin Mellen Press, Lewiston NY, pp 159–187

Allegretti MH (1990) Extractive reserves: an alternative for reconciling development and environmental conservation in Amazonia. In: Anderson AB (ed) Alternatives to deforestation. Steps towards sustainable use of the Amazon rain forest. Columbia University Press, New York, pp 252–264

Angelsen A, Wunder S (2003) Exploring the forest–poverty link: key concepts, issues and research implications. CIFOR Occasional Paper 40, CIFOR, Bogor

Arnold JEM (2001) Forests and people. 25 years of community forestry. FAO, Rome

Arnold JEM, Ruiz Pérez M (2001) Can non-timber forest products match tropical forest conservation and development objectives? Ecol Econ 39:437–447

Belcher B, Ruiz Pérez M, Achdiawan R (2005) Global patterns and trends in the use and management of commercial NTFPs: implications for livelihoods and conservation. World Dev 33(9):1435–1452

Boaz, AA (2004) Case study of tendu leaves (*diospyros melanoxylon*) in Harda District, Madhya Pradesh, India. In: Kusters K, Belcher B (eds) Forest products, livelihoods and conservation. Case studies of non-timber forest product systems, vol 1: Asia. CIFOR, Bogor

Boot RGA (1997) Extraction of non-timber forest products from tropical rain forest. Does diversity come at a price? Netherlands J Agric Sci 45:439–450
Brown K, Rosendo S (2000) Environmentalists, rubber tappers and empowerment: the political and economic dimensions of extractive reserves. Dev Change 31(1):201–227
Brown D, Schreckenberg K, Shepherd G, Wells A (2002) Forestry as an entry point for governance reform. ODI Briefing Paper 2002(1)
Bubb P, May I, Miles L, Sayer J (2004) Cloud forest agenda. UNEP-WCMC, Cambridge, UK
Chomitz KM (2007) At loggerheads? Agricultural expansion, poverty reduction, and environment in the tropical forests. IBRD/World Bank, Washington, DC
East African (2001) Kenya in grave danger over forest excisions. Regional – A joint report Nairobi: Monday 5 November 2001
Fay C, de Foresta H, Sirait M, Tomich TP (1998) A policy breakthrough for Indonesian farmers in the Krui damar agroforests. Agrofor Today 10(2):25–26
Feder G, Feeny D (1991) Land tenure and property rights: theory and implications for development policy. World Bank Econ Rev 5(1):135–153
Hall A (1997) Sustaining Amazonia: grassroots action for productive conservation. Manchester University Press, Manchester
Hall A (2008) Better RED than dead: paying people for environmental services in Amazonia. Philos Trans R Soc B 363:1925–1932
Hobly M (2007) Where in the world is there pro-poor forest policy and land reform? Rights and Resources Initiative, Washington
Jessop B (1998) The rise of governance and the risks of failure: the case of economic development. Int Soc Sci J 50(155):29–45
Kenya Forests Working Group (KFWG) (2002) Misitu News: the newsletter of the Kenya Forest Working Group Sept–Dec 2002
Kooiman J, Bavinck M (2005) The governance perspective. In: Kooiman J, Bavinck M, Jentoft S, Pullin R (eds) Fish for life. Interactive governance for fisheries. Amsterdam University Press, Amsterdam, pp 11–24
Kumar S (2002) Does "participation" in common pool resource management help the poor? A social cost-benefit analysis of joint forest management in Jharkhand, India. World Dev 30:763–782
Kusters K, Ros-Tonen MAF, van der Top G, Dietz T (2001) The potential contribution of non-timber forest product extraction to tropical forest conservation: lessons from a case study of bamboo utilisation in a Sierra Madre community, the Philippines. J Bamboo Rattan 1:77–94
Kusters K, de Foresta H, Ekadinata A, van Noordwijk M (2007) Towards solutions for state vs local community conflicts over forestland: the impact of formal recognition of user rights in Krui, Sumatra, Indonesia. Hum Ecol 35(4):427–438
Lemos MC, Agrawal A (2006) Environmental governance. Ann Rev Environ Resour 31:297–325
Lynch OJ, Talbott K (1995) Balancing acts: community-based forest management and national law in Asia and the Pacific. WRI, Washington, DC
Macqueen DJ, Bila A (2004) Gleanings on governance: learning from a two year process of forest policy support to ProAgri. IIED, London
Mayers J, Vermeulen S (2002a) Power from the trees: how good forest governance can help reduce poverty. WSSD Opinion. IIED, London
Mayers J, Vermeulen S (2002b) Company-community forestry partnerships: from raw deals to mutual gains? IIED, London
Mayers J, Bass S, Macqueen D (2002) The pyramid: a diagnostic and planning tool for good forest governance. IIED, London
Mbugua D (2001) Forest outlook studies in Africa (FOSA): Kenya country report. FAO, Rome
Mugabe J, Krhoda G (1999) The politics of public policy reform in Kenya: the case of environmental adjustment. Paper presented at workshop on Environmental adjustment: opportunities for progressive policy reform in the forest sector? WRI, Washington, April 6, 1999

Nelson F, Agrawal A (2008) Patronage or participation? Community-based natural resource management reform in Sub-Saharan Africa. Dev Change 39(4):557–585

Neumann RP (1997) Primitive ideas: protected area buffer zones and the politics of land in Africa. Dev Change 28:559–582

Pagdee A, Kim Y, Daugherty PJ (2006) What makes community forest management successful: a meta-study from community forests throughout the world. Soc Nat Resour 19(1):33–52

Palis HG (2004) Rattan (*Calamus* spp.) extraction in the Philippines: the case of Manggapin and Kalakwasan watersheds, Palawan. In: Kusters K and Belcher B (eds) Forest products, livelihoods and conservation. Case studies of non-timber forest product systems, vol 1: Asia. CIFOR, Bogor

Pierce A, Shanley P, Laid S (2003) Certification of non-timber forest products: limitations and implications of a market-based conservation tool. Paper presented at the International Conference on Rural Livelihoods, Forests and Biodiversity, 19–23 May 2003, Bonn, Germany

Rosendo S (2007) Partnerships across scales: lessons from extractive reserves in Brazilian Amazonia. In: Ros-Tonen MAF (ed, in cooperation with H van den Hombergh and A Zoomers) Partnerships in sustainable forest resource management: learning from Latin America. Brill, Leiden, Boston, pp 3–35

Ros-Tonen MAF, Wiersum KF (2005) The scope of improving rural livelihoods through non-timber forest products: an evolving research agenda. For Trees Livelihood 15(2):129–148

Ros-Tonen, MAF, van den Hombergh H, Zoomers A (2007) Partnerships for sustainable forest and tree resource management in Latin America: the new road towards successful forest governance? In: Ros-Tonen MAF, van den Hombergh H, Zoomers A (eds) Partnerships in sustainable forest resource management: learning from Latin America. Brill, Leiden, Boston, pp 3–35

Ros-Tonen MAF, van Andel T, Morsello C, Otsuki K, Rosendo S, Scholz I (2008) Forest-related partnerships in Brazilian Amazonia: there is more to sustainable forest management than reduced impact logging. For Ecol Manage 256:1482–1497

Scherr S, White A, Kaimowitz D (2003) A new agenda for achieving forest conservation and poverty alleviation: making markets work for low-income producers. Forest Trends, CIFOR, Washington, DC, Bogor

Schlager E, Ostrom E (1992) Property rights regimes and natural resources: a conceptual analysis. Land Econ 68:249–262

Schneider R (1995) Government and the economy on the Amazon frontier. World Bank Environmental Paper 11, World Bank, Washington, DC

Schwartzman S (1989) Extractive reserves: the rubber tappers' strategy for sustainable use of the Amazon rainforest. In: Browder JO (ed) Fragile lands of Latin America: strategies for sustainable development. Westview, Boulder, pp 150–163

Seymour FJ, Mugabe J (2000) Kenya. In: Seymour FJ, Dubash NN (eds) The right conditions: the World Bank, structural adjustment and forest policy reform. WRI, Washington, DC, pp 113–163

Shackleton S, Campbell B, Wollenberg E, Edmunds D (2002) Devolution and community-based natural resource management: creating space for local people to participate and benefit? Nat Res Persp series 76. ODI, London

Skutsch MM, Banskota K, Trines E, Karky B (2008) Governance for REDD. Policy note 4, Kyoto: think global act local project, Poznan

Sunderlin W, Angelsen A, Belcher B, Burgers P, Nasi R, Santoso L, Wunder S (2005) Livelihoods, forests, and conservation in developing countries: an overview. World Dev 33(9):1383–1402

UNESCAP (n.d.) http://www.unescap.org/huset/gg/governance.htm. Accessed Feb 2009

Verweij P, Schouten M, van Beukering P, Triana J, van der Leeuw K, Hess S (2009) Keeping the Amazon forest standing: a matter of value. Zeist, WWF Netherlands

White A, Martin A (2002) Who owns the world's forests? Forest tenure and public forests in transition. Forest Trends, Washington, DC

World Bank (1989) Sub-Saharan Africa: from crisis to sustainable growth. World Bank, Washington, DC
World Bank (2007) Sustaining economic growth, rural livelihoods, and environmental benefits: strategic options for forest assistance in Indonesia. World Bank, Washington, DC
Zerner C (ed) (2000) People, plants, and justice: the politics of nature conservation. Columbia University Press, New York

Chapter 10
Non-timber Forest Products and Conservation: What Prospects?

Terry C.H. Sunderland, Ousseynou Ndoye, and Susan Harrison-Sanchez

Abstract Non-timber forest products (NTFPs) were hailed as a "silver bullet" to provide the economic incentives to conserve standing forests while contributing to local livelihoods. While the livelihood benefits of NTFPs have been widely acknowledged, the contribution of the NTFP sector to biodiversity conservation is less certain. Despite increasing skepticism of the ability of NTFPs to contribute to conservation, their promotion and development remains a readily implemented tool for many site level conservation projects. However, this chapter dispels certain assumptions related to NTFP sustainability and the links between NTFP extraction systems and conservation. We conclude that the links are generally tenuous to say the least and suggest that perceptions of the relative value of NTFPs in terms of biodiversity conservation need to be revised.

10.1 Introduction

In the late 1980s, non-timber forest products (NTFPs) were mooted as a potential alternative to deforestation and other land conversion activities (Falconer 1990; Plotkin and Famolare 1992). Some NTFPs have a significant market value and it was postulated that the long-term value accruing from the sustainable harvest of these products could override the short-term gain of converting that forest or individual

T.C.H. Sunderland
Center for International Forestry Research (CIFOR), P.O. Box 0113 BOCBD, Bogor 16000, Indonesia
e-mail: t.sunderland@cgiar.org

O. Ndoye
Food and Agriculture Organisation of the United Nations (FAO), P.O. Box 281 Yaoundé, Cameroon
e-mail: Ousseynou.Ndoye@fao.org

S. Harrison-Sanchez (✉)
Life Cycle Associates LLC, 985 Portola Valley Road, Portola Valley, California, 94028, USA
e-mail: sanchez@lifecycleassociates.com

S. Shackleton et al. (eds.), *Non-Timber Forest Products in the Global Context*, Tropical Forestry 7, DOI 10.1007/978-3-642-17983-9_10,

trees to other intensive uses such as timber, agriculture, or plantations (Peters et al. 1989; Godoy and Bawa 1993).

From a biodiversity conservation and livelihood development perspective, it is a compelling concept. Local people live in harmony with nature, deriving their livelihood and subsistence needs from the forest while actively protecting and ultimately sustainably utilising the "subsidy from nature" (Hecht et al. 1988: 25). The concept of the "rainforest harvest" (Prance 1992: 21) was rooted heavily in the extractivist culture of Latin America where rubber and Brazil nuts, amongst other non-timber commodities, are harvested in forested landscapes and sold to established markets, providing material and subsistence needs for local inhabitants. The formal recognition of extractive reserves as a legal entity was a direct result of lobbying from the National Council of Rubber Tappers (CNS), established in 1985 as a result of considerable social unrest related to land use in Amazonia (Richards 1993). The appeal of such extractive systems began to be realised, and in the late 1980s, a number of influential studies were published almost simultaneously, waxing eloquence about the possibilities to save the tropical forests of the world through sustainable use that provided an economic alternative to timber exploitation or other destructive land uses such as agricultural conversion (Gradwohl and Greenberg 1988; Hecht et al. 1988; de Beer and McDermott 1989; Fearnside 1989; Peters et al. 1989). This optimism was further endorsed through the purported potential commercialisation of forest products, including medicinal plants (Farnsworth 1988; Nepstad and Schwartzman 1992; Panayotou and Ashton 1992; Plotkin and Famolare 1992) that would contribute to local livelihoods, and by increasing the value of tropical forests, their conservation would be guaranteed (Godoy and Bawa 1993). The bottom line was that rainforests were perceived to have the ability to pay for themselves (Peters et al. 1989) and the expression "use it or lose it" (Freese 1997: 1; Struhsaker 1998: 930; Putz et al. 2001: 7) entered the parlance of the day. The underlying principle was that if the forest had no immediate value, it would be converted to other, more productive, land uses; hence, the sustainable use of NTFPs became regarded as a direct means of affording protection to standing forest.

However, 20 years later the original optimism surrounding the prospects for NTFPs to contribute substantially to both conservation and development has been significantly tempered (Chap. 2), particularly in the academic community (Arnold and Ruiz-Pérez 2001; Ros-Tonen and Wiersum 2005; Kusters et al. 2006; Belcher and Schreckenberg 2007). Much of this early optimism was based on claims of economic potential (de Beer and McDermott 1989; Peters et al. 1989; Godoy et al. 2000) which have been described as over-simplistic assessments of "value" (Southgate et al. 1996; Sheil and Wunder 2002). These shortfalls have been combined with limited understanding and evaluation of the complexity of political, economic, social, and market-oriented issues surrounding NTFP harvesting and trade (Dove 1995; Lawrence 2003). Increased commercialisation of forest products for promoting conservation and development has also been questioned as a means to contribute to both conservation and development objectives (Marshall et al. 2003), and it is argued that many households barely cover the opportunity costs of collection (Southgate et al. 1996). For example, even for high-value forest products, the majority of income accrues to those who transform the product, usually the wealthier members of the society (Alcorn 1995), local elites who control the market, or the state (Dove 1993; Richards 1993; Struhsaker 1998; Ros-Tonen and Wiersum 2005). Despite this, NTFP development and promotion continue to underpin many conservation efforts in the field, particularly in the implementation of integrated conservation and development projects (ICDPs) (Kiss 2004), and the potential for NTFPs to contribute to forest conservation and poverty alleviation is still being debated in the contemporary literature (see, for example, id21 Forestry Insights, issue 77, May 2009: http://www.id21.org/insights/insights77/index.html).

In this chapter, we review the early prospects for NTFP-based systems for sustainable forest management and whether or not biodiversity conservation goals can be achieved within such systems. In this context, we discuss not only the ecological and biological constraints related to NTFP sustainability but also some of the political and socioeconomic drivers that compromise the ability of NTFPs to contribute significantly to wider biodiversity conservation, particularly in the forested regions of the humid tropics.

10.2 NTFPs and Rural Livelihoods

Before discussing the potential of NTFP harvesting to contribute to tangible biodiversity conservation outcomes, it is important to recognise the valuable contribution that such products make to rural (and urban) livelihoods. Many authors highlight just how important NTFPs are to the local economy and livelihoods (de Beer and McDermott 1989; Nepstad and Schwartzman 1992; Prance 1992, 1998; Colfer 1997; Shanley et al. 2002, 2008; Belcher and Schreckenberg 2007; Paumgarten and Shackleton 2009). In many rural locations, particularly in areas that lack basic infrastructure and market access, the collection of NTFPs provides considerable subsistence support to local livelihoods through the provision of food, medicines, and plants and animals of cultural importance (Chap. 3; Fig 10.1).

Fig. 10.1 The start of a storage basket from *Papyrus* sp., harvested in Bigodi Wetland Sanctuary, and community initiative, Bigodi, Uganda (photo: Claire Shackleton)

When certain products are sold, this provides direct, and often the only means of access to the cash economy (Ros-Tonen and Wiersum 2005). In many instances, the trade in NTFPs has reached such levels that the chain of beneficiaries is long and complex (Belcher and Kusters 2004). There are many examples of such systems: the Brazil nut trade in Amazonia (Ortiz 2002), the trade in *Prunus africana* from African montane forests for the herbal medicinal industry (Ndam and Tonye 2004), the international rattan trade from southeast Asia and Africa (Sunderland and Dransfield 2002), the trade of cardamom (*Amomum villosum*) from the Lower Mekong region (Aubertin 2004; Tu 2004), and the harvest of palm fronds from Guatamala for the floriculture trade (Guariguata et al. 2008) to name but a few (see Chaps. 3–6 for more). In addition, Tabuna (1999) highlights that the strength of the trade in Central African NTFPs has led to substantial marketing of these products for the African diaspora based in Europe. Hence, it is clear that there are substantial livelihood and cash benefits from the harvest and trade in NTFPs, but how does this subsistence or economic value contribute to conservation of the wild resource or biodiversity in general?

10.3 Is the Harvesting of NTFPs Sustainable?

Underpinning the NTFP conservation and development debate is an assumption that NTFP harvest is inherently, or can potentially be, sustainable and ecologically more benign than alternative economic activities (such as timber exploitation) or non-forest utilisation (such as agricultural conversion) (Arnold and Ruiz-Pérez 2001; Putz et al. 2001; Ruiz-Pérez 2005; Chap. 7). Sustainability is a complex concept and there are many definitions of what sustainable means (Ostrom 2009; Tovey 2009). For example, in the case of NTFPs it is necessary to highlight the differences between ecological and economic sustainability, as the two are not always consistent (Hall and Bawa 1993). Over-harvesting of a particular resource may lead to a general decline of wild populations, while continued demand results in a constant market value. In contrast, increasing scarcity may increase the costs of exploitation, driving market prices upwards, leading to a reduction in demand and possible substitution with another, more easily available, product (Homma 1992). However, in economic terms, the effects of unsustainable extraction are more subtle and not so easily detected (Hall and Bawa 1993). This is particularly the case for long-lived trees such as Brazil nuts (*Bertholletia excelsa*), in which low-impact harvesting of fruits can have uncertain impacts on regeneration (Ortiz 2002). Ecologically, harvesting can only be considered sustainable at the species level if it has no long-term deleterious effect on the reproduction and regeneration of the plant or animal populations being harvested (Chap. 7). In addition, harvesting should also not have any discernable adverse effect on other species within the community, or on ecosystem structure or function (Hall and Bawa 1993; Ticktin 2004). There are a number of authors who challenge that there can never be truly sustainable harvest of NTFPs (Hall and Bawa 1993; Redford and Stearman 1993;

Peters 1994; Struhsaker 1998) as the long-term impacts of harvesting may manifest themselves in a variety of ways, not all of which are fully understood or capable of being understood at an acceptable level of probability (Struhsaker 1998). While not immediately harmful, the harvesting of fruits and seeds can decrease the availability of food for frugivores as well as affect the future regeneration of a species (Hall and Bawa 1993). Hunting animals who predate and distribute seeds can lead to regeneration problems, or in the case of pollinator species a reduction in reproductive capacity can lead to, in extreme cases, an "empty" forest (Redford 1992). The harvesting of bark and other woody parts can cause short to long-term senescence and, ultimately, the death of the organism (Peters 1994; Ros-Tonen and Wiersum 2005). This situation is often aggravated by poor monitoring and enforcement of rules when they are in place, thus so-called sustainable practices are rarely adhered to (Ostrom 2009).

For forest products (including timber) sustainability can only be determined by directly measuring the rate of extraction with the rate of harvest (Godoy and Bawa 1993). Unfortunately few natural science studies have directly measured sustainability in this way (Chap. 7). As a result, there are few concrete examples of the measurable sustainable harvest of tropical forest products (Prance 1998). Dynamic data are needed on growth and mortality; data that are woefully incomplete for even the most-studied forest resources (Boot and Gullison 1995; Sunderland et al. 2004). The time period for such data to be useful is in the scale of years to tens of years. Unfortunately, this is a much longer time-frame than is available for most field workers with uncertain funding. Measures of sustainability are also hindered by a lack of appropriate inventory techniques for NTFPs (Wong et al. 2001). With such a diversity of living organisms and range of harvesting techniques and impacts, inventory methods required to assess sustainability of NTFPs are highly complex and hence are rarely implemented. Perhaps unsurprisingly, as a result of this knowledge gap, there are few forestry graduate courses that include formal training on NTFP ecology and management (Guariguata et al. 2008).

10.4 NTFPs, Protected Areas, and Conservation?

10.4.1 The Growth of Protected Areas and the Annexation of Forest Land

Although the assumption that nature is pristine and should be protected from use is commonly challenged (Willis et al. 2004, Chap. 7), protected areas remain the most commonly implemented means of formally conserving biodiversity and have been established in almost every country in the world (Chape et al. 2005; Coad et al. 2009). In response to the current biodiversity crisis, there has been an exponential increase in the number of protected areas and, correspondingly, the area under protection in recent years (Chape et al. 2005). The global network of protected areas now covers 11.5% of the worlds surface area (Rodrigues et al. 2004) with 8.4% of the total protected area falling within categories I–IV of the IUCN's classification (Schmidt et al. 2009); the highest levels of protection. In theory, the majority of protected areas regulate and restrict access, denying millions of rural dwellers usufruct rights from forest lands they often previously relied upon for their livelihoods (Cernea 2005). If low-impact harvesting of NTFPs is seen as the optimum approach to land management in the tropics (Prance 1992), why has there been an exponential increase in protected area coverage and hence an annexation of land available for extractive activities? This contradiction could be explained as follows:

> "Although there has been much discussion suggesting that low-level economic activity would be compatible with biodiversity conservation, it is clear that if the full range of genetic, species and ecosystem diversity is to be maintained in its natural abundance on a given piece of land, then virtually any significant activity by humans must not be allowed" (Redford and Stearman 1993: 252).

This clear advocacy for the protection of "wild nature" (Willis et al. 2004: 402) in the absence of human activity continues to drive, and justify, the growth of protected areas in the tropics, despite the halcyon view that NTFP harvesting could be compatible with achieving conservation objectives; a contradiction within the conservation community that has seldom been acknowledged. In spite of their extensive coverage, protected areas have been relatively poor at conserving the full representative range of biodiversity (Rodrigues et al. 2004), and most of the world's biodiversity remains outside the protected areas, often in complex, multi-functional landscapes occupied by people (Alcorn 1993; Putz et al. 2001; Sayer and Maginnis 2005). It is these landscapes that are perhaps the most valuable for NTFPs and where they make their greatest contribution to rural livelihoods. Yet this requires a transition from natural forest to more intensively, and ultimately less diverse, forested systems, thus resulting in the conversion of biodiversity-rich natural forests (Fig 10.2).

Fig. 10.2 Making charcoal on the border of Bwindi Impenetrable National Park, Uganda (photo: Ross Shackleton)

10.4.2 The Transition from Natural Forests to Agro-Forests

Although highly diverse closed canopy forests can be important sources of forest products, a substantial proportion of NTFPs is harvested in secondary forest, farm fallows, and plantations (Ambrose-Oji 2003; Ros-Tonen and Wiersum 2005), which have been termed "domestic forests" (Michon 2005: 21). The peri-urban agroforests that supply the burgeoning NTFP economy of Belem are a good example of such forests (Shanley et al. 2002) (Fig 10.3). As are the domesticated forests of Indonesia, which supply up to 95% of fruits for the domestic markets as well as 80% of resin from Dipterocarpaceae trees (Michon 2005). Domestic forests are forested areas that contain trees that have been planted and can also include mosaics of natural forest as well as forest fields and fallow lands. Such agroforests are often mixed stands of trees cultivated for commercial purposes, usually NTFPs, rather than timber species.

One of the reasons why there is a strong transition from "nature to culture" (Dove 1995: 194; Levang et al. 2005) is that NTFPs in older, closed canopy forest are often present at very low densities (Peters 1994). As a result of the highly heterogenous nature of tropical forests, where individuals or stands of harvested species may be spread throughout a forested area, harvesters experience extremely low returns. For example, fruit production in the Peruvian Amazon (Phillips 1993) and resin and bark production in Indonesia (LaFrankie 1994) show remarkably low productivity levels per hectare, compared to timber and other land uses. Harvesting from such diverse environments requires an intimate and sophisticated knowledge of the forest, where each productive individual is known and monitored (Browder

Fig. 10.3 Joao Brito in Amazonia extracting medicinal bark from his family forest reserve for the treatment of diabetes. This species and other forest fruit trees are managed near urban markets to supply rising demand for NTFPs (photo: Trish Shanley)

1992a; Phillips 1993). Thus the trend is for greater intensification in managed forests.

Hence, NTFP extraction does not necessarily rely on a biodiverse resource base as there is a heavy reliance on only a very few major resources. Given the fact that such economic value is represented in relatively biodiversity-impoverished and often anthropogenic forests, what scope is there for the conservation of high biodiversity value forest using sustainable methods of NTFP extraction? Experience of the past 20 years probably suggests that the corresponding growth of alternative methods of conservation, such as protected areas (Hutton et al. 2005), is perhaps an indication of the limitations of sustainable NTFP harvesting being able to contribute realistically to biodiversity conservation.

10.5 Socio-Economic and Political Issues

10.5.1 Commercialisation: What Prospects for Conservation?

Underpinning the calls for the promotion and development of NTFPs is an assumption that increasing the commercial value of NTFPs will contribute to an increased appreciation of forests, therefore contributing both to poverty alleviation and forest conservation (Neumann and Hirsch 2000; Ruiz-Pérez 2005). However, historically, private capital or government intervention has played a part in the commodification of NTFPs (Dove 1995), and this process is only viable when there is a firm regional or global market for such products. It should be noted that colonial expansion was often driven by the commercialisation of many forest products (Hobhouse 1999) changing the face of not only the natural world, with large tracts of forest being cleared for the plantation-based production of cocoa, coffee, tea, rubber, etc. but also the introduction of a commodity-based economic system, based on short to medium-term financial returns, that is prevalent today (Brockway 2002).

Once commercialisation takes place and markets expand, elite capture occurs as the wealthiest in the community or region, or those able to invest in the expansion of trade, have adequate access to land and can invest in inputs required for cultivation and processing (Dove 1993; Marshall et al. 2003). Increasing demand for forest products initially leads to increased harvest from the wild, resulting in the loss of economic viability of the wild resource and encouraging the process of cultivation and ultimately, domestication (Homma 1992). Once cultivated systems are in place, the removal of an economically valuable product from the forest economy reduces the value of the standing forest leading to more lucrative, often destructive, land-use alternatives (Homma 1992). More recent assessments of NTFP commercialisation have concurred with these earlier studies and concluded that, while there are certainly local livelihood benefits (Marshall et al. 2003), such schemes are unlikely to be a successful means of ensuring wider biodiversity conservation for the reasons detailed above (Belcher and Schreckenberg 2007).

Underpinning the complexities of commercialisation and therefore the ability of NTFPs to provide an economic incentive to forest conservation is the fact that such a trade is often part of the informal forestry sector or the "hidden harvest" (Scoones et al. 1992: 17). Formalising the harvest and sale of NTFPs such that revenues contribute to the formal forestry sector is an oft-forgotten issue in NTFP research, and their near absence from the policy agenda (Laird et al. 2010).

10.5.2 Tenure, Politics, and Culture

To ensure that any modicum of sustainability in NTFP harvest, land, and resource tenure is critical Without secure tenure, open-access harvesting of an important commodity will almost certainly guarantee resource depletion (Angelsen and Wunder 2003). Where land and resource tenures are uncertain, many local people choose production systems that maximise short-term yield at the cost of long-term sustainability, such as the production of annual crops. This is a particular problem where landless migrants inhabit a forest area, such as Amazonia (Browder 1992b) and Indonesia (Levang et al. 2007), where short-term livelihood strategies take precedence over more sustainable land use practices. In short, it could be argued that more sustainable production systems would be more attractive to rural communities if they could gain *in perpetuo* rights to their land.

Unfortunately, many NTFP production systems operate within open or semi-open access systems of resource tenure, often resulting in mining of a resource if it is of particular economic value. It has been noted that rapid market expansion of products with little or no tenurial security often leads to significant over-harvesting (Alcorn 1995). The case of rattan in Indonesia is a good example of how fast-growing markets and open-access combine to compromise long-term conservation goals (Belcher et al. 2004). However, even if the open-access problems that lead to destructive harvesting were resolved, increases in the value of NTFPs might not benefit the conservation of tropical forests or the livelihoods of their inhabitants (Southgate et al. 1996). The reason given for this partly historical observation is that, as discussed earlier in this chapter, as extractive commodity becomes scarce, cultivation outside the natural ecosystem has been a characteristic response to its subsequent depletion in the wild. However, often such intensification efforts exclude the original resource users with the majority of resultant profits accruing to local élites, commercial concerns, or the state (Dove 1993; Marshall et al. 2003).

One additional tenurial problem relates to the interplay between customary and statutory legislative frameworks. For most of Africa, for example, governments have de jure control of land and can impose often contradictory land use policies over the same areas (e.g., mining concessions being issued within protected areas in Gabon). Overlapping layers of class, education, elite, and statutory "rights" overlying basic traditional tenure systems will affect the way innovations and management options are implemented, and together these relationships will play a direct role in successes or failures related to biodiversity conservation and livelihood improvement, not just for NTFPs but for other natural resources as well.

Political issues related to power and resource control may play an important role in determining sustainable management of forests and their resources than biological factors. The initial establishment of extractive reserves in Brazil was certainly driven more by socio-political issues than ecological considerations (Richards 1993). Yet studies show that extractive reserves and other models of sustainable use require strongly hierarchical political-economic systems (Dove 1995) which often promote inequity (Browder 1992b). Dove (1995) describes how the development of the rubber economies in Amazonia and Brazil has resulted in very different conservation and livelihood outcomes due to dissimilar historical trajectories based on socio-political differences on each region and suggests that the promotion of NTFPs is as much a political challenge as a technological or economic one.

Indigenous NTFP management systems have often imparted controls on the over-harvesting of many products (Redford and Stearman 1993), yet traditional knowledge systems often break down during a transition from subsistence economies to sedentary agriculture and exposure to modern communications and schooling (Ros-Tonen and Wiersum 2005). Conflicts within communities over whether traditional patterns of resource harvesting are able to provide for modern development needs such as health care, education, market access, etc. have taken place throughout the tropics, particularly between the youth and more established traditional institutions, often leading to a breakdown of traditional controls over resource use and management. Mythologising the role of indigenous people in natural resource management or denying them due societal and economic development and possible routes out of poverty, which may result from unsustainable resource exploitation, would be a significant disservice. Indeed, the long-held assumption that local people are inherent conservationists has also been questioned (Redford and Stearman 1993), as a lack of secure land and resource tenure does not encourage resource sustainability, and resource depletion is often a characteristic of NTFP production systems. For example, research conducted in Cameroon concludes that even for economically valuable NTFPs, few management techniques are applied under traditional harvest practices (Malleson 1999).

10.6 Discussion

Despite early optimism that the sustainable harvest of NTFPs was the silver bullet to save the tropical forests, it has become increasingly clear over the past 20 years that the assumptions underpinning this paradigm were based on somewhat simplistic, and generalised, approaches. Although, as we have discussed, NTFPs play a significant role in rural livelihoods, it has been argued that this is often in instances when there are few economic alternatives available (Ros-Tonen and Wiersum 2005; Shackleton et al. 2007). Hence, a major assumption that needs to be questioned is related to the "real" value of NTFPs to local people, and whether such value is reflected in the sustainable management of wild resources. In general, it is often the very poorest who primarily rely on forest products for their livelihoods (Alcorn 1993; Paumgarten and Shackleton 2009). This is because the NTFP sector is characterised by low or medium returns on labour, low capital, and skills requirements and subject to open or semi-open resource access: a reflection of the development characteristics of forested landscapes (Angelsen and Wunder 2003). For many rural communities, in the absence of access to skilled labour, markets, political power or credit, NTFP harvest and trade becomes increasingly important. As such, NTFPs provide safety nets, whereby the benefits provided by forest resources stop rural dwellers from becoming poorer and provide cash income at critical times of the year, particularly during times of low agricultural production (Angelsen and Wunder 2003; Ros-Tonen and Wiersum 2005, Chap. 3). When products become commoditised and highly profitable, few material benefits trickle down to the primary producer (Dove 1993, 1995; Marshall et al. 2003). In the context of extractive reserves in Brazil, Fearnside (1989: 388) writes: "when the value of trade accrues to intermediaries, extractivists remain poor, no matter how much wealth they generate" This scenario is often mirrored elsewhere in the tropics and is the reason why, in some instances, NTFPs are regarded as "poverty traps" (Angelsen and Wunder 2003: 21). Given the often low returns from NTFPs, studies of livelihood trajectories suggest that given alternatives to NTFP production, most people will prefer to practice intensive agriculture and become involved in wage labour, rather than rely on the forest alone (Levang et al. 2007; Paumgarten and Shackleton 2009). This further compromises the potential for the sustainable harvest of NTFPs to contribute directly to biodiversity conservation as agricultural expansion is a leading driver of forest loss.

Again, questioning the assumption that NTFP harvesting is inherently sustainable, there is considerable evidence that long-term harvesting of any non-timber forest product often results in resource depletion (Peters 1994; Ticktin 2004). Responses to depletion include exclusion through statutory or customary controls, other forms of legal control (tariffs and harvesting quotas), cultivation, and ex situ conservation (Hamilton 2008). However, from the perspective of the rural harvester, two local responses to scarcity are also common: (1) increasing the harvest range and (2) substitution (Cunningham 2000). In both the instances, it is important to note the weak link between biodiversity conservation and NTFP harvesting. In essence, the evidence suggests that resources will be utilised until they become scarce, and then either alternative sources are identified or the raw materials are substituted with more readily available materials:

- *Increased harvesting range*: a typical first response to resource scarcity is to increase the harvest range (Cunningham 2000). However, it is commonly found that local harvesters do not factor in the increased opportunity costs of the additional labour needed to collect these resources from a greater distance and that the "*payment received by households [for NTFPs] barely covers the opportunity cost of labour employed in harvesting*" (Southgate et al. 1996: 1).

As the demand for some products continues to expand, this response will undoubtedly lead to further scarcity and local extirpation.

- *Substitution*: in a number of instances, when a preferred species becomes scarce due to over-harvesting, a similar product is utilised in its place. For example, rattan baskets and other products are being replaced in rural African societies with synthetic substitutes due to resource depletion. The same occurs in the case of woodcarving, where indigenous species that have become overexploited are being replaced with fast-growing indigenous or, latterly, exotic species, particularly to supply the thriving Kenyan and Zimbabwean woodcarving industries.

More recent attempts to couple NTFP extraction with environmental and social responsibility through schemes such as fair trade and certification remain somewhat incipient (Shanley et al. 2008) and are hindered by the very issues discussed earlier in this chapter: lack of tenure security, elite capture, insufficient monitoring capacity, poor management capacity, and low levels of organisation among producers. It will take considerable effort and expertise, if not a complete paradigm shift, to make such schemes effective if they are to contribute to both biodiversity conservation and local livelihoods.

The conditions that could foster the sustainable harvest of NTFPs, secure property rights, low population densities, customary rule of use, and simple technologies, all seem to encourage more intensively managed systems *outside* highly biodiverse forested systems. This may be just as well as the growing network of global protected areas is essentially excluding access rights to many forest dwellers who rely on NTFPs for their livelihoods. As opposed to creating more protected areas, biodiversity could perhaps be better conserved through the sustainable use of resources in multi-strata, multi-functional landscapes (Putz et al. 2001). However, the complexities of such land-use planning, management and monitoring required to ensure compliance and for the required, and often intricate, management systems to be in place for such landscapes seem thus far insurmountable, and there remain few examples of holistic landscape scale management that provide optimum outcomes for both conservation and development (Sayer and Maginnis 2005).

Although the academic world has begun to realise the limitations of the NTFP sector and has recently been more sanguine in re-assessing this potential (Chap. 3), NTFP promotion and development remain a mainstay of many site-level projects. This disjunction provides an excellent example of a lack of communication between conservation practice and academia, recently discussed by Sunderland et al. (2009), and the same attempts at NTFP development and promotion continue to be made across the tropics as a tool to support forest conservation, albeit with very questionable outcomes for either conservation or local livelihood development.

In summary, despite the laudable and significant efforts at establishing extractive systems of NTFPs that foster the conservation of highly biodiverse forested systems, insufficient evidence has been offered over the past 20 years to suggest this is a feasible land-use option. Indeed, the fact that the primary approach to biodiversity conservation has focused on the establishment of protected areas that exclude local access for the harvest of forest products suggests that confidence in the ability of sustainable NTFP extraction to meaningfully contribute to conservation is relatively low. NTFP management as a component of complex landscapes, as we have discussed, can certainly play a role in contributing to more diverse production systems, but the potential for NTFPs to provide a singular solution to the on-going biodiversity crisis should be severely tempered. The common tendency to embrace simple solutions to complex problems and pursue what are essentially fads [be it NTFPs, debt-for-nature swaps, green marketing, payments for environmental services and probably, the next big thing, Reduced Emissions from Deforestation and Degradation (REDD)] needs to be replaced with a more holistic, long-term, multi-disciplinary, and equitable approach to integrating human needs and biodiversity conservation.

References

Alcorn J (1993) Indigenous peoples and conservation. Conserv Biol 7:424–426

Alcorn J (1995) Economic botany, conservation and development: what's the connection? Ann Mo Bot Gard 82:34–45

Ambrose-Oji B (2003) The contribution of NTFPs to the livelihoods of the "forest poor": evidence from the tropical forest zone in south-west Cameroon. Int Forest Rev 5:106–117

Angelsen A, Wunder S (2003) Exploring the forest-poverty link: key concepts, issues and research implications. CIFOR Occasional Paper No. 40, CIFOR, Bogor

Arnold JEM, Ruiz-Pérez M (2001) Can non-timber forest products match tropical forest conservation and development objectives? Ecol Econ 39:437–447

Aubertin C (2004) Cardamom (*Amomum* spp.) in Lao PDR: the hazardous future of an agroforest system product. In: Kusters K, Belcher B (eds) Forest products, livelihoods and conservation: case studies of non-timber forest product systems, Vol. 1 – Asia. CIFOR, Bogor, pp 43–61

Belcher B, Kusters K (2004) Non-timber forest product commercialisation: development and conservation lessons. Vol. 1 – Asia. In: Kusters K, Belcher B (eds) Forest products, livelihoods and conservation: case studies of non-timber forest product systems. CIFOR, Bogor, pp 1–22

Belcher B, Schreckenberg K (2007) Commercialisation of non-timber forest products: a reality check. Dev Policy Rev 25:355–377

Belcher B, Rujehan IN, Achdiawan R (2004) Rattan, rubber or oil palm: cultural and financial considerations for farmers in Kalimantan. Econ Bot 58:77–87

Boot RGA, Gullison RE (1995) Approaches to developing sustainable extraction systems for tropical forest products. Ecol Appl 5:896–903

Brockway L (2002) Science and colonial expansion, the role of the British Royal Botanic Gardens. Yale University Press, Newhaven

Browder JO (1992a) Social and economic constraints on the development of market-oriented extractive reserves in Amazon rain forests. Adv Econ Bot 9:33–41

Browder JO (1992b) The limits of extractivism: tropical forest strategies beyond extractive reserves. BioScience 42:174–182

Cernea MM (2005) 'Restriction of access' is displacement: a broader concept and policy. Forced Migr Rev 23:48–49

Chape S, Harrison J, Spalding M, Lysenko I (2005) Measuring the extent and effectiveness of protected areas as an indicator of meeting global biodiversity targets. Philos Trans R Soc B 360:443–455

Coad L, Burgess N, Fish L, Ravillious C, Corrigan C, Pavese H, Granziera A, Besançon C (2009) Progress towards the Convention of Biological Diversity terrestrial 2010 and marine 2012 targets for protected area coverage. Parks 17:35–42

Colfer C (1997) Beyond slash and burn: building on indigenous management of Borneo's tropical rain forests. Adv Econ Bot. Vol. II, New York

Cunningham AB (2000) Applied ethnobotany: people, wild plant use and conservation. Earthscan, London

De Beer J, McDermott M (1989) The economic value of non-timber forest products in Southeast Asia. Netherlands Committee for IUCN, Amsterdam

Dove M (1993) A revisionist view of tropical deforestation and development. Env Conserv 20:17–24

Dove M (1995) Political versus techno-economic factors in the development of non-timber forest products: lessons from a comparison of natural and cultivated rubbers in southeast Asia (and South America). Soc Nat Res 8:193–208

Falconer J (1990) The major significance of 'minor' forest products: the local use and value of forests in the West African humid forest zone. Forests, Trees and People, Community Forestry Note No.6, FAO, Rome

Farnsworth N (1988) Screening plants for new medicines. In: Wilson EO (ed) Biodiversity. National Academy Press, Washington, pp 83–97

Fearnside PM (1989) Extractive reserves in Brazilian Amazonia. BioScience 9:387–393

Freese CH (1997) The "use it or lose it" debate: issues of a conservation paradox. In: Freese C (ed) Harvesting wild species: implications for biodiversity conservation. Johns Hopkins Press, Baltimore, pp 1–48

Godoy RA, Bawa KS (1993) The economic value and sustainable harvest of plants and animals from the tropical forest: assumptions, hypotheses and methods. Econ Bot 47:215–219

Godoy R, Wilkie D, Overman H, Cubas A, Cubas G, Demmer J, McSweeney K, Brokaw N (2000) Valuation of consumption and sale of forest goods from a Central American rain forest. Nature 406:62–63

Gradwohl J, Greenberg R (1988) Saving the tropical forests. Earthscan, London

Guariguata MR, Cronkleton P, Shanley P, Taylor PL (2008) The compatibility of timber and non-timber forest product extraction and management. For Ecol Manage 256:1477–1481

Hall P, Bawa K (1993) Methods to assess the impact of extraction of non-timber forest products on plant populations. Econ Bot 47:234–247

Hamilton AC (ed) (2008) Medicinal plants in conservation and development: case studies and lessons learnt. Plantlife International, Salisbury

Hecht SB, Anderson AB, May P (1988) The subsidy from nature: shifting cultivation, successional palm forests and rural development. Hum Org 47:25–35

Hobhouse H (1999) Seeds of change: six plants that transformed mankind. Papermac, London

Homma AKO (1992) The dynamics of extraction in Amazonia: a historical perspective. Adv Econ Bot 9:23–31

http://www.ecologyandsociety.org/vol11/iss2/art20 Cited 25 November 2008

Hutton JM, Adams WM, Murombedzi JC (2005) Back to the barriers? Changing narratives in biodiversity conservation. Forum Dev Stud 2:341–370

Kiss A (2004) Making biodiversity conservation a land-use priority. In: McShane T, Wells M (eds) Getting biodiversity projects to work. Columbia University Press, New York, pp 98–123

Kusters K, Achdiawan R, Belcher B, Ruiz-Pérez M (2006) Balancing development and conservation? An assessment of livelihood and environmental outcomes of non-timber forest product trade in Asia, Africa and Latin America. Ecol Soc 11(2):20

LaFrankie JV (1994) Population dynamics of some tropical trees that yield non-timber forest products. Econ Bot 48:301–309

Laird SA, McLain RJ, Wynberg R (eds) (2010) Wild product governance: finding policies that work for non-timber forest products. Earthscan, London

Lawrence A (2003) No forest without timber? Int Forest Rev 5:87–94

Levang P, Dounias E, Sitorus S (2005) Out of the forest, out of poverty? For Trees Livelih 15:211–235

Levang P, Sitorus S, Dounias E (2007) City life in the midst of the forest: a Punan hunter-gather's vision of conservation and development. Ecol Soc 12(1):18, www.ecologyandsociety.org/vol21/iss1/art18 Cited 3 January 2009

Malleson R (1999) Community management of non-wood forest resources: a case study from the Korup Forest, Cameroon. In: Sunderland TCH, Clark LE, Vantomme P (eds) Non-wood forest products of Central Africa: current research issues and prospects for conservation and development. FAO, Rome, pp 117–122

Marshall E, Newton AC, Schreckenberg K (2003) Commercialisation of non-timber forest products: first steps in analysing the factors influencing success. Int Forest Rev 5:128–137

Michon G (2005) NTFP development and poverty alleviation: is the policy context favourable? In: Pfund JL, Robinson P (eds) Non-timber forest products between poverty alleviation and market forces. Intercooperation, Bern, pp 20–25

Ndam N, Tonye MM (2004) "Chop, but no broke pot": the case of *Prunus africana* on Mount Cameroon. In: Sunderland TCH, Ndoye O (eds) Forest products, livelihoods and conservation: case studies of non-timber forest product systems. CIFOR, Bogor, pp 37–52

Nepstad DC, Schwartzman S, (eds) (1992) Non-timber products from tropical forests: evaluation of a conservation and development strategy. Adv Econ Bot 9: 1–164

Neumann RP, Hirsch E (2000) Commercialisation of non-timber forest products: review and analysis of research. CIFOR, Bogor

Ortiz EG (2002) Brazil nut (*Bertholletia excela*). In: Shanley P, Pierce AR, Laird SA, Guillen A (eds) Tapping the green market: certification and management of non-timber forest products. Earthscan, London, pp 61–74

Ostrom E (2009) A general framework for analysing sustainability of socio-ecological systems. Science 325:419–422

Panayotou T, Ashton P (1992) Not by timber alone: economics and ecology for sustaining tropical forests. Island, Washington, DC

Paumgarten F, Shackleton CM (2009) Wealth differentiation in household use and trade in non-timber forest products in South Africa. Ecol Econ 68:2950–2959

Peters CM (1994) Sustainable harvest of non-timber plant resources in tropical moist forest: an ecological primer. Biodiversity Support Program, World Wildlife Fund, Washington, DC

Peters CM, Gentry A, Mendelsohn RO (1989) Valuation of an Amazonian rainforest. Nature 339:655–656

Phillips O (1993) The potential for harvesting fruits in tropical rainforests: new data from Amazonian Peru. Biodivers Conserv 2:18–38

Plotkin M, Famolare L (eds) (1992) Sustainable harvest and marketing of rain forest products. Conservation International. Island, Washington, DC

Prance GT (1992) Rainforest harvest: an overview. In: Counsell S, Rice T (eds) The rainforest harvest: sustainable strategies for saving the tropical forests. Friends of the Earth Trust, London, pp 21–25

Prance GT (1998) Indigenous non-timber benefits from tropical rain forest. In: Goldsmith FB (ed) Tropical rain forest: a wider perspective. Chapman and Hall, London, pp 21–42

Putz FE, Blate GM, Redford KH, Fimbel R, Robinson J (2001) Tropical forest management and conservation of biodiversity: an overview. Conserv Biol 15:7–20

Redford K (1992) The empty forest. BioScience 42:422–432

Redford K, Stearman AM (1993) Forest-dwelling native Amazonians and the conservation of biodiversity: interests in common or in collision? Conserv Biol 7:248–255

Richards M (1993) The potential for non-timber forest products in sustainable forest management in Amazonia. Common For Rev 72:21–27

Rodrigues ASL, Andelman SJ, Bakarr MI, Boitani L, Brooks TM, Cowling RM, Fishpool LDC, da Fonseca GAB, Gaston KJ, Hoffmann M, Long JS, Marquet PA, Pilgrim JD, Pressey RL, Schipper J, Sechrest W, Stuart S, Underhill LG, Waller RW, Watts MEJ, Yan X (2004) Effectiveness of the global protected area network in representing species diversity. Nature 428:640–643

Ros-Tonen MAF, Wiersum KF (2005) The scope for improving rural livelihoods though non-timber forest products: an evolving research agenda. For Trees Livelih 15:129–148

Ruiz-Pérez M (2005) Poverty alleviation and forest conservation: the role of non-timber forest products. In: Pfund JL, Robinson P (eds) Non-timber forest products between poverty alleviation and market forces. Intercooperation, Bern, pp 8–13

Sayer JA, Maginnis S (eds) (2005) Forests in landscapes: ecosystem approaches to sustainability. Earthscan, London

Schmidt CB, Burgess ND, Coad L, Belokurov A, Besançon C, Boisrobert L, Campbell A, Fish L, Gliddon D, Humphries K, Kapos V, Loucks C, Lysenko I, Miles L, Mills C, Minnemeyer S, Pistorius T, Ravilious C, Steininger M, Winkel G (2009) Global analysis of the protection status of the world's forests. Biol Conserv. doi:10.1016/j.biocon.2009.04.012

Scoones I, Melnyk M, Pretty JN (1992) The hidden harvest: wild foods and agricultural systems. International Institute for Environment and Development, London

Shackleton CM, Shackleton SE, Buiten E, Bird N (2007) The importance of dry forests and woodlands in rural livelihoods and poverty alleviation in South Africa. Forest Pol Econ 9:558–577

Shanley P, Luz L, Swingland IR (2002) The promise of a distant market: a survey of Belém's trade in non timber forest products. Biodivers Conserv 11:615–636

Shanley P, Pierce A, Laird S, Robinson D (2008) Beyond timber: certification and management of non-timber forest products. CIFOR, Bogor

Sheil D, Wunder S (2002) The value of tropical forest to local communities; complications, caveats and cautions. Conserv Ecol 6(2):9, http://www.consecol.org/vol6/iss2/art9 Cited 5 February 2009

Southgate D, Ritchie MC, Canelos PS (1996) Can tropical forests be saved by harvesting non-timber forest products? CSERGE Working Paper GEC 96-02, p 19

Struhsaker TT (1998) A biologist's perspective on the role of sustainable harvest in conservation. Conserv Biol 12(4):930–932

Sunderland TCH, Dransfield J (2002) Rattan (various spp.). In: Shanley P, Pierce A, Laird S, Guillen A (eds) Tapping the green market; certification and management of non-timber forest products. London, Earthscan, pp 225–239

Sunderland TCH, Harrison ST, Ndoye O (2004) Commercialisation of non-timber forest products in Africa: history, contexts and prospects, Vol. 2 – Africa. In: Sunderland TCH, Ndoye O (eds) Forest products, livelihoods and conservation: case studies of non-timber forest product systems. CIFOR, Bogor, pp 1–24

Sunderland TCH, Sunderland-Groves J, Shanley P, Campbell B (2009) Bridging the gap: how can information access and exchange between conservation biologists and field practitioners be improved for better conservation outcomes. Biotropica 41:559–554

Tabuna H (1999) The markets for Central African non-wood forest products in Europe. In: Sunderland TCH, Clark LE Vantomme P (eds) Non-wood forest products of Central Africa: current research issues and prospects for conservation and development. FAO, Rome, pp 251–264

Ticktin T (2004) The ecological consequences of harvesting non-timber forest products. J Appl Ecol 41:11–21

Tovey H (2009) Sustainability: a platform for debate. Sustainability 1:14–18

Tu DV (2004) Extraction and trade of Cardamom (*Amomum villosum*) from Ba Be National Park, Vietnam. In: Kusters K, Belcher B (eds) Forest products, livelihoods and conservation: case studies of non-timber forest product systems, vol 1, Asia. CIFOR, Bogor, pp 107–116

Willis KJ, Gillson L, Brnic TM (2004) How "virgin" is virgin rainforest? Science 304:402–403

Wong JLG, Thornber K, Baker N (2001) Resource assessment of non wood forest products: experience and biometric principles, vol 13, NWFP Series. FAO, Rome

Part IV
Building on the Opportunities Offered by Non-timber Forest Products

Chapter 11
Regulating Complexity: Policies for the Governance of Non-timber Forest Products

Sarah A. Laird, Rachel Wynberg, and Rebecca J. McLain

Abstract Products from the wild, also known as non-timber forest products (NTFPs), are used as medicines, foods, spices, fibers, and fuel and for a multitude of other purposes. They contribute substantially to rural livelihoods and generate revenue for companies and governments, and their use has a range of impacts on biodiversity conservation. However, throughout the world, NTFPs have been both overlooked and poorly regulated by governments. Inappropriate policies have not only led to over-exploitation but also generated new forms of inequity. Drawing upon cases from around the world, this chapter reviews these experiences and provides information to support new policy approaches toward NTFP regulation and the broader issues of governance associated with these products.

11.1 Introduction

Policies and laws play a central role in regulating trade in non-timber forest products (NTFPs), determining ecological sustainability, and influencing if and how communities benefit from use of these products. However, because NTFPs are harvested, used, and traded by a wide range of groups, in very different ways and contexts (geographical, ecological, economic, political, and cultural, among others) they are difficult to regulate even when great care is taken. Over the past few decades, pressure on policymakers to more effectively regulate NTFPs has increased the attention given to these products, but this new visibility has not

S.A. Laird (✉)
People and Plants International, P.O. Box 251, Bristol, VT 05443, USA
e-mail: sarahlaird@aol.com

R. Wynberg
Environmental Evaluation Unit, University of Cape Town, Rondebosch 7700, South Africa
e-mail: Rachel@iafrica.com

R.J. McLain
Institute for Culture and Ecology, P.O. Box 6681, Portland, OR 97214, USA
e-mail: mclain@ifcae.org

S. Shackleton et al. (eds.), *Non-Timber Forest Products in the Global Context*,
Tropical Forestry 7, DOI 10.1007/978-3-642-17983-9_11,

always been a good thing. As this chapter describes, regulatory measures instituted around NTFPs in recent decades were often tagged onto timber-centric forestry laws, were poorly informed, and had inadequate resources allocated for oversight and implementation. Consequently, in the end, they created new opportunities for corruption and exploitation and often, in conjunction with other bodies of law like agriculture and land tenure, provided perverse incentives to overharvest NTFPs. In many cases, policy interventions also criminalised NTFP extraction, further marginalising harvesters while generating new forms of inequity (Alexiades and Shanley 2005). Customary law and local institutions better suited to regulating NTFPs were also often undermined by efforts to establish statutory control over NTFPs (Arnold and Ruiz Pérez 2001; Michon 2005).

Numerous works have been published about NTFPs over the past two decades, including descriptions of their use, harvest and conservation, analyses of the factors influencing successful commercialisation (e.g., Neumann and Hirsch 2000; Sunderland and Ndoye 2004; Alexiades and Shanley 2005; Kusters et al. 2006; Marshall et al. 2006) and "how-to" manuals for inventorying and monitoring NTFPs or measuring their economic value (e.g., Peters 1996; Cunningham 2001; Shanley and Medina 2005; Stockdale 2005). While many of these works touch on policy issues, NTFP policy is not their primary focus. Other publications have a strong focus on policy (Dewees and Scherr 1996; Jones et al. 2002; Shanley et al. 2002; Michon 2005; McManis 2007; Wynberg and Laird 2007; Cunningham et al. 2009), but tend to be either geographically or topically narrow, i.e., dealing with a single or few species or types of products.

This chapter draws upon case studies from more than a dozen countries and complements the existing NTFP literature by providing a comparative analysis of a broad spectrum of experiences with NTFP policy and law from around the world. By including cases from postindustrial societies as well as the more commonly studied context of developing economies, it emphasises the truly global importance of these products, and highlights similarities in issues and lessons that emerge with NTFP regulation.

11.2 Why and How NTFP Laws and Policies are Developed

NTFP policies and laws are usually a complex, and often confusing, mix of measures developed over time, with poor coherence or coordination. They rarely resemble an overall policy framework. Many policy instruments are enacted as ad hoc responses to a crisis (e.g., perceived over-exploitation of a species) or an overly optimistic view of potential tax revenue should informal activities be made more formal. Rarely does regulatory activity follow from a careful and systematic assessment of the range of opportunities and threats associated with species, ecosystems and livelihoods and a strategic approach to regulating the NTFP sector as a whole is uncommon.

11.2.1 Reactive Policymaking

Reactive policymaking is often an inevitability associated with the NTFP commercial production cycle. The tendency for NTFP laws to be drafted in response to a real or perceived overharvesting crisis is widespread, especially when use of a species changes from local trade and subsistence use to large-scale commercial trade. Booms and busts in NTFP commercial cycles also result from consumer fads, scientific research that supports or undermines markets, and health concerns (Chap. 2). In the botanical and herb industry, for example, griffonia (*Griffonia simplicifolia*), kava (*Piper methysticum*), ephedra (*Ephedra sinica*), and cat's claw (*Uncaria tomentosa*) are just a few examples of species that have experienced increased sales in recent decades, followed by market crashes after media reports raised concerns about safety and efficacy (Alexiades 2002; Nalvarte Armas and de Jong 2005; Pierce and Burgener 2010). Health concerns associated with raw material supplies in the food sector often trigger reactive policy responses, as in the case of aflatoxins found in Brazil nuts sold in Europe and North America (Cronkleton and Pacheco 2010), with Chinese matsutake mushrooms harvested in Yunnan and sold in Japan (Menzies and Li 2010), and with palm hearts in Brazil and Bolivia (Fantini et al. 2005; Stoian 2005a).

Despite the risks associated with reactive and iterative NTFP policymaking, such interventions can also have strengths. For example, the succulent plant *Hoodia*'s entry into the weight-control market in 2001 led to a surge in demand for raw material that required southern African governments to respond rapidly by introducing a stringent permit system and, in some cases, prohibiting wild harvesting. A few years later, an increase in the availability of cultivated material reduced pressure on wild populations, and governments responded in turn with a less severe permitting system (Wynberg 2010) (Fig. 11.1).

Fig. 11.1 Wild-harvested *Hoodia gordonii*, Western Cape, South Africa (photo: David Newton)

11.2.2 Opportunistic Policymaking

Government action is often triggered when politically powerful groups lobby for regulation to increase their control over NTFP production and trade. For example, the Rooibos Tea Control Scheme established by the apartheid state of South Africa in 1954 was promoted by and benefited the "white" farming elite, rather than the mostly "coloured" farmers who had traditionally gathered rooibos tea from the wild. The scheme was a statutory, one-channel marketing system set up to regulate the production and marketing of indigenous rooibos (*Aspalathus linearis*) tea and to support the sector, including subsidies for affiliated producers, research, and the provision of extension services (Hayes 2000; Wynberg 2006).

Governments are also quick to act when a species or set of products appear to show great economic promise, part of which they might capture through royalties, taxes, or other means. In Cameroon, the government instituted new taxes on medicinal plants in the 1990s in response to a widespread belief that these NTFPs were "green gold" (Laird et al. 2010). In India, tendu (*Diospyros melanoxylon*), which provides as much as 74% of Orissa state's total earnings from forests, was nationalised in several states in the 1960s and 1970s due to its high value and the interest of government bodies in benefiting from its trade (Lele et al. 2010).

11.2.3 Information Requirements for Drafting Effective Policies

A common problem with NTFP law and policy is limited understanding on the part of policymakers about the products, people, and activities they seek to regulate. Unlike timber or agricultural crops, NTFPs include a broad range of species with extremely different ecologies and cultural and livelihood roles, and equally diverse market chains, end products, and consumers (Arnold and Ruiz Pérez 2001; Arnold and Ruiz Perez 1996; Peters 1996; Shanley et al. 2002; Alexiades and Shanley 2005). For most species, there remain enormous gaps in understanding, including those widely used such as Brazil nuts, devil's claw (*Harpagophytum* spp.), and eru (*Gnetum* spp.) (Chap. 7).

Solid background information is critical to policy formulation. For example, because NTFPs are an extremely diverse array of species, with a wide range of ecological niches, policymakers cannot assume that intensification of harvesting will have similar impacts in all cases (Chap. 7). Marula (*Sclerocarya birrea* subsp. *caffra*) is widespread and common, fruits abundantly, is planted in homesteads, is retained in fields, and is usually well managed in the southern African region. These circumstances suggest a resilience that does not require immediate government intervention, but rather calls for monitoring of populations in areas with heavy harvesting rates (Shackleton et al. 2003; Wynberg and Laird 2007) (Fig. 11.2). Vesi (*Instsia bijuga*) in Fiji, on the other hand, is slow growing, occurs in low densities, is scattered in distribution, and does not disperse well, all of which are characteristics that make it vulnerable to overharvesting. In addition, *Intsia bijuga* is

Fig. 11.2 Workers for a local NGO project squeezing juice from *Sclerocray birrea* (marula) fruits for sale, Limpopo Province, South Africa (photo: Myles Mander)

experiencing commercial pressure from the tourist trade, new technology has increased harvesting rates, and cultural changes have eroded customary laws and beliefs that hold *Intsia bijuga* to be a sacred species. This combination of factors has led to a sustainability crisis that, unlike the case of marula, requires legislative and policy attention (Areki and Cunningham 2010).

11.2.4 Consultations Associated with Laws and Policies

Consultations with stakeholders are probably the most important way to gather information and to set priorities and objectives for policy. However, in most countries, NTFP harvesters and producers are drawn from the least powerful members of society and typically have little say in policymaking (Hecht et al. 1988; Shanley et al. 2002; Shackleton and Shackleton 2004; Alexiades and Shanley 2005; Wynberg and Laird 2007). Because such groups are rarely consulted during policy design, their needs seldom drive the policymaking process. Technical experts and even nongovernmental organisations (NGOs) (which may not be representative of producers and harvesters, but can provide important assistance) often have more significant input into the design and drafting process than those directly involved in the harvest or trade of products. The consultations that do take place for NTFP law and policy are often with larger and more powerful business interests.

One reason for the limited involvement of harvesters in the policy process is the dearth of producer organisations or institutional vehicles through which their views and concerns can be expressed, and a lack of organisational capacity to do so. Even in recent decades, Brazil nut measures were drafted and passed in Bolivia without public consultation. It was only in the late 1990s that small Brazil nut producers finally forced their views into the public arena, in part by being better organised (Cronkleton and Pacheco 2010). In the United States, Canada, and the United Kingdom, some effort has recently gone into including harvesters, buyers, and processors in proposed regulatory reforms, either through the formation of industry-specific task forces or through public hearings (Dyke and Emery 2010; McLain and Lynch 2010; Mitchell et al. 2010). In southern Africa, the nonprofit trade association PhytoTrade Africa plays an important role in enabling the voice of marginalised producers to be heard (PhytoTrade Africa 2006) (Chap. 4).

11.2.5 The Few Strategic Exceptions

A few governments have developed NTFP law and policy in a more strategic manner. This includes undertaking research and building ecological, economic, social, and cultural understanding of species, incorporating comprehensive consultations with stakeholders and developing a strategy for the resulting legal framework.

In the past decade, for example, Namibia has taken a proactive and progressive approach toward NTFP policy and regulation, recognising that these products provide vital income and livelihoods for communities in an environment characterised by extreme aridity and few economic opportunities (Bennett 2006; Cole and Nakamhela 2008; Nott and Wynberg 2008; Wynberg 2010). Much of this has been done through the multistakeholder Namibian Indigenous Plant Task Team, which promotes collaborative approaches and effective regulation and facilitates development of the local natural products industry (Nott and Wynberg 2008).

Finland is also a notable exception to the rule of government neglect of NTFPs. The Finnish government has supported scientific research on wild berries for decades, including studies of their cultural and economic importance, as well as biological and ecological research (Kanga 1999). At the same time, it has actively promoted berry and mushroom harvesting as an economic activity and cultural practice. Indeed, rather than discouraging harvesting as many countries have done, the government has developed programs to promote harvesting and related industries. These include a berry crop forecasting system and income-tax relief favourable to harvesters, providing them with the information and incentives they need to participate more effectively in NTFP industries (Richards and Saastamoinen 2010).

11.3 The Policies

11.3.1 Policies and Laws that Directly Address NTFPs

A number of laws and policies directly address NTFPs, often to conserve or sustainably manage resources, and in some cases to improve rural livelihoods or promote broader economic growth in a region. These measures tend to focus on species in commercial trade, or form part of national efforts to protect endangered or indigenous species or regulate international trade under the Convention on International Trade in Endangered Species of Wild Fauna and Flora (CITES). The majority of measures directly addressing NTFPs, however, are found in natural resource law, in particular forestry laws. A range of other measures explicitly regulate specific aspects of NTFP trade and use, including quality control, safety and efficacy standards, transportation, taxation, and trade (Pierce and Burgener 2010).

11.3.1.1 The Inclusion of NTFPs in Forestry Laws of the 1990s

In most countries, forestry laws historically focused almost exclusively on timber resources and paid limited or no attention to NTFPs. Moreover, the subsistence and commercial value of NTFPs was totally disregarded when timber management plans were designed and logging operations undertaken. In recent decades, however, NTFPs have been incorporated into forestry laws as a response to changing international policy trends. In many cases, this resulted from the direct pressure of international agencies, such as large conservation organisations and finance institutions, including the World Bank, to diversify forest management and make it more sustainable (Laird et al. 2010). As a result, in the 1980s and 1990s, many countries integrated a wider range of objectives into forest policies, including forest health and biodiversity conservation, ecosystem functions, and long-term sustainability, as well as broader economic values such as tourism, recreation, and NTFPs.

However, initial efforts to address NTFPs in these new forestry laws were poorly formulated and rarely implemented. The scope and definition of the products covered remained unclear, and few specific actions were stipulated (e.g., Fiji Islands 1992; Republic of Cameroon 1994; República de Bolivia 1996a). When actions were prescribed, they usually focused on permits, quotas (often set in arbitrary ways), management plans, and royalties or taxes – an approach lifted directly from the timber sector, and one that proved entirely inappropriate for the diverse, complex, and perhaps less lucrative NTFP sector.

More usefully, some forestry laws of this time included NTFPs in timber norms, requiring their consideration in management plans and logging operations in order to minimise negative impacts on locally valuable products (Chap. 8). In many countries, the logging of high-value NTFP species for timber has proved their greatest threat. In Brazil in recent years, national and state governments have passed laws prohibiting the logging of high-value NTFP species (Kluppel et al. 2010), and in Bolivia prohibitions on felling Brazil nut trees arrived in 2004 as part of a decree addressing property conflicts (Cronkleton and Pacheco 2010). But the track record for implementing such policies is often poor (e.g., Ortiz 2002; Pierce and Burgener 2010).

In the past 10–15 years, a number of countries have begun to fine-tune well-intentioned forest policies passed in the 1990s to reflect the socioeconomic, ecological, and cultural realities of NTFP use. This has resulted in a number of specific improvements in the ways these products are regulated, including rethinking the use of costly and complex inventories and management plans for NTFPs and revising quota and permitting systems (Areki and Cunningham 2010; Cronkleton and Pacheco 2010; Kluppel et al. 2010; Laird et al. 2010). There is still a long way to go, and NTFPs continue to have low priority in most forestry departments and curricula, but the trend in several countries is toward greater understanding and better-elaborated regulatory frameworks.

11.3.1.2 Quality Control, Safety, and Efficacy

Quality control and proof of safety and efficacy are increasingly important in developed country markets. This means that NTFP producers may be required to institute sophisticated procedures for tracking materials that end up as botanicals, personal care and cosmetic products, and food and beverages. Food safety legislation has often proved a formidable obstacle to international trade of NTFPs (Iqbal 1993; Brown 2005; Burgener 2007; Pierce and Burgener 2010). However, governments tend to act quickly when these obstacles arise; unlike environmental and social justice concerns, health concerns often get their attention, and pressure from influential commercial players involved in the trade can be great. For example, in the 1990s, when the EU and the USA set maximum acceptable levels of aflatoxins that threatened the Brazil nut trade, the Bolivian government jumped into action, passing a series of measures that created norms for Brazil nut classification, sanitation practices, and aflatoxin sampling, drawing upon the Food and Agriculture Organisation's Codex Alimentarius (Soldán 2003, in Cronkleton and Pacheco 2010). These steps allowed Bolivian Brazil nuts to maintain access to international markets.

11.3.1.3 Transportation

Transportation laws can have direct and indirect impacts on NTFPs. Most significant for all natural resources, including NTFPs, is the opening of previously remote forest areas following road building. More specific to the case of NTFPs is the use of transportation law to monitor trade. The State of Washington in the USA relies heavily on transportation permits as a mechanism for monitoring and tracking the harvesting of floral greens and other NTFPs; these permits also play an important role in identifying thefts of products from state and private land (McLain and Lynch 2010). In Brazil, a 1993 regulation required a license to transport any forest product. This included essential oils, medicinal plants, and the seedlings, roots, bulbs, vines, and leaves of native plants, many of which were not regulated in any other way. Because the law was so broad, and local harvesters and traders could not easily acquire the necessary license, they either could not participate in commercial trade or did so illegally. This measure was amended in 2006, in response to these problems (Kluppel et al. 2010).

11.3.1.4 Taxation, Including "Unofficial Taxation"

Governments sometimes tax the NTFP trade to gain revenue from what is perceived as a lucrative business, but this often negatively impacts the sector. In Cameroon, new taxes instituted in the 1990s on the medicinal plant export business resulted in the near collapse of that sector, and a blossoming of bureaucracy and opportunities for corruption (Laird et al. 2010; Ndoye and Awono 2010). In Bushbuckridge, South Africa, the government charges kiaat (*Pterocarpus angolensis* – African or wild teak) harvesters and craftsmen a fee per running meter of wood to promote responsible use of this valuable material. In reality, however, reports of harassment and corruption (e.g., government rangers taking wood or issuing incorrect receipts) are common. As a result, craftsmen and harvesters usually choose to bypass the system (Shackleton 2010) (Fig. 11.3). Some governments, however, use tax structures as a way of providing incentives to the NTFP sector. In Finland, for example, to encourage and support harvesters, and to offer the sector an incentive, the government makes picking income exempt from tax (Richards and Saastamoinen 2010).

"Unofficial taxation" (i.e., bribery) is a very real cost of doing business in many countries. Bribes are tolerated, and even encouraged, by some governments, and they work like any other policy stick to change behavior. In a number of countries, roadblocks are set up by government officials to "control" the transport of goods from rural to urban areas, check required documents, bleed profits from traders, and have knock-on effects for harvesters (Arquiza et al. 2010; Ndoye and Awono 2010; Sunderland et al. 2010). In The Philippines, one study showed that unofficial payments, or "standard operating procedures" (SOPS), significantly impact the already meager NTFP livelihoods of indigenous peoples (Arquiza et al. 2010).

Bribery can be a good indicator not only of problems with broader governance, but also with NTFP policies and laws. Bureaucratic and confusing NTFP measures can leave communities and government authorities unclear about proper procedures, providing government officials an opportunity to request additional "unofficial payments" (Arquiza et al. 2010; Laird et al. 2010; Ndoye and Awono 2010).

Fig. 11.3 Locally produced Kiaat (*Pterocarpus angolensis*) carvings for sale at a popular tourist destination (viewpoint) in Mpumalanga Province, South Africa (photo: Sheona Shackleton)

Inappropriate and burdensome measures can also make unofficial payments or bribes preferable to following regulations.

11.3.2 Policies and Laws that Indirectly Impact NTFPs

In addition to laws that explicitly address NTFPs, there are a myriad of measures that may not mention the term, and yet impact their use, management, and trade as much as, or more than, those that do (Dewees and Scherr 1996). The high impact of these measures is largely because the role of NTFPs in subsistence and local livelihoods is often poorly understood and rarely considered when drafting other measures. Laws tend to be drafted along sectoral lines that do not take into account other land uses and the complex and interconnected nature of activities.

Laws and policies with an indirect impact on NTFPs include agricultural policies, land tenure and resource rights, intellectual property, and labour law. In addition, a range of natural resource laws have a significant impact on NTFPs, including the forestry laws discussed above, mining (Novellino 2010) and protected area and conservation laws that discourage or forbid NTFP harvesting (e.g., Baird and Dearden 2003; Jaireth and Smyth 2003; Dowie 2005).

11.3.2.1 Agricultural Policies

Agricultural policies can impact NTFPs in a range of ways. They might discourage or promote farming practices that are linked to NTFP harvests and associated livelihoods. For example, in the 1990s, an international policy movement identified swidden (slash and burn) agriculture as a major cause of tropical deforestation. Although this was unproven and controversial, the impact of restricting practices associated with swidden agriculture was significant, including on NTFPs. In the case of the Batak in Palawan, these policy restrictions led to a surge in NTFP harvesting and trade to buy food to supplement low agricultural production (Novellino 2010). Agricultural policies can also include subsidies and other incentives to cultivate NTFPs, with both positive and negative impacts on rural livelihoods and species. The cultivation of rooibos tea in South Africa, for example, is promoted by a regulatory framework that encourages the clearing of natural biodiversity for rooibos plantations, and discourages wild collection of this species (Wynberg 2006).

Agricultural policies can also be a vehicle for land and resource rights reform, with significant consequences for NTFPs. For example, the 1996 Agrarian Reform Law (República de Bolivia 1996b) in Bolivia initially appeared to have little relevance for the Brazil nut economy, but its impact was dramatic because it sought to resolve the complex and contradictory property rights system of the country (Cronkleton and Pacheco 2010). Agricultural policies can also impact NTFPs through their effect on the supply of labour available to harvest products. In Finland, the loss of domestic price supports for agricultural products following the country's accession to the EU in 1995 accelerated rural economic restructuring and the out-migration of many rural residents to urban areas. To overcome the resulting labour shortage during the berry season, Finnish berry companies have increasingly turned to the use of immigrant labour, thereby creating further changes in the NTFP economy (Richards and Saastamoinen 2010).

11.3.2.2 Land Tenure and Resource Rights

NTFPs are harvested under a wide range of landownership systems, including communal, private, and various tiers of state control, and under different access regimes, from strict prohibitions on use through to open access. Four basic kinds of rights typically underpin such systems: use, transfer, exclusion, and enforcement (Neumann and Hirsch 2000). The many combinations of rights and forms of ownership mean that NTFP tenure systems are complex. However, clear land tenure and resource rights are fundamental to the success of any NTFP policy measure seeking equity and sustainability (Chaps. 9 and 12). These rights do not necessarily take the form of government titles, something often not possible in vast rural areas, but there must be a working understanding between stakeholders. When such understanding is not in place, conflicts over NTFP resources are common (e.g., Arquiza et al. 2010; Cronkleton and Pacheco 2010; Laird et al. 2010; Novellino 2010).

In some cases, land tenure may be secure, but resource rights are not. In Mexico, most forests are collectively owned, and while local communities have some autonomy in the management of their natural resources, the state sporadically exerts control over their use. For example, agave extraction has been regulated for hundreds of years through local institutions within the *ejido* and indigenous community structure. These have been responsible for regulating access, management practices, and the distribution of benefits based on history and traditional knowledge of the species. Norms and agreements are established by general assembly and are continually modified or replaced in a dynamic process that responds to new situations and to tensions of environmental, socioeconomic, cultural, or technological origin. Even with such a dynamic and sophisticated system, however, the Environmental Protection Agency now often fines local harvesters when they do not present a legal harvesting permit (Granich et al. 2010).

In Yunnan, China, changing land and resource rights have created opportunities for greater local control and a more effective policy framework for matsutake mushroom harvests. During most of the latter half of the twentieth century, China's forests were under state ownership. In the 1980s, however, forests were divided into state, collective, and household holdings. In Yunnan, forests under the new tenure arrangements continued to be managed largely for timber until 1998, when logging was banned as a flood prevention measure. These developments coincided with expansion in demand for the region's matsutake, a product that previously had little value and for which rights of tenure and usufruct were in flux. This state of flux and the resulting flexibility in tenure arrangements left space for villages to develop codes of conduct for access to local matsutake grounds and the monitoring of harvest practices. Local regulation has had the added benefit of fostering adaptive management, since villages can adjust to new conditions more quickly and easily than higher levels of government (Menzies and Li 2010).

The security of resource rights may also depend on the commercial value of an NTFP. This is illustrated in India, where the state owns all NTFPs and grants usufruct rights for collection, as well as transport and sale. In theory, the state is involved in resource rights to protect and benefit collectors, but in practice the distribution of income from these resources is considered highly inequitable, and government is interested only in those species with high commercial value like tendu (*Diospyros melanoxylon*). Political devolution has recently transferred rights over many NTFPs to local communities, but these are primarily products of low commercial value. The state retains control over more lucrative NTFPs (Lele et al. 2010).

Resource rights are undergoing change alongside broader views of property rights in many developed countries of the North. In Sweden and Finland, for example, the centuries-old principle of "everyman's right" (see Box 5.1) to harvest wild berries and mushrooms is being tested by the seasonal in-migration of large numbers of non-Nordic pickers, raising public concerns about immigration and tax policies, labour practices, and benefit sharing (Richards and Saastamoinen 2010); in England and Scotland, tension exists between customary rights to roam and the codified versions of those rights (Dyke and Emery 2010); and in Canada, in a reversal of trends in many other countries, as part of asserting aboriginal rights and title, First Nations are demanding the return of their right to regulate access to NTFPs (Mitchell et al. 2010).

When intact, customary law can play an important role in ensuring sustainable and equitable use of NTFPs. Arquiza et al. (2010) describe landownership vested in Philippine communities, each with its own rattan territory and many with strong customary laws that promote sustainable rattan management. Communities with a poorly defined sense of collective ownership and no traditional institutions tend to have weaker enforcement and manage resources less sustainably. Similarly, in the case of marula (*Sclerocarya birrea* subsp. *caffra*) in southern Africa, Wynberg and Laird (2007) found that where tenure is secure, customary laws are strong, and local capacity exists to manage the resource base and deal with the pressures of commercialisation, customary law achieves a balance between sustainable resource use and livelihood needs. However, when customary laws are weak and insecurities persist with land tenure and resource rights, significant conflicts arise around resource management, and government intervention is often necessary. In Fiji, 83% of the total land area is under customary tenure (native lands) as a result of British colonial policy that prohibited the sale of land to colonial settlers. However, even with secure land tenure and resource rights, dramatic social, cultural, technological, economic, and other changes have strained customary and local laws and have led to significant sustainability problems for *Intsia bijuga* (Areki and Cunningham 2010).

In many countries, customary and statutory laws play complementary roles, but it is common for new statutory laws to weaken effective customary systems. For example, in Bolivia, small producers maintained strong de facto control over the resource base for decades through a customary system of tree tenure. Access rights were based on rubber trails and later, when Brazil nuts became important, on access to Brazil nut trees and related infrastructure. All these activities operated in a statutory policy vacuum until 1995. At that time the government superimposed another layer of rights over the region's forests by allocating timber concessions. Conflicts were further exacerbated when well-intentioned efforts to modify the 1996 Agrarian Reform Law to expand the size of land grants to communities instead undermined customary tree tenure arrangements. Land reform gave smallholders formal recognition of their tenure rights, but by basing it on control of contiguous territory (allocating each family 500 ha), it undermined effective traditional tenure arrangements and access rights based on key resources (once rubber, and now Brazil nut trees) (Stoian 2005b; Cronkleton and Pacheco 2010).

11.3.2.3 Intellectual Property Rights

Policies relating to intellectual property rights (IPRs) can also have a significant impact on NTFP harvest and trade. The Agreement on Trade-Related Aspects of Intellectual Property Rights (TRIPS) of the World Trade Organisation has created a global regime for IPRs, the result of which is that many NTFPs are increasingly included in patents and other forms of IPRs (Dutfield 2002). This has important implications for the broader trade in and use of these products, since IPRs can create barriers against nonaffiliated companies entering the market (Gebhardt 1998). If narrowly applied, IPRs need not restrict the trade or commercialisation of products by other companies or groups, but there are a number of cases where this has occurred. For example, the 1997 patenting of active components of *Hoodia* and the specification of a particular extraction technique have directly inhibited trade in *Hoodia* extracts over the past decade (Wynberg et al. 2009; Wynberg 2010).

The pharmaceutical, crop protection, and seed industries, in particular, use patents to protect innovations, and plant breeders' rights (or plant patents in the USA) serve the same function in the horticultural industry. To a lesser extent, patents and other IPRs are also used in industries that rely on whole-plant material, such as the botanical medicine and personal care and cosmetic industries. These products contain multiple compounds and therefore do not lend themselves easily to patent protection, but other areas of product development, such as manufacturing and processing techniques, formulations, dosage forms, and unique release characteristics, enable IPRs to be secured. IPRs are clearly a complex, difficult, and expensive way for small-scale producers to ensure benefits from NTFPs, although trade organisations such as PhytoTrade Africa (see Chap. 4) are increasingly looking toward using intellectual property tools to protect small producers and enhance their competitiveness.

Increasingly, geographical indications, or appellations of origin, are used as an intellectual property mechanism to protect regional products and the communities associated with them. This is done through labels on products identifying the country, region, or locality from which they originate, and that yields the particular qualities or reputation associated with the products (Commission on Intellectual Property Rights 2002). Because geographical indications are anchored to a region and are a means to identify and market products easily, they can play a role in protecting traditional and cultural practices, as well as local economies associated with non-timber and other products. However, if poorly applied, geographical indications can also result in the disenfranchisement of local groups (Granich et al. 2010).

11.3.2.4 Labour

Labour policies, and those like immigration that directly affect labour supplies, can have significant impacts on NTFPs and those whose livelihoods depend on them. These impacts are particularly evident in the case studies from the global North, where many countries have experienced significant rural restructuring over the past two decades. In the north-western USA in the 1990s, for example, floral greens harvesters were transformed from self-employed sole proprietors or microfirms with relatively independent access to floral greens harvesting sites to predominantly de facto wage labourers heavily dependent on the floral greens companies not only for access to harvesting sites but also for the transport needed to get to those sites (McLain and Lynch 2010). In the UK and Finland, rural restructuring has also been accompanied by an influx of immigrants to harvest NTFPs, but most of these have legal authorisation to be in those countries and wage laborer conditions analogous to those in the USA have not developed.

Insider–outsider conflicts about accessing, harvesting, and trading NTFPs are significant and occur consistently around the world. NTFPs are an important, and sometimes the most easily accessed, source of cash for rural communities. "Outsiders" often enter communities' lands to harvest products without permission, use destructive methods, and take more than wild populations can support, disregarding local and customary laws and controls (Lynch and Alcorn 1994; Michon 2005; Wynberg and Laird 2007; Laird et al. 2010; Novellino 2010). This dynamic is played out from northern Europe to South Africa, and from Palawan to Canada to Bolivia. Migrants might harvest for their own use, but most often they exploit an available commercial opportunity, sometimes under contract with companies. The government of Sweden sought to ease tensions between local and migrant harvesters of wild berries by eliminating tax advantages for migrants (Richards and Saastamoinen 2010). In some cases, so-called "outsiders" have resided in a region for generations (e.g., Cronkleton and Pacheco 2010). Policymakers must tread carefully when dealing with this potential minefield. Both insiders and outsiders require support, but in very different ways, and measures should take into account, and guard against inflaming, this common form of conflict.

It is also important for policymakers to consider the many different types of labour involved in the harvest, trade, and processing of NTFPs. Harvesters and producers typically receive a small fraction of the final value of NTFPs (e.g., Padoch 1988; Hersch-Martinez 1995; King et al. 1999; Biswas and Potts 2003; Schreckenberg 2004; Arquiza et al. 2010). In general, profits from NTFPs increase with greater processing and as the value chain progresses, as does political power (Southgate et al. 1996; Neumann and Hirsch 2000; Schreckenberg 2004; Alexiades and Shanley 2005; Cronkleton and Pacheco 2010). Existing inequities and power imbalances in the value chain should be understood by policymakers in order to create laws that benefit all stakeholders, and do not set them against each other.

11.4 Common Features of NTFP Policy and Legal Frameworks

11.4.1 The Tension Between Broad Policy Prescriptions and the Need to Limit the Scope of Laws

Measures regulating NTFPs must carefully balance a wide range of objectives. These might include the protection of species under threat, the promotion of sustainability, the distribution of greater benefits to harvesters and producers, quality control, the generation of government revenues through taxation, and support for local businesses. A law heavily weighted to serve a single goal and one category of products (e.g., commercially traded medicinal plants and increased tax revenues) might create obstacles for achieving objectives associated with different kinds of NTFPs or stakeholders (e.g., improved livelihoods from local trading or subsistence use of the same species).

As described, the majority of laws that specifically regulate NTFPs do so in response to perceived threats to a species, and the result is often a narrow scope: species-based measures or those regulating a category of products, rather than umbrella measures for a wide range of NTFPs. In some cases, this may be the most effective response. However, this type of measure runs the risk of producing unintended consequences if it lumps locally traded and subsistence NTFPs into a regulatory framework designed for commercially traded species.

There is an inherent tension in the objectives and scope of NTFP laws: on the one hand, there exists a need for broad measures that address a range of species and, on the other, measures must be focused to be effective and meaningful, and avoid unintended consequences. How to focus and narrow the scope of laws is a challenge, however, and requires significant understanding.

11.4.2 The Tendency Toward Overwhelming Bureaucracy and Reporting Requirements Inappropriate for Small-Scale Producers

NTFP regulations are often unnecessarily bureaucratic. Regulations lifted from industrial timber production that include permitting, fees and management plans have proven unworkable. Even regulations tailored to NTFPs can be cumbersome, and often favour large-scale commercial exploitation over small-scale NTFP harvesters or producers. In one area of Mexico, for example, it is easier to obtain authorisation to log timber than to extract mushrooms (Granich et al. 2010). In the Philippines, the Department of Environment and Natural Resources established community-based forest management agreements to allow communities to manage forests for NTFPs, but the bureaucratic obligations that came with these agreements proved insurmountable for most indigenous communities (Arquiza et al. 2010; Novellino 2010). In Cameroon, complex bureaucratic requirements create obstacles for both large- and small-scale traders, and have driven much of the commercial trade in medicinal plants underground (Laird et al. 2010).

Most policies assume that communities are literate, have technical skills or funds to pay experts, and can easily find cash to pay for permits. This is rarely the case. Additionally, the logic underlying elaborate regulations eludes most harvesters and producers because they offer little or no benefit in return for increased cost and effort, sometimes criminalise NTFP extraction, and open the door to corruption and exploitation at the hands of government officials. Ill-conceived and bureaucratic requirements associated with government interventions are unlikely to change, however, and this is an important reason why "less is often more" when it comes to NTFP regulation (Wynberg and Laird 2007).

11.4.3 Poor Coordination of Laws and Policies Resulting in Inconsistency, Conflicting Mandates, and Confusion About Jurisdiction

NTFP laws and policies tend to be poorly integrated with existing federal, provincial, or state laws and are rarely coordinated with customary law. A comprehensive policy framework for NTFPs that addresses laws and policies acting at different levels requires time, funds, research, and comprehensive consultations with stakeholders. This level of investment in NTFP law and policy is extremely rare. The result is legal frameworks that are inconsistent and confusing, and a lack of clarity about which laws and government departments have jurisdiction over these products and activities.

For example, the NTFP policy environment in South Africa is characterised by a plethora of inefficient and sometimes contradictory national and provincial laws. These laws are only sporadically implemented, are often incompatible with each other, and are largely unknown by local communities. The laws then interface with customary systems that have eroded to varying degrees as a result of colonial and apartheid administration, but often offer the most effective regulation for NTFPs (Wynberg and Laird 2007; Shackleton 2010).

11.4.4 Inconsistent and Often Underfunded Policy Implementation

It is difficult to interest governments in effective NTFP law and policy because NTFPs fall into institutional and sectoral cracks, and are usually part of informal or loosely organised trade, or are consumed for subsistence. Moreover, most producers are politically and economically marginalised and there is little political will to address their needs. When governments do engage with this sector and draft laws, it is common for implementation, monitoring, and compliance to be poor since resources and capacity are rarely allocated to what are perceived as minor products (Tomich 1996; Wynberg and Laird 2007; Areki and Cunningham 2010; Laird et al. 2010). In Fiji, for example, the government recently sought to regulate the NTFP sector more effectively through the 2007 National Forest Policy and the Endangered and Protected Species Act of 2002. Despite good intentions, however, implementation has been weak: few traders know of the laws, and monitoring and enforcement is nonexistent (Areki and Cunningham 2010).

Sometimes a lack of implementation results when government departments compete with each other or their mandates conflict or overlap. As a result, no institution delegates the resources or staff needed to implement NTFP regulations (Antypas et al. 2002). In Cameroon, the 1994 Forestry Law (Republic of Cameroon 1994) set up an NTFP Subdirectorate within the then Ministry of Environment and Forests. This new body was provided with a civil servant to oversee activities, but had no budget and extremely limited power compared to the timber interests residing in the same ministry. Financial returns from taxes and fees on NTFPs went to other departments and ministries (Laird et al. 2010). It is often the case that revenue streams, which could strengthen and build capacity within government to effectively regulate and manage NTFPs, are diverted to other, more powerful, entities in government. In the Western Ghats in India, for example, royalties collected on uppage (*Garcinia gummi-gutta*) went to the state treasury, with no allocation for conservation of the resource, and state efforts focused on policing the movement of material in order to collect royalties, rather than monitoring harvest and trade to ensure sustainability (Lele et al. 2010).

Unimplemented policy measures can be worse than no measures. In some cases, they weaken traditional structures that might better promote sustainable management or equity in trade; even cursory government regulation of NTFPs can undermine community institutions and control over resources (Arnold and Ruiz Pérez 2001; Michon 2005). Confusion, conflict, and corruption can also result when laws are unclear or unenforced, making the lives of producers, harvesters and traders more difficult and encouraging unsustainable harvests of species (Arquiza et al. 2010; Laird et al. 2010; Ndoye and Awono 2010).

11.5 Conclusions and Recommendations

A few catchphrases emerge repeatedly in NTFP policy cases from around the world – "less is more", "carrots not sticks", "leave well enough alone", "the best-laid plans" – all suggesting a sector that has endured poorly directed and formulated policy. The need for better information, simplification, clarity, and consistency in NTFP policy frameworks is repeatedly stated. Although the state of NTFP law and policy is not encouraging at present, it is possible that recent interest in laws and policies regulating NTFPs will yield more strategic, better-informed, and effective policy frameworks. Following are some recommendations to help move in this direction.

(a) The extent of commercialisation and the heterogeneity of NTFP resources, markets, and stakeholders should be reflected in policies and laws.

- The extent of commercialisation should have a strong bearing on the nature of regulations. Laws should recognise the different types of NTFP use, including subsistence, local trade, commercial trade, and recreation. For example, subsistence use should not be regulated except in cases where there are clear risks of overharvesting, but government attention should be paid to internationally traded industrial-scale NTFPs.
- NTFP measures should be flexible and adaptive to accommodate shifts in market demand, safety concerns and other common disruptions to NTFP trade.
- Market access is as important as market prices for small-scale producers. Policies that support certification and other efforts to set producers apart from competitors are most effective when the administrative costs of such systems do not exceed their benefits.
- Processors and traders often control NTFP sectors, with small-scale producers having limited power over the commercial trade, including prices. Policymakers can help reduce monopolistic tendencies in NTFP markets, but should do this in a way that supports all stakeholders along the value chain and does not set them against each other.
- Although commercial uses of NTFPs are often based on traditional uses, the relationship between the two grows weaker as commercial demand increases and products move outside the original cultural and geographical context of their use. However, it remains important that traditional knowledge holders provide consent for and benefit from the commercial use of their knowledge, and measures should be instituted to achieve this.

(b) NTFPs are part of land-use systems that include a range of activities, many with significant impacts on NTFPs. NTFP regulations should reflect these inter-connected patterns of land and resource use.

- NTFP laws and policies must take into account the most pressing threats to species and the ecosystems within which they are found. It is often the case that forest degradation and destruction resulting from commercial agriculture, logging, mining, and other land uses cause far more damage to NTFPs than overharvesting.
- Governments should regulate timber and NTFPs in very different ways given the enormous differences in how they are harvested and used, and their role in local economies and cultures. However, timber regulations should minimise the negative impacts of logging on locally and commercially valuable NTFPs.
- Prior to drafting regulations, policymakers should understand the relationship between NTFPs and agriculture, the importance of NTFP harvest timing for subsistence and cash income and other critical features of these systems.
- Given current and future shifts in the geographic distribution of plant species, climate change mitigation and adaptation strategies and policies need to address NTFP harvesting and trade alongside other land-use activities.

(c) Power and other social relations must be factored into law and policy formation.

- The power dynamics between stakeholders should be understood prior to policy formulation and implementation. Policies should avoid criminalising harvesting activities, and further marginalising producers.
- The potential for tensions between "insiders" and "outsiders" to arise must be allowed for in policy measures and addressed in consultations with stakeholders. Where conflict exists, facilitators trained in conflict resolution are likely to be needed to help formulate equitable and viable policies.
- The capacity of local and indigenous people needs to be built so that communities can organise, navigate overly bureaucratic NTFP permitting procedures, and assert their rights against more powerful players.
- In many countries, entrenched corruption and abuse of power on the part of governments and their circle of patronage means that new measures will stall. Small producers, who lack political or economic power, can easily lose out if measures are drafted in a way that primarily promotes the interests of the elite.

(d) Information requirements for effective laws and policies should be carefully considered before regulations are developed.

- Policymakers require a vast range of information about NTFPs when drafting laws, including: the ecology and management of species, harvesting

practices, key stakeholders, and the socioeconomic costs and benefits along the value chain. Capacity building, broad research, and data-collection efforts should be ongoing, but when governments have limited resources, they should focus on threatened species and those that are intensively traded.
- The greatest threats to NTFPs generally come from degradation or destruction of habitats, but the overharvesting of NTFPs can be a significant problem, as CITES and national endangered species lists make clear. Policymakers should, however, be cautious about concluding that overharvesting is the main threat to NTFPs or that concerns about unsustainable sourcing necessarily mean there is a crisis at hand.

(e) Policy development must incorporate comprehensive, ongoing, and iterative stakeholder consultations.

- Laws and policies should grow from extensive consultations with the full range of affected stakeholders, including harvesters and producers, traders, companies, and government departments. The participation of diverse groups is particularly important for species that are heavily traded and thus involve strong economic interests.
- Intermediary organisations such as producer and harvester groups, trade associations, and NGOs should be supported to help strengthen consultations, and ensure these voices are heard in policy processes.

(f) Capacity should be built in government, trader, and producer communities to enable the development and implementation of effective NTFP policies and laws.

- Government capacity to develop and implement NTFP laws and policies is notoriously underfunded and marginalised, due in part to the lack of importance given to these "minor" forest products. Capacity and technical skills should be developed in government departments.
- Producers, traders and their support organisations need greater capacity to engage with government on the development of effective laws and policies. Creative approaches should also be explored to involve producer communities and traders in monitoring resource use and assisting with policy implementation.

(g) Many seemingly unrelated areas of law can significantly affect NTFP management, use, and trade and should be considered while developing NTFP policy and legal frameworks.

- A range of laws directly and indirectly impact NTFPs. Governments should identify the socioeconomic and environmental effects of such laws on NTFPs when developing a policy framework, and should seek to mitigate the negative impacts of these seemingly unrelated bodies of law. Governments must be careful to build on or complement traditional resource rights, minimise paperwork, and avoid duplication of existing laws.

- It is vital that access and ownership rights to resources and land be clarified when developing regulatory frameworks for NTFPs. Governments should ensure that laws provide an enabling environment for traditional knowledge protection and local NTFP industries and producers.
- Laws governing labour relations, quality control, and food safety need to ensure that they do not exclude producers or products that may not qualify.

(h) The impact of regional and international policies on NTFPs must be examined as national, state, and provincial NTFP policy frameworks are developed.

- Policymakers need to consider how regional and international policies on NTFPs interact in order to minimise negative, unintended consequences for NTFP harvesting and trade.
- Countries that share commercially traded species should collaborate to develop regional policies for their management, use, and trade.
- International treaties such as CITES are important tools to regulate trade in endangered species but need to be used with caution to ensure that trade restrictions are appropriate, targeted, and effective and that the negative effects of regulation on livelihoods are minimised.
- National, state, and provincial policies regarding trade and benefit sharing from the commercial use of biodiversity are typically not coordinated. Governments should attempt to integrate these bodies of law when developing policy frameworks for NTFPs.

(i) Policy frameworks should be strategic, comprehensive, and coordinated across government departments.

- Care should be taken to consider the wide range of issues that converge upon and can distort the effects of NTFP policy and law. Most NTFP laws are built incrementally and lack an overall strategy or clear objectives.
- Governments should aim to synchronise laws affecting NTFPs, avoid duplication, and ensure the mandates of government departments do not overlap.
- Governments should examine NTFP laws with a view to eliminating permits and procedures that are inappropriate and burdensome for small-scale producers and bring no clear management or livelihood benefits.
- Unintended consequences often result from policies regulating NTFPs and from those found outside the sector. Policies based on theoretical frameworks and assumptions originating outside a region are particularly likely to lead to unanticipated outcomes when they interact with local political, cultural, economic, and ecological conditions.

(j) NTFP policies work best when based on incentives ("carrots") rather than penalties ("sticks").

- "Sticks", such as permits, quotas, taxes, and restrictions on trade are often employed to regulate NTFPs, particularly in a perceived overharvesting

crisis. However, "carrots" in the form of incentives and supportive legal frameworks, such as government support for producer, trade, and processing groups; market access and premium prices via certification; tax breaks; and outreach and education on new policies and laws usually work best for this category of products. In some cases, particularly when there is sudden and high commercial demand, both approaches are necessary.
- Revenue generated by the state from royalties, taxes, or the sale of NTFPs should be channeled to conservation and sustainable management of NTFPs, supporting the sector, and building government capacity on NTFPs.

(k) Less is often more: NTFP regulation should be approached with a light hand.

- Governments should approach NTFP regulation with a light hand and in ways that reflect the financial, ecological, and social costs and benefits of such actions, the government's implementation capacity, and the likelihood of compliance.

(l) Existing customary and local laws are often better suited to this diverse set of products and activities.

- Where land tenure and resource rights are secure, customary laws are still strong, and local capacity exists to manage the resource base and deal with commercial pressures, customary laws often provide a more nuanced approach to regulation, integrating unique local cultural, ecological, and economic conditions in ways that better suit this category of products.
- In cases where customary law has broken down to a significant degree, or outside commercial pressure has intensified well beyond the carrying capacity of traditional measures, governments can offer important and necessary complementary levels of regulation, something often requested by local groups. Interventions should be crafted to include local-level institutions and management systems, where these are effective.
- Governments should explore NTFP policy frameworks that integrate and coordinate customary and statutory law and governance systems. This requires real commitments of time, money, research, and extensive stakeholder consultation.

Acknowledgments This chapter draws from an edited volume titled *Wild Product Governance: finding policies that work for non-timber forest products*, edited by Sarah A. Laird, Rebecca J. McLain and Rachel P. Wynberg, and published by Earthscan in 2010 (http://www.peopleandplants.org). We are grateful to the contributors of that book for providing rich material from which to develop a synthesis. Financial and institutional support was provided by People and Plants International, the Environmental Evaluation Unit (University of Cape Town), the Institute for Culture and Ecology, the United Nations University and the Center for International Forestry Research. The Christensen Fund also provided invaluable support to the research and writing process. Additional financial support was given to Rachel Wynberg by South Africa's National Research Foundation (NRF) although any opinion or findings expressed in this book are those of the authors alone.

References

Alexiades MN (2002) Cat's claw (*U. guianensis* and *U. tomentosa*). In: Shanley P, Pierce AR, Laird SA, Guillen A (eds) Tapping the green market: certification and management of non-timber forest products. People and Plants Conservation Series. Earthscan, London

Alexiades MN, Shanley P (2005) Forest products, livelihoods and conservation; case studies of non-timber forest product systems, vol 3 – Latin America. CIFOR, Bogor, Indonesia

Antypas A, McLain RJ, Gilden J (2002) Federal non-timber forest products policy and management. In: Jones ET, McLain RJ, Weigand J (eds) Non-timber forest products in the United States. University Press of Kansas, Lawrence, KS

Areki F, Cunningham AB (2010) Fiji commerce, carving and customary tenure. In: Laird SA, McLain, RJ, Wynberg RP (eds) Wild governance – finding policies that work for non-timber forest products. Earthscan, London, pp 229–242

Arnold JEM, Ruiz Pérez M (1996) Current issues in non-timber forest products research. Center for International Forestry Research, Bogor, Indonesia

Arnold JEM, Ruiz Pérez M (2001) Can non-timber forest products match tropical forest conservation and development objectives? Ecol Econ 39:437–447

Arquiza YD, Guerrero MCS, Gatmaytan AB, Aquino AC (2010) From barter trade to Brad Pitt's bed: NTFPs and ancestral domains in the Philippines. In: Laird SA, McLain RJ, Wynberg RP (eds) Wild governance – finding policies that work for non-timber forest products. Earthscan, London, pp 155–182

Baird IG, Dearden P (2003) Biodiversity conservation and resource tenure regimes: a case study from northeast Cambodia. Environ Manage 32:541–550

Bennett B (2006) Foreign direct investment in South Africa: how big is southern africa's natural product opportunity and what trade issues impede sectoral development? Regional Trade Facilitation Programme, Pretoria

Biswas T, Potts J (2003) Grounds for action: looking for a sustainable solution to the coffee crisis. Bridges 7:17–18

Brown HCP (2005) Governance of non-wood forest products and community forests in the humid forest zone of Cameroon. PhD thesis, Cornell University, Ithaca, NY

Burgener M (2007) Trade measures – tools to promote the sustainable use of NWFP? An assessment of trade-related instruments influencing the international trade in non-wood forest products and associated management and livelihood strategies. Non-Wood Forest Products Working Document no 6, FAO, Rome

Cole D, Nakamhela U (2008) Review and clarification of procedures for the issuing of permits for the collection and export of biological resources in Namibia. Unpublished report, Centre for Research Information Action in Africa and Southern African Development and Consulting (CRIAA SA-DC), Windhoek, Namibia

Commission on Intellectual Property Rights (2002) Integrating intellectual property rights and development policy. Report of the Commission on Intellectual Property Rights, September, London. http://www.iprcommission.org/papers/pdfs/final_report/CIPR_Exec_Sumfinal.pdf. Accessed 16 May 2008

Cronkleton P, Pacheco P (2010) Changing policy trends in the emergence of Bolivia's Brazil nut sector. In: Laird SA, McLain RJ, Wynberg RP (eds) Wild governance – finding policies that work for non-timber forest products. Earthscan, London, pp 15–42

Cunningham AB (2001) Applied ethnobotany, people, wild plant use and conservation. Earthscan, London

Cunningham AB, Garnett S, Gorman J, Courtenay K, Boehme D (2009) Eco-enterprises and *Terminalia ferdinandiana*: best laid plans and Australian policy lessons. Econ Bot 63:16–28

Dewees PA, Scherr SJ (1996) Policies and markets for non-timber tree products. EPTD discussion paper no. 16, Environment and Production Technology Division, International Food Policy Research Institute, Washington, DC

Dowie M (2005) Conservation refugees: when protecting nature means kicking people out, *Orion Magazine*, November/December, pp 16–27, abridged version available at http://www.orionsociety.org/index.php/articles/article/161. Accessed 7 June 2009

Dutfield G (2002) Sharing the benefits of biodiversity: Is there a role for the patent system? J World Intellect Prop 5:899–932

Dyke A, Emery M (2010) NTFPs in Scotland: changing attitudes to access rights in a reforesting land. In: Laird SA, McLain RJ, Wynberg RP (eds) Wild governance – finding policies that work for non-timber forest products. Earthscan, London, pp 135–154

Fantini AC, Guries RP, Ribeiro RJ (2005) Palm heart (*Euterpe edulis* Mart.) in the Brazilian Atlantic rainforest: a vanishing resource. In: Alexiades MN, Shanley P (eds) Forest products, livelihoods and conservation; Case studies of non-timber forest product systems, vol 3 – Latin America. CIFOR, Bogor, Indonesia, pp 83–110

Fiji Islands (1992) Forest Decree No. 31 of 1992, A Decree Relating to Forest and Forest Produce, Government of the Sovereign Democratic Republic of Fiji

Gebhardt M (1998) Sustainable use of biodiversity by the pharmaceutical industry? Int J Sustain Dev 1:63–72

Granich CI, Purata SE, Edouard F, Pardo MFS, Tovar C (2010) Overcoming barriers in collectively managed ntfps in mexico. In: Laird SA, McLain RJ, Wynberg RP (eds) Wild governance – finding policies that work for non-timber forest products. Earthscan, London, pp 205–228

Hayes PB (2000) Enhancing the competitiveness of the rooibos industry. M.Phil thesis, University of Stellenbosch, Stellenbosch, South Africa

Hecht SB, Anderson AB, May P (1988) The subsidy from nature: shifting cultivation, successional palm forests, and rural development. Hum Organ 47:25–35

Hersch-Martinez P (1995) Commercialisation of wild medicinal plants from southwest Puebla, Mexico. Econ Bot 49:197–206

Iqbal M (1993) International trade in non-wood forest products: an overview, FAO Misc/93/11 Working Paper. Food and Agriculture Organisation, Rome

Jaireth H, Smyth D (eds) (2003) Innovative governance: indigenous peoples, local communities, and protected areas. World Conservation Union, Ane Books, New Delhi

Jones ET, McLain RJ, Weigand J (eds) (2002) Non-timber forest products in the United States. University of Kansas Press, Lawrence, KS

Kanga K (1999) Trade of main wild berries in Finland. Silva Fennica 33:159–168

King SR, Meza EN, Carlson TJS, Chinnock JA, Moran K, Borges JR (1999) Issues in the commercialisation of medicinal plants. HerbalGram 47:46–51

Kluppel MP, Ferreira JCP, Chaves JH, Hummel AC (2010) Case study A: in search of regulations to promote the sustainable use of NTFPs in Brazil 43. In: Laird SA, McLain RJ, Wynberg RP (eds) Wild governance – finding policies that work for non-timber forest products. Earthscan, London, pp 43–52

Kusters K, Achdiawan R, Belcher B, Ruiz Pérez M (2006) Balancing development and conservation? An assessment of livelihood and environmental outcomes of non-timber forest product trade in Asia, Africa, and Latin America. Ecol Soc 11(2):20

Laird SA, Ingram V, Awono A, Ndoye O, Sunderland T, Fotabong EL, Nkuinkeu R (2010) Bringing together customary and statutory systems: the struggle to develop a legal and policy framework for NTFPs in Cameroon. In: Laird SA, McLain RJ, Wynberg RP (eds) Wild governance – finding policies that work for non-timber forest products. Earthscan, London, pp 53–70

Lele S, Pattanaik M, Isoni O (2010) NTFPs in India: rhetoric and reality. In: Laird SA, McLain RJ, Wynberg RP (eds) Wild governance – finding policies that work for non-timber forest products. Earthscan, London, pp 85–111

Lynch OJ, Alcorn JB (1994) Tenurial rights and community-based conservation. In: Western D, Wright M (eds) Natural connections: perspectives in community-based conservation. Island, Washington, DC, pp 373–392

Marshall E, Schreckenberg K, Newton AC (2006) Commercialisation of non-timber forest products: factors influencing success: lessons learned from Mexico and Bolivia and policy implications for decision-makers. Issue 23, UNEP-WCMC Biodiversity Series

McLain RJ, Lynch K (2010) Managing floral greens in a globalised economy: resource tenure, labour relations and immigration policy in the Pacific Northwest, USA. In: Laird SA, McLain RJ, Wynberg RP (eds) Wild governance – finding policies that work for non-timber forest products. Earthscan, London, pp 183–198

McManis CR (2007) Biodiversity and the law: intellectual property, biotechnology and traditional knowledge. Earthscan, London

Menzies NK, Li C (2010) One eye on the forest, one eye on the market: multi-tiered regulation of matsutake harvesting, conservation and trade in north-western Yunnan Province. In: Laird SA, McLain RJ, Wynberg RP (eds) Wild governance – finding policies that work for non-timber forest products. Earthscan, London, pp 243–264

Michon G (2005) NTFP development and poverty alleviation: is the policy context favourable. In: Pfund JL, Robinson P (eds) Non-timber forest products: between poverty alleviation and market forces. Intercooperation, Switzerland

Mitchell DA, Tedder S, Brigham T, Cocksedge W, Hobby T (2010) policy gaps and invisible elbows: NTFPs in British Columbia. In: Laird SA, McLain RJ, Wynberg RP (eds) Wild governance – finding policies that work for non-timber forest products. Earthscan, London, pp 113–134

Nalvarte Armas W, de Jong W (2005) Potentials and perspective of cat's claw (*Uncaria tomentosa*). In: Alexiades MN, Shanley P (eds) Forest products, livelihoods and conservation; Case studies of non-timber forest product systems, vol 3, Latin America. CIFOR, Bogor, Indonesia, pp 281–298

Ndoye O, Awono A (2010) Case study B: policies for *Gnetum* spp. trade in Cameroon: overcoming constraints that reduce benefits and discourage sustainability. In: Laird SA, McLain RJ, Wynberg RP (eds) Wild governance – finding policies that work for non-timber forest products. Earthscan, London, pp 71–76

Neumann RP, Hirsch E (2000) Commercialisation of non-timber forest products: review and analysis of research. Centre for International Forestry Research, Bogor, Indonesia

Nott K, Wynberg R (2008) Millennium challenge account Namibia compact, vol 4: Thematic analysis report – indigenous natural products. Namibia strategic environmental assessment, task order under the Project Development, Project Management, Environmental and General Engineering ID/IQ Contract no. MCC-06–0087-CON-90, Task Order No. 02, ARD, Washington, DC

Novellino D (2010) From indigenous customary practices to policy interventions: the ecological and sociocultural underpinnings of the non-timber forest trade on Palawan Island, the Philippines. In: Laird SA, McLain RJ, Wynberg RP (eds) Wild governance – finding policies that work for non-timber forest products. Earthscan, London, pp 83–198

Ortiz EG (2002) Brazil nut. In: Shanley P, Pierce A, Laird S, Guillen A (eds) Tapping the green market: certification and management of non-timber forest products. Earthscan, London

Padoch C (1988) The economic importance and marketing of forest and fallow products in the Iquitos region. In: Denevan WM, Padoch C (eds) Swidden fallow agroforestry in the Peruvian Amazon. Advances in Economic Botany 5. New York Botanical Garden, Bronx, NY, pp 74–89

Peters CM (1996) The ecology and management of non-timber forest resources. World Bank Technical Paper No 322, World Bank, Washington, DC

PhytoTrade Africa (2006) PhytoTrade Africa's approach. http://www.phytotradeafrica.com/about/approach.htm. Accessed 7 June 2009

Pierce A, Burgener M (2010) Laws and policies impacting trade in NTFPs. In: Laird SA, McLain RJ, Wynberg RP (eds) Wild governance – finding policies that work for non-timber forest products. Earthscan, London, pp 327–342

Republic of Cameroon (1994) Law No 94/01 of 20 January 1994 to Lay Down Forestry. Yaoundé, Republic of Cameroon, Wildlife and Fisheries Regulations

República de Bolivia (1996a) Ley Forestal (Forestry Law) No 1700 of 11 July 1996. El Congreso Nacional, La Paz, Bolivia

República de Bolivia (1996b) Ley del Servicio Nacional de Reforma Agraria (National Agrarian Reform Law) No 1715 of 18 October 1996. El Congreso Nacional, La Paz, Bolivia

Richards RT, Saastamoinen O (2010) NTFP policy, access to markets and labour issues in Finland: impacts of regionalisation and globalisation on the wild berry industry. In: Laird SA, McLain RJ, Wynberg RP (eds) Wild governance – finding policies that work for non-timber forest products. Earthscan, London, pp 287–308

Schreckenberg K (2004) The contribution of shea butter (*Vitellaria paradoxa* CF Gaertner) to local livelihoods in Benin. In: Sunderland T, Ndoye O (eds) Forest products, livelihoods and conservation: case studies of non-timber forest product systems, vol 2 – Africa. Centre for International Forestry Research, Bogor, Indonesia

Shackleton S (2010) Case study D: overregulation and complex bureaucratic procedure: a disincentive for compliance? The case of a valuable carving wood in Bushbuckridge, South Africa. In: Laird SA, McLain RJ, Wynberg RP (eds) Wild governance – finding policies that work for non-timber forest products. Earthscan, London, pp 287–308

Shackleton CM, Shackleton SE (2004) The importance of non-timber forest products in rural livelihood security and as safety nets: a review of evidence from South Africa. S Afr J Sci 100:658–664

Shackleton CM, Botha J, Emanuel PL (2003) Productivity and abundance of *Sclerocarya birrea* subsp. *caffra* in and around rural settlements and protected areas of the Bushbuckridge lowveld, South Africa. For Trees Livelihood 13:217–232

Shanley P, Medina G (2005) Frutíferas e plantas úteis na vida Amazônica. CIFOR/Imazon, Belém, Brazil. http://www.cifor.cgiar.org/publications/pdf_files/Books/BShanley0501.pdf

Shanley P, Pierce AR, Laird SA, Guillen A (2002) Tapping the green market: certification and management of non-timber forest products. People and Plants Conservation Series. Earthscan, London

Soldán MP (2003) The impact of certification on the sustainable use of Brazil Nut (*Bertholletia excelsa*) in Bolivia. Non-Wood Forest Products Programme, Food and Agriculture Organisation of the United Nations. http://www.fao.org/forestry/foris/pdf/NWFP/Brazilnuts.pdf. Accessed 2 October 2008

Southgate D, Coles-Ritchie M, Salazar-Canelos P (1996) Can tropical forests be saved by harvesting non-timber products? A case study for Ecuador. In: Adamowicz WL (ed) Forestry, economics and the environment. CAB International, Wallingford, UK, pp 68–80

Stockdale M (2005) Steps to sustainable and community-based NTFP management. Non-Timber Forest Products-Exchange Programme for South and Southeast Asia, The Philippines

Stoian D (2005a) What goes up must come down: the economy of palm heart (*Euterpe precatoria* Mart.) in the Northern Bolivian Amazon. In: Alexiades MN, Shanley P (eds) Forest products, livelihoods and conservation; Case studies of non-timber forest product systems, vol 3: Latin America. CIFOR, Bogor, Indonesia, pp 111–134

Stoian D (2005b) Harvesting windfalls: the Brazil nut (*Bertholletia excelsa*) economy in the Bolivian Amazon. In: Alexiades MN, Shanley P (eds) Forest products, livelihoods and conservation; Case studies of non-timber forest product systems, vol 3, Latin America. CIFOR, Bogor, Indonesia, pp 83–110

Sunderland TCH, Ndoye O (eds) (2004) Forest products, livelihoods and conservation: case studies of non-timber forest product systems, vol 2: Africa. Center for International Forestry Research, Bogor, Indonesia

Sunderland T, Asaha S, Balinga M, Isoni O (2010) Case study C: regulatory issues for Bush Mango (*Irvingia* spp.) trade in southwest Cameroon and southeast Nigeria. In: Laird SA, McLain RJ, Wynberg RP (eds) Wild governance – finding policies that work for non-timber forest products. Earthscan, London, pp 77–84

Tomich TP (1996) Market, policies and institutions in NTFP trade: nothing is perfect. In: Leakey RRB, Temu AB, Melnyk M, Vantomme P (eds) Domestication and commercialisation of non-timber forest products in agroforestry systems. Non-Wood Forest Products no 9, FAO, Rome

Wynberg RP (2006) Identifying pro-poor best practice models of commercialisation of southern African non-timber forest products. PhD thesis, Graduate School of Environmental Studies, University of Strathclyde, Glasgow

Wynberg RP (2010) Navigating a way through regulatory frameworks for *Hoodia* use, conservation, trade and benefit sharing. In: Laird SA, McLain RJ, Wynberg RP (eds) Wild governance – finding policies that work for non-timber forest products. Earthscan, London, pp 113–134

Wynberg RP, Laird SA (2007) Less is often more: governance of a non-timber forest product, marula (*Sclerocarya birrea* subsp. *caffra*) in southern Africa. Int For Rev 9:475–490

Wynberg R, Silveston J, Lombard C (2009) Value adding in the southern African natural products sector: how much do patents matter? In: The economics of intellectual property in South Africa. World Intellectual Property Organisation, Geneva

Chapter 12
Building a Holistic Picture: An Integrative Analysis of Current and Future Prospects for Non-timber Forest Products in a Changing World

Charlie Shackleton, Sheona Shackleton, and Patricia Shanley

Abstract This final chapter seeks to synthesise key discussions and conclusions from the preceding chapters. Each chapter deals with a specific dimension of NTFP use and management, but when read together, offers a revealing overview of the discourses, debates, and dilemmas associated with the use and promotion of NTFPs over the past two to three decades. Here, we capture this bigger picture through the development of an integrated understanding of these issues and debates. While unpacking broadly applicable lessons and generalisations, we also attempt, given the varied profiles and contexts of NTFPs, to go further through questioning how an integrated understanding stands up to scrutiny as local and global circumstances change. This is particularly pertinent as the key foundations underlying the policy and functional value of NTFPs change. Lastly, we consider some of what we perceive to be the key areas that need investigation or resolution over the coming decade.

12.1 Introduction

The preceding chapters have together presented a comprehensive appraisal of the role and importance of non-timber forest products (NTFPs) in livelihoods, economies, and biodiversity conservation throughout the world. Although the emphasis has been on rural situations and the developing world, evidence is also provided of the contribution of NTFPs in urban settings and in developed countries. Each chapter deals with a specific dimension of NTFP use and management, but when read together they offer a revealing overview of the discourses, debates, and

C. Shackleton and S. Shackleton (✉)
Department of Environmental Science, Rhodes University, Grahamstown 6140, South Africa
e-mail: c.shackleton@ru.ac.za, s.shackleton@ru.ac.za

P. Shanley
Center for International Forestry Research and Woods and Wayside International, 211 Ridgeview Road, Princeton, NJ 08540, USA
e-mail: p.shanley@cgiar.org

S. Shackleton et al. (eds.), *Non-Timber Forest Products in the Global Context*, Tropical Forestry 7, DOI 10.1007/978-3-642-17983-9_12,

dilemmas associated with the use and promotion of NTFPs over the past two to three decades. With contributions from ecologists, economists, anthropologists, botanists, geographers, conservationists, and policy analysts, this final chapter offers an interdisciplinary synthesis that will be useful for experienced actors operating in the field of NTFPs, as well as newcomers seeking insights into the complexities and subtleties of the debates, approaches, and policy options.

While each chapter presents its own conclusions, it is the integration across the chapters that facilitates the development of a deeper understanding. Promoting an integrated understanding of issues surrounding NTFPs is the primary purpose of this concluding chapter. This is difficult, however, as a vast array of plants and animals are lumped under the NTFP umbrella. The sheer magnitude of species, their extensive geographic ranges, divergent socioeconomic contexts, and the idiosyncratic nature of the many species termed an NTFP make it challenging to develop valid generalisations (Pierce 2002). Given such varied profiles and contexts of NTFPs, we attempt to go further through questioning how an integrated understanding stands up to scrutiny as local and global circumstances change. This is particularly pertinent as the key foundations underlying the policy and functional value of NTFPs change with growing urbanisation, globalisation, and environmental change. Lastly, we consider some of what we perceive to be the key issues that need investigation or resolution over the coming decade.

12.2 Building a Holistic Picture

12.2.1 NTFPs in Livelihoods

As a starting point it is instructive to revisit the underlying value of NTFPs. Throughout human history, people have devised ingenious ways of living with the natural resources available to them; serving cultures from widely diverse climates, landscapes, and vegetative zones (Chap. 2). Demands for exotic NTFPs and the subsequent spice trade were responsible for some of the earliest and most distant trade networks circumnavigating the earth. Harrowing voyages and intense conflicts took place to gain possession of spices such as nutmeg which, at the time, was worth its weight in gold.

Today, NTFPs continue to be used on an enormous scale for subsistence and food security, with much of this use remaining unchanged over generations and occurring without any form of external intervention; it is simply part of the way people live (Fig. 12.1). In addition, NTFPs remain a major component of local, regional, and global trade networks (Table 3.6, Chaps. 3 and 4), sustaining millions of families and linking urban and rural areas across the world (Chap. 6). While many markets evolved endogenously, more recently governments and development agencies have sought to promote the NTFP trade (Sect. 12.3), particularly at a global level, as a means to address poverty and to incentivise natural resource management and conservation.

Fig. 12.1 An everyday NTFP (winnowing basket) from the community-run Bigodi Wetland Sanctuary, Uganda. This community-based initiative combines conservation and tourism, particularly bird watching, with the sustainable use of wild resources (photo: Claire Shackleton)

The research reported in this book has shown that, despite a wide range of methodological approaches and divergent perspectives within the growing volume of work on NTFP valuation, NTFPs constitute an extremely important component of local livelihoods (Chap. 3). Contributions span the spheres of direct provisioning, income generation, cultural and spiritual needs, and safety nets in adverse times. It was recognition of this that first raised the expectation that it might be possible to "lift" poor people out of poverty through NTFP promotion and commercialisation (signifying accumulation of sufficient capital to move out of the World Bank's economic definition of poverty) (Chap. 3). There are now sufficient studies to show that the use and marketing of NTFPs can assist some households move out of poverty, but that they play their most significant role in preventing a deepening of poverty for many, many more (Chap. 3). The fact that NTFPs offer a locally accessible, free resource which serves to reduce vulnerability of some of the world's most marginalised populations is one which needs greater appreciation, especially in the context of global environmental change and the complex linkages between ecosystem health and human well-being (MA 2005).

It is also important to recognise that NTFPs not only have economic value but are often the backbone of cultural traditions among communities worldwide. Customs, beliefs, rituals, and traditions in areas as diverse as art, medicine, food, hunting, religion, and marriage often include or revolve around a selection of specific forest resources (Chap. 5). This aspect of NTFPs is critical for maintaining social capital and conserving culture and tradition, and may provide as important an incentive for natural resources management and conservation as monetary benefits. Certainly, the cultural role of NTFPs is one that is often neglected. Reductionist models and a primarily economic lens can blind researchers and policy makers to the substantial, but mainly unquantifiable, direct use and cultural value of forest products.

12.2.2 NTFPs in Conservation and Development

At the turn of the new millennium it was widely recognised that poverty throughout the world was rife and would require concerted and coordinated global efforts to make any positive impacts (WRI 2005). A similar outlook applied to what has been described as the biodiversity crisis (Laurance 2007). Previously the domain of researchers and activists on the margins, environmental issues began to gain credence in national politics and agendas throughout the world. Highest among these were climate change and biodiversity loss. However, it was widely perceived that the global poverty and biodiversity crisis were incompatible; development and poverty alleviation could only be achieved at the expense of biodiversity, and protection of biodiversity and conservation at the expense of improved human well-being and development. Given that levels of poverty and biodiversity are highest in the developing world, and in rural areas, this dilemma was most acute in these regions. Due to the shorter time scale to incur benefits, policy decisions frequently favoured development and economic exploitation over biodiversity (Edwards and Abivardi 1998).

During the late 1980s and 1990s, NTFPs were touted as one means of possibly reconciling this dilemma (Plotkin and Famolare 1992), namely a route to promoting and improving local livelihoods with only limited land transformation and biodiversity loss. At that time, the potential of NTFPs was still inadequately documented and poorly understood, but, in theory, the fundamental basis of the argument was sound and became attractive to a range of lobbies, particularly NGOs promoting local economic development and those linked to agencies concerned with biodiversity conservation (e.g., WWF, IUCN). They embraced the argument and hoped to accumulate sufficient evidence of its veracity and practicality to encourage industries to trade with rural communities engaged with NTFP production. Over time, in some countries, the arguments were also taken to the governmental level with the intent of amending and improving local and national policies which impact NTFP collectors on a wide scale. Consequently, there was a burgeoning of development initiatives in this area accompanied by diverse research that aimed to inform practice (Box 12.1). The results and findings from this work are mixed and are discussed in more detail in the next section.

12.2.3 Sustainability

Having established that NTFPs contribute meaningfully to the livelihoods of many, indeed most, rural people in developing countries (Sect. 12.2; Chaps. 3–6), it is necessary to ascertain whether this is at expense of biodiversity and the future of the particular species used. The fundamental question is – can livelihood improvement go hand in hand with improved natural resource management and biodiversity conservation?

Box 12.1 The Growth in NTFP Studies Over the Last Two Decades

Charlie Shackleton and *Dylan McGarry*, Department of Environmental Science, Rhodes University, Grahamstown, South Africa, c.shackleton@ru.ac.za; armadylan@gmail.com

Although NTFPs have been part of the daily lives of rural people for millennia, they remained effectively hidden to the eyes of forest researchers and managers until the late twentieth century. A few academic papers toward the end of the 1980s served to ignite a small, but latterly accelerating interest in every facet of NTFPs in terms of management, ecology, economics, governance, and contribution to local livelihoods.

We sought to display this growing interest and consequently undertook a small numerical literature search. We used the global Science Direct and Scopus search engines, using the terms "non-timber forest product", "minor forest product", and "nonwood forest product". The period of the search was from 1990 until the end of 2009. However, the returns for 2009 were lower than 2008, which we suspect was a consequence of not all 2009 papers being abstracted in those databases by the time of our search in March 2010. We excluded papers that dealt with laboratory testing of the properties of NTFPs, and those dealing with joint forest management or community-based natural resources management generally. A total of 398 papers were returned. Interestingly, although the field of NTFP research is over two decades old, almost half were published only in the last 4 years. There has been a steady increase in the output over the 20-year period covered (Fig. 12.2 and Fig. 12.3).

Examining the data by region of study shows that Asia dominates as a reporting area for NTFP studies. India had the most papers. South America (mainly Brazil, Peru, and Colombia) and North America (Canada, Mexico, and USA) followed. Undoubtedly there is a great deal more material from South America, but is not abstracted in English language databases. North Africa and the Middle East were the most poorly represented.

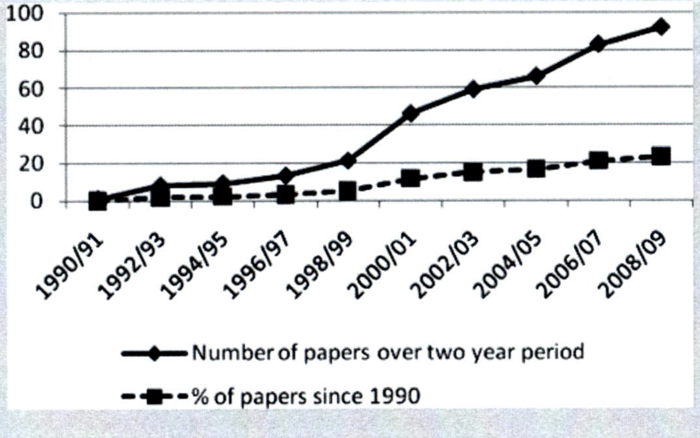

Fig. 12.2 Number of papers on NTFPs published in two-yearly intervals since 1990

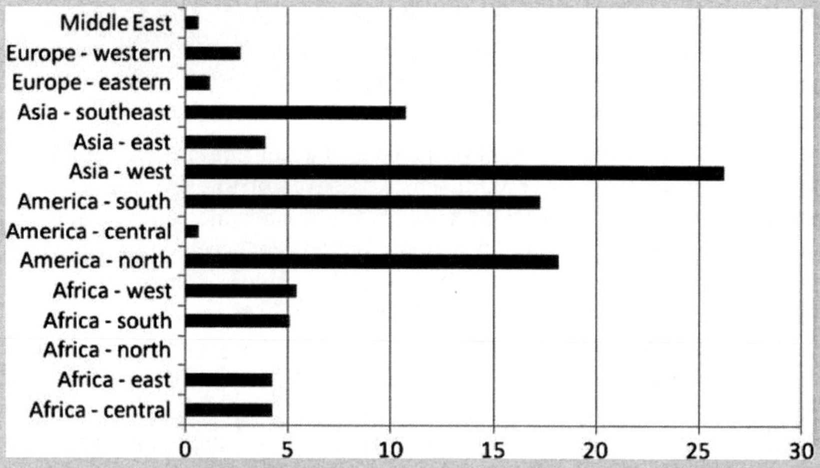

Fig. 12.3 The proportion of NTFP papers published between 1990 and 2009 by region, as abstracted in Science Direct and Scopus databases

When contemplating the sustainability of NTFP harvest, first it is important to recall that the majority of NTFP harvesting is invisible, undocumented, and used to meet subsistence and local needs. It is also worth noting that the scale of harvesting for local and household needs is frequently nondestructive. In terms of commercially harvested NTFPs the situation is highly context-dependent. Unfortunately, in contrast to the wealth of studies in NTFP value chains and value to livelihoods, there is far less work on quantifying the impacts of use on biodiversity (of the species harvested as well as the broader system) and determination of sustainable harvesting levels (Chap. 7).

As reflected above and in Chap. 1, and commented on further by Guariguata et al. (Chap. 8), there are literally tens, if not hundreds, of thousands of NTFP species. The basic biology of most of these has not been studied. The very detailed work in determining ecologically sustainable levels of harvest requires time; estimates spanning 1 or 2 years (perhaps three for the average PhD study) are insufficient (Chap. 8). Some impacts, such as nutrient declines or changes in competitive interactions between species, will only manifest over substantially longer time periods (Chap. 10). Even when the information is procured for a specific species, it cannot be unquestioningly extrapolated throughout the range of that species due to differences in biophysical conditions which may change various factors such as its growth rate, response to harvesting, and exposure to exacerbating pressures (such as browsing, fire, diseases, or predators). Secondly, the information dates rapidly. As human population pressures grow or livelihood options change, what may have been an ecologically sustainable harvesting system may become unsustainable. Thirdly, biological systems are complex and prone to unpredictable changes. Thus, there may be changes in the broader system other than human population pressures which affect the potential for sustainable harvests from one year to the next (e.g., drought, floods, disease), and slowly through time (e.g., salinisation of soils, climate change).

Thus, ascertaining the ecological sustainability of NTFP use is not just a case of balancing supply and demand because both of these are highly variable, and so assessments must be context-specific in time and place. There is a dire need for a great deal more effort in understanding and developing models around harvesting impacts and ecologically sustainable off-take levels. However, given the substantial commitment of time and funds required to effectively research NTFPs and the tens of thousands of species lacking study, it would be necessary to prioritise research to focus on species under threat (long-lived, slow growing, rare), those identified by rural and urban users to be of critical local and regional use, and those involved in large-scale commercial trade which are nondomesticated. Conservation managers, policy makers, and researchers need to be able to place species along a gradient of urgency of research need. Scant governmental resources should be committed for species under threat or of particular health, nutritional, or cultural interest.

Drawing on what limited evidence there is, although very few with strong, long-term data sets, the results of sustainability studies parallel those from the livelihood and market chain studies, i.e., there is high variability. Some NTFP systems are ecologically unsustainable, caused by harvesting systems that have negative impacts on the target species and even the wider ecological system. Sunderland et al. (Chap. 10) argue that this is the most common scenario. This is the case for species that are filling a spike in commercial demand, or for those which are harvested where institutions and social and cultural norms have weakened. But along the spectrum of management there are innumerable harvest systems that, although often based on incomplete datasets, demonstrate ecological sustainability. Indeed, the principal use of NTFPs – at the local level for subsistence use – often exhibits sustainability given their relatively small-scale demand at community level. All biological resources have some theoretical level of ecologically sustainable harvest. The management and harvesting challenge is often whether or not governance systems recognise and respect these (Chap. 9) and the opportunity costs of doing so.

Making use of the quadrant depiction of Shaanker et al. (2004) for the analysis of winners and losers between livelihoods and biodiversity, it is possible to explore why in some situations harvesting is sustainable and in others not (Fig. 12.4).

		Livelihoods	
		Win	Lose
Biodiversity	Win	Achievement of positive livelihood and conservation/ sustainable natural resource management outcomes.	Protected areas that exclude harvesting.
	Lose	Positive outcomes for livelihoods, but high resource impacts.	Overexploitation and long term degradation. Loss of livelihood opportunity.

Fig. 12.4 Assessing the winners and losers in NTFP harvest systems

If particular products lie in the win–lose quadrant, it is necessary to question why and what might be required to shift these into win–win situations, as well as prevent current win–lose situations from degrading into lose–lose ones. It is here that current NTFP research is grappling.

The array of factors that potentially influence whether or not an NTFP is harvested sustainability and so continues contributing to livelihoods is vast. Most studies examine one or two factors in isolation. A few take on the herculean task of examining several, but these are still a subset of the broader array. But most point to aspects of governance systems as being crucial in shaping a particular situation or product falls, and into which other quadrant it may move in time.

For instance, it is strongly argued in diverse literature on forest management that without secure resource/land tenure rights for forest users, the goal of sustainable forest management will likely remain unattainable in the majority of situations. Yet despite recognition of this as a primary factor in good governance, it tends to be the one that is most often lacking (RRI and ITTO 2009, Chap. 9). Lack of property rights for forest resources undermines the governance and sustainable management of NTFPs, acting as a disincentive for long-term investment and protection and, at the same time, rendering poor forest-dependent people more vulnerable (Chap. 9). Secure property rights are therefore fundamental in shifting to the win–win block in Fig. 12.4. Data collected by RRI and ITTO (2009) have shown that increasing areas of forest around the world are coming under the ownership of local communities, and that this is likely to bode well for the future of these forests and the NTFPs they contain.

However, providing secure tenure, while an underlying necessity, will not on its own guarantee good governance or sustainable resource management. The situation is considerably more complex than that and many other factors may play a part. For example, other factors mentioned in Chap. 9 that are critical for pro-poor forest and NTFP governance include: local decision making and participatory processes, the rule of law which if not enforced can result in overregulation of poor peoples' access while powerful forest users (often harvesting valuable resources such as timber) face few restrictions, enabling market opportunities for poor resource users that provide incentives for good management, and effective alliances and partnerships between local people and other government and private actors. Furthermore, recognition of customary law and building on it for effective governance can assist in ensuring sustainability (Chap. 10). In many countries, customary and statutory laws play complementary roles, but it is common for new statutory laws to weaken effective customary systems. This may cause confusion and contribute to the erosion of governance and management systems (Chap. 10). Of course, for all of the above to happen there must be enabling policies in place that support devolution and decentralisation, communal tenure rights, the rights of indigenous people, local decision-making processes, capacity building, etc. This issue is provided substantive coverage in Chap. 10.

Another possible solution to the lose–lose scenario that has been mooted is the domestication and cultivation of NTFPs, often in an agroforestry context (Leakey et al. 2004; Michon 2005). This involves a move from wild harvesting to deliberate cultivation (which may mean different actors), something not discussed specifically in this book. Like everything else we have considered regarding NTFPs, domestication is also a complex process involving both biological and social elements. Similar to any intervention, it has been successful for some species and in some contexts, but not in others. There are cases where previously wild harvested products have been domesticated to the extent they would be better classified as agricultural crops than NTFPs (e.g., rubber). In other situations, harvesters continue to harvest the wild resource as no real motivation may exist to promote cultivation unless the product is highly marketable or scarce in the wild. Cultivation of tree NTFP species has been encouraged within agroforestry systems and these have assisted in providing an additional (often rather than an alternative) source of selected NTFPs. A frequently expressed concern with domestication is that it may play into the hands of the wealthy, i.e., those with land assets, or in the worst case scenario into the hands of the private sector, thus displacing poor local people from traditional income-generating activities (Wynberg 2004).

12.3 Research and Methodological Concerns

The research highlighted in Box 12.1 and reported in this book has provided valuable insights into the importance of NTFPs for human well-being, their harvesting and governance, the characteristics, uses and trade chains of particular species, and whether and how NTFPs can play a lead role in conservation and development. While the long-standing significance of NTFPs for rural people and their dependence on them has been confirmed, the initial euphoria around using NTFPs for poverty elimination and biodiversity conservation has been tempered. Not unsurprisingly a wide range of outcomes were documented, reflecting the disciplinary biases of researchers, the methods employed, the local and regional contexts in which livelihoods were operating, the agro-ecological potential of the site, the prevailing market structures, and the overarching policy and regulatory environment. This diversity of results and outcomes inevitably led to greater questioning of the worth of NTFPs, as well as the conditions in which they are used and managed. Conceptual divides between disciplines, conclusions based on inadequate or inaccurate data, and methodological drawbacks have further contributed to debates and confusion about NTFPs. This book has served to bring these debates and various perspectives together in a single volume.

Recently there has been increased inquiry regarding some of the methods used in NTFP research. Indeed, the diversity of methods employed is an area for concern, because these undermine the ability to draw comparisons between different studies (Gram 2001; Wong et al. 2001; Chauhan et al. 2008; Menton et al. 2010). To that end, the current cross-continental work under the Poverty and Environment Network (PEN) coordinated by CIFOR is useful (http://www.cifor.cgiar.org/pen/). However, it is also prudent to recognise that global studies are fraught with methodological challenges given the difficulty of generating truly comparative data across species, continents, climates, cultures, and economies. Thus, broad-based conclusions based on international case studies need to be treated with caution, given the complexity and diversity of NTFPs, the seasonal nature of their use, and the specifics of the contexts in which they operate. For example, a key aspect of many NTFP valuation studies is quantifying subsistence value of NTFPs, however, this can often require multiple years of ecological and socioeconomic inquiry which the majority of research projects cannot afford (Chap. 3). The trend toward shorter-term field work, questionnaires as opposed to observation and snap shot, rapid appraisals disregarding seasonal and annual fluctuations of NTFPs has the potential to grossly misrepresent their use and value contribute little regarding their management or cultural significance. Accurate representation of NTFPs, particularly long-lived and culturally significant species, requires longitudinal data, yet funding and project cycles rarely provide for this.

Of further concern relating to methods is the number of studies that rely on a single approach, i.e., a once-off interview asking respondents to recall amounts harvested or consumed (in units alien to local users) during the last year. Such an approach commonly under-reports quantities harvested. For example, (1) children's use of forest goods, which is often substantial, is generally not captured at all; (2) recall of quantities for less frequently used resources is weak; (3) use of some resources can be highly variable from month to month or year to year, and therefore enquiring about "last year" can be imprecise; and (4) many surveys do not break down aggregate measures into a series of component questions, which are easier to comprehend and answer, and provide more reliable estimates. Furthermore, the perspective obtained may be biased according to the person interviewed and the interviewer (Fisher et al. 2010). These concerns require that most studies would benefit from a range of approaches over a period of time to elicit the required information at an acceptable level of quality.

This point is illustrated in a recent study by Menton et al. (2010) in which they undertook a comparative analysis of diary approaches for measuring quantities of forest products used versus recall methods recorded via a questionnaire. They found that 33% of product level estimates showed a threefold difference between methods. For products used in high quantities, survey methods tended to result in lower values than diaries and vice versa for less frequently used products. This suggests that many of the quantitative measures that exist for NTFP use could be underestimates since these are mainly based on surveys.

Furthermore, the significance of contextual setting requires further investigation, in particular agro-ecological potential and economic options available (Shackleton et al. 2007a). Much of the research on NTFPs is conducted in moist tropical systems, where the potential for agriculture production is significantly higher than in dry savannas or other semi-arid systems. Therefore, it may be feasible for investments in small-scale agriculture to have greater returns in reducing poverty in moist tropical systems. Similarly, in regions with high absorptive capacity for relatively unskilled labour in the formal or informal markets, NTFPs may not be the first choice for poverty alleviation. But, in countries or regions with low absorptive capacities for unskilled labour, NTFPs may well have a comparative advantage. In countries or sites with both low agricultural potential and a surplus of unskilled labour NTFPs can be expected to be particularly important. As has been shown throughout the developing world, the poor rarely invest in a single livelihood strategy, and NTFPs are a common component within the suite of strategies employed. This very fact requires greater attention from government and development agencies, many of which remain unaware and uninformed, despite research generated over the last two decades.

12.4 NTFPs in a Changing World

The local and global contexts in which NTFPs are managed, harvested, used, or sold are constantly changing. Indeed, it is widely argued that we are presently experiencing a period of major global environmental and societal change, linked to increasing uncertainty regarding the future and greater fragility and instability in the complex social–ecological systems that form our world. A recent paper by Rockström et al. (2009) reviews the state of the Earth's nine life support systems and attempts to quantify their boundaries. Their research suggests that humanity has already transgressed the boundaries for three of these systems (climate change, rate of biodiversity loss, and changes in the global nitrogen cycle), and we are fast approaching the boundaries of others – the consequences of which are largely unknown. The Millennium Ecosystem Assessment (2005) indicated similar trends and highlighted the risk of pushing many of our systems beyond their thresholds. In terms of socioeconomic systems, the recent global economic recession is an illustration of how interconnected the world is and how a change in one region can result in unanticipated consequences for people and economies throughout the world (Meltzer 2009). Other changes such as improved communications technology may provide increased opportunities for forest product commercialisation and livelihood benefits from this.

These changes have implications for use of NTFPs and indicate the need for a dynamic view in terms of the systems as well as associated policies and interventions. Some of the broad-scale changes that are likely to impact distribution, access, and use of NTFPs are outlined below.

12.4.1 Land Transformation

The greatest impact on NTFPs is not overharvesting but land uses such as logging, agro-industries, and mining, which rapidly transform entire landscapes. Land transformation remains the primary driver of biodiversity loss internationally (Millennium Ecosystem Assessment 2005). Conversion of largely natural forests or extensive rangelands to agricultural fields or plantations is the most widespread manifestation. Land transformation invariably results in diminished flows, or complete loss of benefits, from a wide range of NTFPs. Land transformation will generally have severe negative impacts on incomes of forest-reliant households, consumption patterns, and livelihood security. The relative trade-offs from the diminished supply, or loss of NTFPs, relative to the benefit streams from the transformed land, are crucial in the final analysis of livelihood gains or losses. But the conservation outcomes are invariably negative, with the possible exception of one or two species favoured by the transformation (e.g., cultivated NTFP species such as *Hoodia* in southern Africa).

The first exposé of the comparative economic potential of NTFPs was revealed (Peters et al. 1989) in evaluating the logging of high-value timber in moist tropical forests. But the pressure of commercial logging and transformation of tropical forests to plantations and agriculture continues unabated. Hence, the continued need to make visible the still hidden contribution of NTFPs to livelihoods and biodiversity conservation. Yet, the analyses in Chaps. 8 and 10 suggest that NTFPs have been insufficient to turn the tide of deforestation. In areas of Latin America and Asia, logging serves as a catalyst of broad land use change, frequently followed by agro-industries or fire. The synergistic effect of these land uses often suppresses regeneration of valuable NTFPs, leaving an impoverished landscape.

Given the consequences to livelihood security of land transformation for forest-reliant peoples, there is a dire need for longitudinal studies on NTFPs use and markets and elucidation of the drivers of change. In the meantime, governance structures, decision-makers, and planners need to appreciate that local land use options are not static, and so their policies and strategies for NTFPs and security of local livelihoods must be built around adaptive management and social learning.

12.4.2 Urbanisation

Early research on NTFPs tended to focus on remote forested locations, relatively intact environments, and forest-dependent people. During the past decade, additional attention has been paid by researchers to the use of NTFPs in urban and peri-urban environments and the active flux of products between forests and cities (Stoian 2005; Kilchling et al. 2009, Chap. 6, Fig. 12.5). The critical socioeconomic and cultural role of NTFPs in cities has expanded from the tropical to the temperate zone with rising demand for forest goods in both developed and developing countries. Furthermore, the increasing mobility of human populations globally has enhanced trade of forest goods as people's customs and traditions follow them (Padoch et al. 2008). Thus, rather than reducing demand, for many NTFPs it is simply a change in the locus of demand, and may even stimulate increased demand (Chaps. 4 and 6).

Fig. 12.5 Forest fruits and medicinal plants are purchased by all classes of society on street corners and in shops and open air markets in Belem, Brazil (photo: Trish Shanley)

12.4.3 Communication and Information Technology

The information and communication technology revolution is over two decades old and even remote countries and regions have felt its effects. Cellphone coverage in developing countries is increasing at a remarkable rate, as has e-mail and internet access. To the best of our knowledge, there is little analysis of the impacts of such new technologies on NTFP use and trade, but all the authors have first-hand experiences of the differences it has made in the lives of traders and rural communities in which we operate. The recent work of Overå (2006) provides some insights of how communication technology (most notably cellphones) has reduced transaction costs and increased profitably, networks and trust among small traders operating between remote regions and urban centers in Ghana. The most palpable effects of increased access to and use of communication technology relate to (1) increased access to information that underpin market negotiations and transactions (such as consumer preferences, international prices, design trends, transport availability and prices, etc.), (2) increased communication between suppliers and marketers which can potentially increase trust, and (3) potentially increased market demand as new markets are discovered further afield from the local or traditional ones.

Modern communication technology means that while the frequency of communication is increased, the costs of business are decreased as the transacting parties do not need to rely on face-to-face meetings, which are far more expensive in travel costs as well as the time for the personnel involved. Overå (2006) comments that the increased frequency of communication helps build trust between the transacting parties. This does not apply solely to producers and buyers. For example, wood-carvers in South Africa need a permit to fell carving timber. The process requires that they locate a suitable tree in the wild and then contact the relevant official, with whom they make an appointment to accompany them to measure the tree and calculate the required fee. In the past this used to take several visits to first locate the official in the office (as he might be out with another woodcarver), which usually costs several days and repeated taxi fares to and from the office (Shackleton and Shackleton 2004). These arrangements are now done by cellphone, saving time and money (A. Shabangu pers comm. – former chairperson of the Mhala Woodcarvers Association). If costs are reduced, then goods are potentially cheaper and hence more competitive on international markets. Similarly in the state of Para, Brazil, radio programs have brought market intelligence to remote villagers, and cellphones assist traders to arrange with whom, how, and when their products will be transported to the market.

The benefits do not accrue solely from increased communication between producers and sellers, but also between networks at each end of the market chain. Thus, producers can communicate with other producers in different regions, to source surplus production, check prices, identify reliable buyers, share transport, plan joint harvesting trips, etc. Similarly, buyers can communicate more readily with networks of other buyers with whom they have a working association. Intermediaries can network both, such as Phytotrade Africa (http://www.phytotradeafrica.com), which is a business development and trade organisation, networking rural producers of forest products with markets and business partners in Europe. This need not apply only to cross-border trade but also to increasing regional trade with countries, especially large ones of the developing world, such as Brazil, South Africa, India, and China, where the costs of in-country travel are high due to the large distances.

12.4.4 Global Climate Change

The implications of global climate change for NTFPs and local communities are profound and will have a threefold impact. First will be increased livelihood vulnerability, especially of rural communities and small-scale farmers in drier ecosystems, and coastal fishing communities. As presented in Chap. 3, when already marginalised and remote communities are faced with hardships, many frequently turn to NTFPs as a fall back option. Thus, global climate change impacts are likely to increase the need for the traditional safety net function of NTFPs. This will only be possible if adequate habitats and stocks of NTFPs are maintained. Perversely, global climate change may, in some regions, accelerate rates of land transformation as farmers seek to cultivate larger areas to compensate for declining yields or more frequent losses of their entire yield through diseases, drought, or floods, which would reduce stocks or species available as a safety net.

The second repercussion of global climate change will be the increased vulnerability of some NTFP species as their environmental envelope changes. This may affect the distributions of NTFP species as well as the growth rates. There may also be synergistic negative impacts on species from the double stress of climate change and harvesting. Forest pest and disease outbreaks are also expected to increase as a result of changing rainfall patterns (IUFRO 2010), and this may further impact some NTFP species. Thus, for some regions and local communities, the supply of

NTFPs will diminish, or new indigenous and exotic species will come in as potential substitutes. These consequences will be particularly marked for species with cultural uses as cultural beliefs and practices are built up over time and adapt more slowly than direct subsistence requirements. Over the last few years there have been many niche models depicting expected changes in vegetation types and distributions of specific species (e.g., Wiens et al. 2010; Wiens and Bachelet 2010), but none have focused explicitly on NTFPs. This requires redress, albeit hampered by the limited biological knowledge pertaining to most NTFPs species. The effects of global climate change are also expected to favour some already aggressive invasive species (e.g., Bradley et al. 2010), which will add further threat to some NTFPs species; but the extent and degree currently remain unknown.

The third outcome of the global climate change is that arguments for conservation and reforestation of ecosystems will become more compelling (Hannah 2010). Current initiatives to reduce emissions from deforestation and degradation (REDD and REDD+) through sustainable forest management, for instance, could benefit the stocks of certain NTFPs. However, the manner in which such programs are implemented will be crucial for local livelihoods. If local communities are included and are partners in the management of and benefits from the conservation efforts, then it may become a win–win solution. But if, as is unfortunately often the case (Brockington and Igoe 2006), local communities are excluded, then promotion of forest conservation under REDD and other conservation projects is likely to have negligible or negative impacts on local livelihoods.

12.4.5 Globalisation and International Policies

At a national and international level, policies, aid, and trade agreements are important indirect drivers of change having both positive and negative impacts on ecosystem goods and services and their links to livelihoods. At an international level globalisation is a significant process having numerous impacts in many different spheres.

Regarding the effects of globalisation and international trade policies on NTFPs, limited information exists other than some evidence on the positive benefits of promotion and growth of global markets for NTFPs. But impacts may not always be obvious or positive and may involve trade-offs, especially for the poor. A report by UNEP-WCMC (2007) stressed how intensification of production and reliance on global markets may not always be of benefit to the poor. The changes associated with globalisation tend to "reduce the capacity of the local area to meet the needs of the local population, increasing dependency on the vagaries of markets" (UNEP-WCMC 2007). Global markets tend to seek the lowest priced supply, so that producers are vulnerable to being undercut and losing market share. Because of this, producers often try to maximise short-term gains, usually leading to accelerated rates of environmental degradation and eventually leaving local people in the position where they have neither the capacity to produce commodities for sale nor the local resource base on which to fall back.

In Mozambique, pressures on forests are growing due to Chinese interests in hardwoods (a trend also mentioned in tropical Africa) (Shackleton et al. 2010). Such commercial timber harvesting often occurs at the expense of the poor as they lose access to a host of important ecosystem services including NTFPs and receive few returns from timber sales. Moreover, the recent economic recession suggests the need for market diversification and to seek opportunities for NTFP commercialisation within local, national, regional, and global markets to spread risk (Shackleton et al. 2007a, b). Globalisation and global markets should not be seen as the silver bullet, certainly on their own. As is wise advice in an increasingly uncertain world, diversification is key – diversification of markets, NTFP products, and value chain pathways – as a route to building resilience to change.

There are many other global and local level changes affecting livelihoods, NTFPs, and the interface between the two. It is not possible to speculate on them all. The above provides ample illustration of the dynamic contexts in which livelihoods and NTFPs occur. Adaptation and change are integral aspects of sustainability, with use of NTFPs being a vital ingredient in facilitating adaptability. Consequently, researchers and policy makers need to develop a more nuanced view of NTFPs that takes into account their ever-changing contexts and if and how these contribute to vulnerability and/or resilience. Change and vulnerability can perhaps provide a new lens to consider the importance of NTFPs. It is expected that such ecosystem goods may become increasingly crucial in building resilience in social–ecological systems (IUFRO 2010).

12.5 The Next Chapter in the NTFP Story

12.5.1 Expected Trends

During the next decade, in both developed and developing countries, temperate and tropical regions, and urban and rural environments, use of NTFPs is anticipated to rise, with continuing trends observed during the past two decades. In developed nations, consumer preferences have shifted strongly toward organic and traditional products which have become firmly entrenched in middle class markets. Products exhibiting a rise in trade include scientifically confirmed herbal remedies, artisanal goods, naturally sourced cosmetics, and healthcare products. The sizeable growth of the middle class in Asian countries, such as India and China, would be expected to generate substantial new markets for NTFPs. Underlying motivations for increased use of forest goods among people include self-interest in health; concern with quality, fascination with the exotic; and a renewed interest in where, by whom, and how products are made.

In the future herbal remedies may not be a preferred option, but become the only affordable one. In times of economic hardship, rising unemployment, and reduction of state services, households turn to wild plants for food and healthcare out of need. This is presently witnessed in conflict areas throughout the globe as well as among refugee populations (Pierce and Emery 2005). In African countries afflicted with high rates of HIV, for example, households increasingly rely on traditional medicines and foods as part of their coping strategies (Kengni et al. 2004). As forests become degraded and conflict over natural resources becomes more acute, the knowledge base of elders and written documentation of traditional healing and management practices may resume significance. The impacts of climate change on conventional agricultural systems are also expected to increase use and reliance on wild resources (see Sect. 12.4.4). For example, a study of responses to climate induced risk in Mozambique revealed increased consumption of wild foods (as an asset conservation strategy) and the sale of wild products in nearby urban centers as two key coping mechanisms (Osbahr et al. 2008).

Where and when will use of NTFPs be expected to decrease? Trade in niche, high end, luxury products may decline due to the lack of a sufficiently strong consumer base coupled with global economic downturn. Furthermore, use and/or sales of specific NTFPs which are wholly overharvested and where efforts toward domestication are embryonic or nonexistent will also drop. Trade in vulnerable, slow-growing species, which occur in low densities with low reproductive capacity, will decline as the species become rarer and/or extinct. Examples include the trade to China of wildlife parts from endangered African mammals, wild sourced swift nests from Kalimantan, and select "conflict of use" NTFPs in Latin America.

Thus, the future scenario is both optimistic and pessimistic in terms of the ability of NTFPs to continue to deliver livelihood benefits to local people and ensure biodiversity and forest conservation. However, due to the importance of NTFPs to the livelihoods of millions of poor rural and urban people, research and practical and policy support will be needed. Some suggestions for areas requiring greater attention are outlined below.

12.5.2 Moving Forward: Facilitation, Capacity Building, and Increasing the Representation of Marginalised Groups

In considering how to improve the use, sustainable management, and trade of NTFPs, it is important to recall that for centuries, people throughout the world have capably devised ways to extract, process, transport, and trade NTFPs without any intervention at all. Often, what is needed is not intervention, but simply allowing collectors' access to resources and markets (Laird et al. 2010). Policies which do not inhibit the use of land and/or resources and which do not favour large agro-industries over small holders are some of the most fundamental and crucial factors in supporting NTFP use. However, given the tendency of NGOs, research institutions, and governments to favour and implement "interventionist" and/or "project" approaches to both conservation and development, it is useful to understand if, what, and how interventions assist in helping to realise the full potential of NTFPs.

Professionals in the fields of development and research routinely offer policy recommendations based on the assumption that "intervention" is necessary and will lead to a "trickle down" of benefits for households classified as poor. However, in some cases, supporting locally crafted practices of NTFP collectors and customary resource management arrangements can be more appropriate than outside intervention. In addition, as case studies within Chap. 11 of this book attest, caution is advised with any policy change or intervention which will impact the lives of marginalised, often "invisible" populations.

Governmental and NGO capacity to effectively work with NTFPs and harvesters is nascent and needs to be built among foresters, policy makers, NGOs, and researchers. Cross-sectoral collaboration among governmental agencies involved in fields such as forestry, health, culture, agricultural, and education is also essential for successful facilitation of NTFP use, management, and trade. Investments in NTFPs as a vehicle for poverty alleviation and development can be included in the basket of options considered by central and regional governments alongside other strategies such as subsidies, public works programs, local economic development initiatives, and investments in agricultural infrastructure. NTFPs, like other, often part-time, livelihood options in which the rural poor engage, i.e., small-scale agriculture, livestock, petty trading, unskilled labour, or low-level employment, deserve concomitant government attention and resources.

Although not the domain of most governmental officials or researchers, one major area of weakness for NTFP collectors is lack of representation. Given their weak political position throughout the globe, attention and action is needed to organise disparate collectors. Select grass roots movements have proven successful in organising forest-reliant people for the purpose of exchanging information, improving sustainable management practices and trade, and working toward improved representation at the national level (see Box 12.2). In southern Africa, PhytoTrade (Southern African Natural Products Trade Association) has developed sustainable and ethical supply chains for wild harvested food, cosmetic, and medicinal plants, involving 20,000 collectors from eight countries (Nemarundwe et al. 2009). Resin tappers in Cambodia are expanding from small village associations into a federation which is providing practical information on trade and product development as well as pressing for policy change to protect forests in which resin tapping trees occur. In Brazil, the National Council of Rubber Tappers has successfully lobbied for the creation of extractive reserves, thus securing over 52 reserves and land rights for tens of thousands of forest-reliant people.

Box 12.2 Development from the Ground Up: Forest Honey in Asia

Jenne de Beer, International Association of Ethnobiology, Darrell Posey Fellow, and South and South East Asian NTFP Exchange Program, Manila, Philippines, ntfp7@yahoo.co.uk

What types of facilitation do collectors of NTFPs need to harvest resources sustainably, process products efficiently, and market profitably? The South and Southeast Asian NTFP Exchange Programme (NTFP-EP) has been working for over a decade in seven countries in support of local communities

(continued)

that depend on non-timber forest resources for their livelihoods.[1] This network of People's Organisations and NGOs is driven by local demand and aims at offering bottom-up support and exchange between NTFP collectors on how to initiate and sustain successful livelihood ventures. The effort focuses on a broad range of NTFPs, including wild fruits, tree resins, materials for craft making, and forest honey.

The case of honey is useful to demonstrate key, generic lessons learned. Wild gathered forest honey, primarily produced by the bee species, *Apis dorsata*, has long been a sought after delicacy in much of Asia. However, consumers have been concerned with quality issues such as adulteration, and the resource base has been under threat due to forest degradation. In order to address the situation, starting in the late 1990s, several pilots with forest honey collector groups were initiated within the NTFP-EP network. Recently, these pilots have grown into full maturity and expanded well beyond the original sites. The introduction of sustainable harvesting practices and improved hygienic handling techniques has led to the achievement of high-quality standards region-wide. Furthermore, network partners have begun taking pride in guaranteeing the purity of the product. Strict internal rules have been established, while traceability of the source of origin has been made relatively easy.

Results from these initiatives are impressive, and local media have helped to spread the word to consumers. Markets have opened and income derived from honey and wax collection has, in many cases, increased significantly.[2] Finally, organised honey collectors play an increasingly active role in forest conservation.[3]

Select lessons learned[4] from building capacity among honey collectors include:

- *Aim for a holistic approach.* From the start, simultaneously address livelihood, conservation, and sustainable management, as well as issues regarding land tenure

(continued)

[1]The Philippines, Indonesia, Malaysia, Cambodia, Vietnam, India, and Bangladesh.

[2]For an example see Pavel Partha and Rumaisa Samad "In Solidarity with Sunderban Honey Wisdom" in: Voices from the Forest 18, April 2010.

[3]See for an example: JMHI, Dian Niaga, NTFP-EP 'Forest Honey and Forest Conservation: What is the Link? Report of the national forest honey workshop', Jakarta, October 2008.

[4]For more detailed observations and guidelines, particular in relation to community enterprise development see: Jenne de Beer and Ma. Christina S Guerrero "Lessons learned from Experience" in Yasmin Arquiza (ed.) From Seeds to Beads; Tales, Tips and Tools for Building a community-based Enterprise. NTFP-EP, Quezon City, 2008.

- *Avoid quick fixes*. Instead, assure long-term commitment over extended periods
- Ensure that the communities' aspirations and concerns are fully taken into consideration. The work should begin as and remain the community's own initiative
- Eventual self-reliance of the grass root partners involved requires finding simple and culturally appropriate solutions

Finally, avoid becoming a gate keeper. Empower community organisations by providing them with direct links to top notch expertise, trade contacts, and policy makers.

In the Brazilian case, where forest-reliant communities have successfully organised on issues related to NTFPs, the participation of women was critical to success. Although women have demonstrated a vital role in the use in and trade of NTFPs worldwide, they generally possess relatively scant political power and often have negligible opportunities for entry into decision-making roles. Given their crucial household responsibility in the use and trade of NTFPs, and their tendency to perceive forests as related to the health and welfare of their families, organisation of NTFP harvesters will likely be more successful with increased women's participation. Notably, the organisation, education, and lobbying efforts needed to facilitate the use and trade of NTFPs require a combination of elements which are difficult to encounter singly much less together: sustained commitment on the part of collectors and facilitators; flexibility of NGOs to listen to local needs; collaboration and transparency among governmental sectors, long-term time frames; and a low level of steady financial support.

12.5.3 Future Research and Policy Needs and Challenges

On the positive side, statistics indicate that there have been a rising number of publications on NTFPs over the last two decades indicating increased interest in NTFPs on the part of researchers. However, it is less common to find examples where research has helped lead to a change in policies and/or practices that assist NTFP collectors. In spite of scattered exceptions, NTFP collectors remain poorly represented, their products continue to be absent from national statistics, and collectors receive few state services. Not only invisible to policy makers, forest product collectors may have also become less visible to and/or consulted by researchers, who have a growing tendency to "design their studies from behind a computer". NTFP studies that are insufficiently grounded in real life needs and scenarios can lead to less irrelevant, financially wasteful, and inappropriate investigations (J. de Beer personal communication).

Furthermore, donors frequently reinforce the latest funding trends rather than more essential investigations. Over the last two decades there has been ongoing interest in market-based conservation efforts, principally focused on international markets (Laird et al. 2010). However, these are often inappropriate, of high risk, and and/or impossible for remote forest collectors to attain (i.e., carbon). The greatest and most consistent value for local communities is usually found in subsistence use and local trade in NTFPs, yet far fewer funders, researchers, or policy makers have given sustained attention to local use and trade and/or in securing access to and protecting the safety net functions of forests (e.g., Shackleton et al. 2007a, b). Nor do donors and researchers generally commit or invest in the long-term time frames necessary to build the trust of forest-reliant people and to understand the ever-changing landscape dynamics necessary to gain even a partial understanding of these vital issues.

An additional bias in the fields of conservation and development toward intervention frequently drives the establishment of new laws or actions before a solid understanding of the problems and issues they are meant to address is built (Laird et al. 2010). In addition, substantial attention has been given by researchers, funders, and policy makers to sustainable extraction of NTFPs. However, forest degradation and destruction resulting from commercial agriculture, logging, and mining frequently cause far more damage to NTFP populations than overharvesting. Given limited resources, primary focus is needed on threatened species and those that are intensively traded (Laird et al. 2010).

Positive exceptions at the local and national levels exist and effective initiatives need attention so that lessons can be learned. For example, the Finnish government collects detailed annual statistics on trade in berries and other NTFPs, making visible the significant economic and socio-cultural value of NTFP collection (Richards and Saastamoinen 2010). In the Philippines the national government is recognising ancestral domain and the significant local capacity to manage forest resources based on customary laws which often provide a more nuanced approach to regulation (Novellino 2010). In eastern Amazonia, the state of Para has recently decreed that forest product collectors will be entitled to the same rights, such as retirement benefits, as agricultural workers.

In addition to legislation, actions which celebrate the cultures and traditions linked to forest products can fortify efforts to protect species and landscapes. In India, China, and the UK, formal healthcare systems recognise and value herbal healing traditions. Revitalisation of forest-based customs and pride in traditional crafts and foodstuffs has grown within communities (Fig. 12.6) as well as been instilled through rising international and regional demand for native crafts and natural foods and medicines. Government support of craft traditions throughout Latin America, Africa, and Asia is assisting in elevating the status of both forest goods and collectors. As researchers and governments grapple with issues such as environmental degradation, adaptation to climate change, declining state services, and increased incidence of infectious diseases, cursory attention has been drawn to NTFPs as one potential means to mitigate environmental harm and socioeconomic ills. NTFPs currently serve crucial safety net functions and contribute to the livelihoods of billions of people on earth. In an

Fig. 12.6 Women benefit from being able to manufacture NTFP goods at home. Grass (*Festuca costata*) broom makers, Bushbuckridge, South Africa (photo: Sheona Shackleton)

uncertain future, their role is likely to expand; thus governmental agencies, researchers, NGOs, and donors may be well served by devoting more serious and sustained attention to forest goods, services, and peoples.

12.6 Concluding Remarks and Key Emerging Messages

This book is a first in providing an analytical global overview of the many aspects that need consideration when researching and promoting the use and management of NTFPs. Beginning with a discussion around what is meant by an NTFP, we moved on to considering the importance of these products for poor people; their role in livelihoods; the cultural benefits they bring; the factors affecting their management, use, and governance; their marketing and economic importance as globally traded products; and where they are situated in a changing world context. We explored the evolution of sentiments regarding the potential of NTFPs in promoting options for sustainable multi-purpose forest management, income generation, poverty alleviation, and biodiversity conservation. Based on critical analysis of debates and discourses, we employed a systems approach to providing an integrated, balanced, and realistic perspective on the benefits and challenges associated with NTFPs and their use. At the same time, we were outspoken in our analysis where we believed this necessary, drawing on our own extensive experiences on the ground. While absorbing all of this information, a number of diverse issues began to emerge that we judge to be important, but somewhat neglected;

these key "take home" messages are summarised in Box 12.3. We hope that they will stimulate the reader to look at NTFPs from a deeper, more integrated and renewed perspective.

Box 12.3 Key Emerging Messages

Patricia Shanley and *Sheona Shackleton*, Center for International Forestry Research and Woods and Wayside International, Princeton, NJ 08540, USA and Department of Environmental Science, Rhodes University, Grahamstown, South Africa, p.shanley@cgiar.org; s.shackleton@ru.ac.za

Theme	Message
Thinking broadly and locally about NTFPs	
Multiple values	NTFPs have multiple values and contribute a range of livelihood assets or capitals. More attention needs to be given to their vast importance in subsistence use and local trade. The role of NTFPs is best understood not solely in narrow economic terms or on internationally traded goods. The focus on internationally traded commodities has skewed research efforts, conclusions, funding, and policy.
Culture matters	Given that they are not quantifiable, cultural values which, to varying degrees, are of critical importance in the daily lives, and Culture Matters customs of people worldwide have generally been ignored with negative repercussions for funding, research direction, and policy. The cultural connections of people to specific species and their landscapes can be a powerful motivator for conservation.
Recognise local efforts	Terms such as "multiple use" and "agroforestry" describe practices that are regularly employed by rural people throughout the world. It is important to recognise and build on land management systems that are already undertaken by local people, rather than to reinvent it as an academic exercise.
Women are critical	In many cultures, women's task of treating the health needs of the family drives them to see and utilise the landscape in terms of its nutritional and medicinal benefits. Capitalising on these sensibilities can give momentum to conservation movements and/or sustainable management practices.
NTFPs are part of a diverse livelihood portfolio	NTFPs usually form one activity that rural households undertake to sustain their livelihoods. There is a need to situate NTFP use within the broader livelihood portfolio and income sources to fully understand their role and importance.
Seeing the bigger picture	Despite decades of research, a recurrent blind spot regarding NTFPs is an understanding of the socioeconomic impact that declining access and reduced abundance of NTFPs has on markets and households. As NTFPs become less accessible, families that independently met some of their basic health, nutritional, and income needs through use of forest goods no longer can. A deeper understanding of the true value of forests is necessary to prevent further degradation of social and ecological systems and the subsequent ample costs to governmental agencies

(*continued*)

Theme	Message
Getting NTFPS on the agenda: policy and practice issues	
Make NTFPs more visible: record the statistics	Most countries are unable to provide FAO with reliable statistics regarding the national and international volume of trade in NTFPs. Modifications to censuses and to natural resource transportation accounting practices could assist in filling this enormous gap in knowledge. Incentives to quantify NTFP values would help to improve the visibility and understanding of these critical but still hidden products.
Enhance forestry extension and education	Forestry extension/education is often insufficient for timber and frequently nonexistent for NTFPs. Forestry training is needed which integrates management of timber and NTFPs and which takes into account both economic and livelihood concerns.
Cross-sectoral collaboration	Cross-sectoral collaborations are needed for effective work on NTFPs. Ministries of agriculture, education, health, culture, technology, etc. can each play a significant role in the use and management of forest resources. Health care, in particular, can be an important catalyst for sustainable practices through fostering an understanding of landscape health as a reflection of human health.
Green marketing	Additional attention and consumer education is warranted on issues of sustainability of trade in NTFPs. Efforts in research, education, and policy are needed to ensure that vulnerable tree species or species with high livelihood importance are not extracted and marketed as certified and/or "green" timber.
Identifying research problems and gaps	
Conceptual bias	The literature regarding NTFPs reflects a conceptual divide between researchers who spend time in rural communities with local people (anthropologists, botanists, and ecologists) and those from a policy or economics perspective who generally have less direct contact with the environment or people they study. This disciplinary divide can influence discourse regarding the value of NTFPs.
Conflict of use	Large-scale logging, ranching, and agriculture are expanding in many regions yet insufficient attention is being paid to the impact of these industries on species composition and livelihoods. Placing blame on collectors for overexploitation and research emphasis on sustainable off-take has taken attention away from large-scale regional deforestation issues.
Which NTFPs?	Research is needed on species which are vulnerable to trade (such as long-lived primary forest species) and also on those species which can live in harsh, pioneer environments. Universities, research organisations, communities, and local cooperatives can help to fill in knowledge gaps by proactively identifying widely used and traded species, pooling their knowledge bases and filling gaps. Widely marketed barks, roots. and latexes can represent particularly vulnerable parts of plants and need particular attention.

References

Bradley BA, Wilcove DS, Oppenheimer M (2010) Climate change increases risk of plant invasion in the Eastern United States. Biol Inv 12:1855–1872

Brockington D, Igoe J (2006) Eviction for conservation: a global overview. Conserv Soc 4:424–470

Chauhan KV, Sharma AK, Kumar R (2008) Non-timber forest products subsistence and commercial uses: trends and future demands. Int Forest Rev 10:201–216

Edwards PJ, Abivardi C (1998) The value of biodiversity: where ecology and economy blend. Biol Conserv 83:239–246

Fisher M, Reimer JJ, Carr ER (2010) Who should be interviewed in surveys of household income? World Dev 38:966–973

Gram S (2001) Economic valuation of special forest products: an assessment of methodological shortcomings. Ecol Econ 36:109–117

Hannah L (2010) A global conservation system for climate change adaptation. Conserv Biol 24:70–77

IUFRO (International Union of Forest Research Organisations) (2010) Making African forests fit for climate change. A regional view of climate change impacts on forests and people and options for adaptation. Policy brief. IUFRO, Austria

Kengni E, Mbofung CMF, Tchouanguep MF, Tchoundjeu Z (2004) The nutritional role of indigenous foods in mitigating the HIV/AIDS crisis in West and Central Africa. Int Forest Rev 6:149–160

Kilchling P, Hansmann R, Seeland K (2009) Demand for non-timber forest products: Surveys of urban consumers and sellers in Switzerland. For Policy Econ 11:294–300

Laird SA, McLain RJ, Wynberg RP (2010) Wild Product Governance: finding policies that work for non-timber forest products. Earthscan, London

Laurance WF (2007) Have we overstated the tropical biodiversity crisis? Trends Ecol Evol 22:65–70

Leakey RRB, Tchoundjeu A, Smith RI, Munro RC, Foundoun JM, Kengue J, Anegbeh PO, Atangana AR, Waruhiu AN, Asaah E, Usoro C, Ukafor V (2004) Evidence that subsistence farmers have domesticated indigenous fruits (*Dacryodes edulis* and *Irvingia gabonensis*) in Cameroon and Nigeria. Agrofor Syst 60:101–111

Meltzer AH (2009) Reflections on the financial crisis. Cato J 49(1):25–30

Menton MC, Lawrence A, Merry F, Brown ND (2010) Estimating natural resource harvests: conjectures? Ecol Econ 69:1330–1335

Michon G (2005) Domesticating Forests: how farmers manage forest resources. CIFOR, ICRAF, Bogor Indonesia. Millennium Ecosystem Assessment (2005) *Ecosystems and human well-being: synthesis*. Island, Washington, DC

Nemarundwe N, Le Breton G, Chibaya I (2009) Linking rural producers of natural products to global markets for poverty alleviation in Southern Africa. XIII World Forestry Congress Buenos Aires, Argentina, 18–23 October

Novellino D (2010) From indigenous customary practices to policy interventions: the ecological and sociocultural underpinnings of the non-timber forest trade on Palawan Island, the Philippines. In: Laird SA, McLain RJ, Wynberg RP (eds). Wild Governance – Finding Policies that Work for Non-timber Forest Products, Earthscan, London, 83–198

Osbahr H, Twyman C, Adger N, Thomas D (2008) Effective livelihood adaptation to climate change disturbance: scale dimensions of practice in Mozambique. Geoforum 39:1951–1964

Overå R (2006) Networks, distance, and trust: telecommunications development and changing trade practices in Ghana. World Dev 34:1301–1315

Padoch CE, Brondizio S, Costa M, Pinedo-Vasquez R, Sears R, Siqueira A (2008) Urban forest and rural cities: multi-sited households, consumption patterns, and forest resources in Amazonia. Ecol Soc 13(2):2, [online] URL: http://www.ecologyandsociety.org/vol13/iss2/art2/

Pierce AR (2002) American ginseng (*Panax quinquefolius*). In Shanley P, Pierce A & Laird S (eds) Tapping the green market: certification and management of non-timber forest products. Earthscan, London, pp 172–182

Pierce AR, Emery MR (2005) The use if forests in times of crisis: ecological literacy as a safety net. For Trees Liveli 15:249–252
Peters CM, Gentry A, Mendelsohn RO (1989) Valuation of an Amazonian rainforest. Nature 339:655–656
Plotkin M, Famolare L (eds) (1992) Sustainable harvest and marketing of rain forest products. Conservation International. Island, Washington, DC
Richards RT, Saastamoinen O (2010) NTFP policy, access to markets and labour issues in Finland: impacts of regionalisation and globalisation on the wild berry industry. In Laird SA, McLain RJ, Wynberg RP (eds) Wild Governance – Finding Policies that Work for Non-timber Forest Products, Earthscan, London, 287–308
Rights and Resources Initiative (RRI) and International Tropical Timber Organisation (ITTO) (2009) Tropical forest tenure assessment: Trends, challenges and opportunities. RRI and ITTO, Washington, DC
Rockstöm J, Steffen W, Noone K, Person Å, Chapin FS, Lambin E, Lenton TM, Scheffer M, Folke C, Schellnhuber HJ, Nykvist B, de Wit CA, Hughes T, van der Leeuw S, Rodhe H, Sörlin S, Snyder PK, Constanza R, Svedin U, Falkenmark M, Karlberg L, Corell R, Fabry V, Hansen J, Walker B, Liverman D, Richardson K, Crutzen P, Foley J (2009) Planetary boundaries: Exploring the safe operating space for humanity. Ecol Soc 14(2):32, [online] URL: http://www.ecologyandsociety.org/vol14/iss2/art32/
Shaankar RU, Ganeshaiah KN, Krishnan S, Ramya R, Meera C, Aravind NA, Kumar A, Rao D, Vanaraj G, Ramachandra J, Gauthier R, Ghazoul J, Poole N, Chinnappa Reddy BV (2004) Livelihood gains and ecological costs of non-timber forest product dependence: assessing the roles of dependence, ecological knowledge and market structure in three contrasting human and ecological settings in south India. Env Conserv 31(3):242–253
Shackleton CM, Shackleton SE, Gambiza J, Nel E, Rowntree K, Urquhart P, Fabricius C, Ainslie A (2010) Linking ecosystem services and poverty alleviation in the arid and semi-arid lands of southern Africa. Nova Publishers, New York
Shackleton CM, Shackleton SE, Buiten E, Bird N (2007a) The importance of dry forests and woodlands in rural livelihoods and poverty alleviation in South Africa. For Policy Econ 9:558–577
Shackleton SE, Shackleton CM (2004) The *Pterocarpus angolensis* DC. based woodcraft industry in the Bushbuckridge district, South Africa. In: Sunderland T, Ndoye O (eds) Forest products, livelihoods and conservation: case studies of Non-timber forest product systems Vol 2 – Africa. CIFOR, Bogor, pp 203–228
Shackleton SE, Shanley P, Ndoye O (2007b) Viable but invisible: Recognising local markets for non-timber forest products. Int Forest Rev 9(3):697–712
Stoian D (2005) Making the best of two worlds: rural and peri-urban livelihood options sustained by non-timber forest products from the Bolivian Amazon. World Dev 33:1473–1490
UNEP-WCMC (2007) Biodiversity and poverty reduction: The importance of biodiversity for ecosystem services. United Nations Environmental Programme World Conservation Monitoring Centre (UNEP-WCMC), Cambridge
Wiens JA, Bachelet D (2010) Matching the multiple scales of conservation with the multiple scales of climate change. Conserv Biol 24:51–62
Wiens JA, Stralberg D, Jongsomjit D, Howell CA, Snyder MA (2010) Niches, models, and climate change: Assessing the assumptions and uncertainties. Proc Natl Acad Sci 106:19729–19736
Wong JLG, Thornber K, Baker N (2001) Resource assessment of non-wood forest products: experience and biometric principles. Non-wood Forest Products 13, FAO, Rome, 109 pp
WRI (2005) The wealth of the poor: managing ecosystem to fight poverty. WRI, Washington, DC, p 255
Wynberg R (2004) Achieving a fair and sustainable trade in devil's claw (*Harpagopytum* sp.). In: Sunderland T, Ndoye O (eds) Forest products, livelihoods and conservation. Case studies of non-timber forest product systems. Volume 2: Africa. Centre for International Forestry Research (CIFOR), Bogor

Index

S. Shackleton et al. (eds.), *Non-Timber Forest Products in the Global Context*, Tropical Forestry 7, DOI 10.1007/978-3-642-17983-9,

D

E

F